METHODS IN CELL BIOLOGY

VOLUME XII

Yeast Cells

Contributors to This Volume

W. BANDLOW

P. BAUER

JOAN H. CAULTON

MARJORIE CRANDALL

D. R. CRYER

JULIAN DAVIES

JOHN H. DUFFUS

R. ECCLESHALL

D. J. FENNELL

B. J. KILBEY

BIRTE KRAMHØFT

ANTHONY W. LINNANE

H. B. LUKINS

C. V. LUSENA

J. MARMUR

H. PONTA

U. PONTA

JOHN R. PRINGLE

ANTHONY H. ROSE

GERALD M. RUBIN

DANIEL SCHINDLER

P. R. STEWART

ANDRES WIEMKEN

D. WILKIE

D. H. WILLIAMSON

E. WINTERSBERGER

L. J. ZEMAN

ERIK ZEUTHEN

Methods in Cell Biology

Edited by

DAVID M. PRESCOTT

DEPARTMENT OF MOLECULAR, CELLULAR AND
DEVELOPMENTAL BIOLOGY
UNIVERSITY OF COLORADO
BOULDER, COLORADO

VOLUME XII

Yeast Cells

1975

ACADEMIC PRESS • New York   San Francisco   London

A Subsidiary of Harcourt Brace Jovanovich, Publishers

ACADEMIC PRESS, INC.
111 Fifth Avenue, New York, New York 10003

United Kingdom Edition published by
ACADEMIC PRESS, INC. (LONDON) LTD.
24/28 Oval Road, London NW1

LIBRARY OF CONGRESS CATALOG CARD NUMBER: 64-14220

ISBN 0–12–564112–5

PRINTED IN THE UNITED STATES OF AMERICA

CONTENTS

5. DNA-Dependent RNA Polymerases from Yeasts
H. Ponta, U. Ponta, and E. Wintersberger

6. The Isolation of Yeast Nuclei and Methods to Study Their Properties
John H. Duffus

7. Isolation of Vacuoles from Yeasts
Andres Wiemken

8. *Analytical Methods for Yeasts*
P. R. Stewart

9. *Methods for Avoiding Proteolytic Artefacts in Studies of Enzymes and Other Proteins from Yeasts*
John R. Pringle

10. *Induction of Haploid Glycoprotein Mating Factors in Diploid Yeasts*
Marjorie Crandall and Joan H. Caulton

11. *Mutagenesis in Yeast*
B. J. Kilbey

12.   Induction, Selection, and Experimental Uses of Temperature-Sensitive and Other Conditional Mutants of Yeast
John R. Pringle

13.   In Vivo and in Vitro Synthesis of Yeast Mitochondrial DNA
L. J. Zeman and C. V. Lusena

14.   Isolation of Mitochondria and Techniques for Studying Mitochondrial Biogenesis in Yeasts
Anthony W. Linnane and H. B. Lukins

15. Separation and Some Properties of the Inner and Outer Membranes of Yeast Mitochondria
W. Bandlow and P. Bauer

16. The Use of Fluorescent DNA-Binding Agent for Detecting and Separating Yeast Mitochondria DNA
D. H. Williamson and D. J. Fennell

17. Cytoplasmic Inheritance and Mitochondrial Genetics in Yeasts
D. Wilkie

LIST OF CONTRIBUTORS

Numbers in parentheses indicate the pages on which the authors' contributions begin.

W. BANDLOW, Institute for Genetics, University of Munich, München, Germany (311)

P. BAUER, Institute for Genetics, University of Munich, München, Germany (311)

JOAN H. CAULTON,[1] Department of Microbiology, Indiana University, Bloomington, Indiana (185)

MARJORIE CRANDALL,[2] Department of Microbiology, Indiana University, Bloomington, Indiana (185)

D. R. CRYER, Department of Biochemistry, Albert Einstein College of Medicine, New York, New York (39)

JULIAN DAVIES, Department of Biochemistry, University of Wisconsin, College of Agricultural and Life Sciences, Madison, Wisconsin (17)

JOHN H. DUFFUS, Department of Brewing and Biological Sciences, Heriot-Watt University, Edinburgh, Scotland, United Kingdom (77)

R. ECCLESHALL, Department of Biochemistry, Albert Einstein College of Medicine, New York, New York (39)

D. J. FENNELL, National Institute for Medical Research, Mill Hill, London, England (335)

B. J. KILBEY, Institute of Animal Genetics, Edinburgh, Scotland (209)

BIRTE KRAMHØFT, The Biological Institute of the Carlsberg Foundation, Copenhagen, Denmark (373)

ANTHONY W. LINNANE, Department of Biochemistry, Monash University, Clayton, Victoria, Australia (285)

H. B. LUKINS, Department of Biochemistry, Monash University, Clayton, Victoria, Australia (285)

C. V. LUSENA, Division of Biological Sciences, National Research Council of Canada, Ottawa, Ontario, Canada (273)

J. MARMUR, Department of Biochemistry, Albert Einstein College of Medicine, New York, New York (39)

H. PONTA, Max-Planck-Institut für Molekulare Genetik, Berlin-Dahlem, Germany (65)

U. PONTA, Max-Planck-Institut für Molekulare Genetik, Berlin-Dahlem, Germany (65)

JOHN R. PRINGLE, Institute of Microbiology, Swiss Federal Institute of Technology, Zürich, Switzerland (149, 233)

ANTHONY H. ROSE, School of Biological Sciences, Bath University, Bath, England (1)

GERALD M. RUBIN,[3] Medical Research Council Laboratory of Molecular Biology, Cambridge, England (45)

DANIEL SCHINDLER, Department of Biochemistry, University of Wisconsin, College of Agricultural and Life Sciences, Madison, Wisconsin (17)

P. R. STEWART, Department of Biochemistry, Faculty of Science, Australian National University, Canberra, Australia (111)

[1] *Present address:* Department of Zoology, Indiana University, Bloomington, Indiana.

[2] *Present address:* T. H. Morgan School of Biological Sciences, University of Kentucky, Lexington, Kentucky.

[3] *Present address:* Department of Biochemistry, Stanford University, Stanford, California.

ANDRES WIEMKEN, Department of General Botany, Swiss Federal Institute of Technology, Zürich, Switzerland (99)

D. WILKIE, Department of Botany and Microbiology, University College, London, England (353)

D. H. WILLIAMSON, National Institute for Medical Research, Mill Hill, London, England (335)

E. WINTERSBERGER, Physiologisch-Chemisches Institut der Universität Würzburg, Würzburg, West Germany (65)

L. J. ZEMAN,[4] Division of Biological Sciences, National Research Council of Canada, Ottawa, Ontario, Canada (273)

ERIK ZEUTHEN, The Biological Institute of the Carlsberg Foundation, Copenhagen, Denmark (373)

[4]*Present address:* Department of Biological Chemistry, Harvard Medical School, Boston Massachusetts.

PREFACE

In the absence of firsthand personal instruction, researchers are often reluctant to adopt new techniques. This hesitancy probably stems chiefly from the fact that descriptions in the literature do not contain sufficient detail concerning methodology; in addition, the information given may not be sufficient to estimate the difficulties or practicality of the technique or to judge whether the method can actually provide a suitable solution to the problem under consideration. The presentations in the volume are designed to overcome these drawbacks. They are comprehensive to the extent that they may serve not only as a practical introduction to experimental procedures but also to provide, to some extent, an evaluation of the limitations, potentialities, and current applications of the methods. Only those theoretical considerations needed for proper use of the method are included. Special emphasis has been placed on inclusion of much reference material in order to guide readers to early and current pertinent literature.

Volume XII of *Methods in Cell Biology* continues with the presentation of techniques and methods in cell research that have not been published or have been published in sources that are not readily available. Much of the information on experimental techniques in modern cell biology is scattered in a fragmentary fashion throughout the research literature. In addition, the general practice of condensing to the most abbreviated form materials and methods sections of journal articles has led to descriptions that are frequently inadequate guides to techniques.

This volume continues coverage begun in Volume XI of research methods that have been developed with yeast cells. Yeast cells are becoming increasingly useful in the analyses of problems in cell biology, and the aim here is to bring together into one compilation complete and detailed treatments of various research procedures used with this cell type. Originally, we had planned to cover yeast methods in a single volume, but the subject has proven to be too large. Volumes XI and XII are therefore entirely devoted to yeast cell methodology.

DAVID M. PRESCOTT

METHODS IN CELL BIOLOGY

VOLUME XII

Yeast Cells

Chapter 1

Growth and Handling of Yeasts

ANTHONY H. ROSE

School of Biological Sciences,
Bath University,
Bath, England

I. Introduction

By definition, yeasts are fungi that exist predominantly in the single-celled state. It follows from this definition that many if not all of the advantages enjoyed by experimenters in growing and handling other single-celled microbes such as bacteria—the ability to obtain small isolated clones or colonies on solidified media, to grow cultures and handle suspensions that contain a homogeneous suspension of cells, and to separate by rapid filtration even fairly large volumes of culture and suspension, all on both a laboratory and commercial scale—are shared by microbiologists and biochemists who experiment with yeasts. Not surprisingly, with such a broad definition of yeasts as that stated above, there are several genera of fungi which, although

1

they contain species able to grow in the yeast state, are not considered true yeasts simply because they do not exist predominantly as single cells. Fungi that are almost, but not quite, yeasts are commonly encountered in genera which include *Blastomyces*, *Geotrichum*, *Histoplasma*, *Paracoccidioides*, and *Sporotrichum*, not a few of them being actual or potential pathogens for man and animals. This is a very grey area of microbial taxonomy, and for many fungi there can never be a hard and fast demarcation. For a trenchant commentary on the meaning of the term "yeast" and the problems that arise from using it, the reader should turn to the admirable little volume by Phaff *et al*. (1966).

The currently accepted authority on the classification of yeasts (Lodder, 1970) recognizes 349 species which are classified into 39 genera. These genera, and their taxonomic relationships to the fungi, are listed in Table I. New species are continually being proposed, and some are accepted by the community of yeast workers. Correspondence concerning new species of yeasts is conducted the *Yeast News Letter*, which is edited by Herman J. Phaff of the University of California at Davis. This biannual publication is recognized as the official organ of communication among microbiologists interested in yeasts, according to a decision made at the Third International Symposium on Yeasts held at Delft and The Hague, The Netherlands, in June 1969.

Although there are 349 recognized species of yeasts, the bulk of the data reported in the literature on yeast physiology, biochemistry, genetics, and industrial use, is concerned with a relatively small number of species, notably members of the genera *Saccharomyces* and *Candida*. Species of *Saccharomy-*

TABLE I

LIST OF YEAST GENERA SHOWING THEIR TAXONOMIC RELATIONSHIP TO THE
MAIN CLASSES OF FUNGI[a]

Fungal class	Yeast genera
Ascomycetes	*Citeromyces, Coccidiascus, Debaryomyces, Dekkera, Endomycopsis, Hanseniaspora, Hansenula, Kluyveromyces, Lipomyces, Lodderomyces, Metschnikowia, Nadsonia, Nematospora, Pachysolen, Pichia, Saccharomyces, Saccharomycodes, Saccharomycopsis, Schizosaccharomyces, Schwanniomyces, Wickerhamia, Wingea*
Basidiomycetes	*Bullera, Leucosporidium, Rhodosporidium, Sporidiobolus, Sporobolomyces*
Deuteromycetes (fungi imperfecti)	*Brettanomyces, Candida, Cryptococcus, Kloeckera, Oosporidium, Pityrosporum, Rhodotorula, Schizoblastosporion, Sterigmatomyces, Torulopsis, Trichosporon, Trigonopsis*

[a]After Lodder (1970).

ces, particularly *S. cerevisiae* and *S. uvarum* [the latter still is referred to as *S. carlsbergensis* despite the recommendations of Lodder (1970)] are used in the manufacture of beers, wines, and ciders, while species of *Candida*, particularly *C. utilis* and *C. lipolytica*, are sources of animal food and fodder. A few yeasts are troublesome pathogens, notably strains of *Candida albicans*. But pride of place in the yeast community must go to *S. cerevisiae*. Pioneer work on extracts of this yeast by the Buchner brothers gave birth to the science of biochemistry, and the volume of data published on strains of this species is massive. It follows that many of the techniques used for growing and handling yeasts were developed with strains of *S. cerevisiae* and may not always be applicable to other yeast species.

The literature on yeasts is truly voluminous and of long standing. For an introduction to these microbes, the reader cannot do better than to turn to the text by Phaff *et al.* (1969). Authoritative articles on the biology, physiology, and technology of yeasts are to be found in the texts edited by Rose and Harrison (1969, 1970, 1971).

II. Isolation and Maintenance of Yeasts

A. Isolation Procedures

Since yeasts can grow to produce discrete colonies on solidified media, they can easily be isolated from materials that contain low proportions of other microbes (bacteria and filamentous fungi) by plating a sample of the material on a rich organic medium such as malt extract–agar or Wickerham's (1951) malt extract–yeast extract–glucose–peptone (MYGP) medium. Details of these media, both of which are available commercially, are given in Section III.

More often than not, however, yeasts must be isolated from materials that contain large proportions of bacteria and/or filamentous fungi, particularly materials that have been sampled from natural environments. Under these conditions the isolation medium must be one that selectively encourages growth of all yeasts present in the material, rather than the bacteria and filamentous fungi. Environmental conditions that encourage the growth of yeasts at the expense of other microbes can also be employed. Yeasts in general grow well over a fairly wide range of pH values (see Section III,B,3) and, by lowering the pH value of the malt extract or MYGP medium to 3.5–4.0, growth of bacteria and filamentous fungi can to some extent be repressed. Acidification of the media is best effected by adding dilute hydrochloric or phosphoric acid to the molten medium. Van der Walt (1970) counsels against the use of organic acids for acidification, since several of these compounds have been found to inhibit the growth of some yeasts. At the same time, Campbell (1972) advocates the use of lactic acid. Moreover, propionic acid (0.25% at pH 4.0) is particularly useful, in that it is fungistatic but does not have untoward effects on the growth of most yeasts. This compound is incorporated into breads for this purpose.

Media have also been made selective for yeast growth by including chemical compounds that inhibit the growth of bacteria and filamentous fungi but have no effect on yeasts. These compounds include basic dyes, such as methylene blue and crystal violet, and the fungistatic agent diphenyl, although some doubts have been expressed about use of the last-mentioned compound since it can inhibit the growth of some yeasts.

But in recent years preference has been given to incorporating antibiotics into media for the isolation of yeasts. Broad-spectrum antibacterial antibiotics, such as chloramphenicol and the tetracyclines, are extensively used for this purpose. Streptomycin has also been used, but Richards and Elliott (1966) caution against using this compound. Unfortunately, antifungal antibiotics are also active against yeasts, and so cannot be used in yeast isolation procedures.

Choosing suitable growth conditions can also favor the growth of yeasts at the expense of bacteria and filamentous fungi. However, this is possible to only a limited extent, since both yeasts and filamentous fungi in general prefer aerobic conditions and a temperature in the range 20°–30°C. Wickerham (1951) devised a rather ingenious method for recovering yeasts from material that contains filamentous fungi. A batch of malt extract or MYGP medium, adjusted to a pH value of 3.5–4.0, is inoculated with the material, and the culture incubated on a rotary shaker at 25°C for 48 hours. The filamentous fungi, which are prevented from sporulating, grow in the form of large pellets and are greatly outnumbered by the multiplication of individual yeast cells. Separation of the yeast cells is then effected by direct plating.

So far, we have described methods that can be used to isolate yeasts in general from materials that contain bacteria and filamentous fungi. These methods have been devised to ensure that no one group or genus of yeasts is prevented from growing under the selective conditions employed. Different problems arise when the aim of the isolation procedure is to obtain particular species or types of yeasts from a material. This situation is commonplace in industries, such as brewing, that use cultured strains of yeasts which can become infected with other species or strains of yeasts. For this purpose the physiological, and particularly the nutritional, peculiarities of yeast species are exploited.

Several yeasts (such as species of *Hansenula*) can use nitrate as the sole source of nitrogen. By including nitrate in the isolation medium, instead of the more commonly used ammonium salt, growth of nitrate-utilizing yeasts can be encouraged. For this purpose, use is often made of the commercially available yeast carbon base (see Section III,A, p. 11) which can be supplemented with potassium nitrate. Another example of the exploitation of a specific nitrogen requirement of a yeast is the use of creatinine in the isolation of the pathogenic yeast *Cryptococcus neoformans* (Shields and Ajello, 1966). *Torulopsis pintolopesii* and *T.* (= *Candida*) *bovina* are auxotrophic for choline (Curey *et al.*, 1960), and so must be isolated using defined media supplemented with this compound. Likewise *Pityrosporum ovale*, which is auxotrophic for an unsaturated fatty acid (Shifrine and Marr, 1963). Other examples of unusual nutritional requirements of yeasts, which can be exploited in isolation and enrichment procedures, are the need for high partial pressure of carbon dioxide in the medium for growth of *Saccharomycopsis guttalata* (Richle and Scholer, 1961), and the abnormally high concentrations of thiamine (10 mg per liter) required for growth of *Brettanomyces* spp. (van der Walt, 1961).

Growth temperature can also be used to encourage the growth of desired strains of yeast in isolation procedures. Psychrophilic yeasts, such as *Leucosporidium scottii*, are easily isolated by incubating plates at 0°–4°C for up to 2 weeks. Some yeasts are thermoduric or thermophilic, and so can be isolated with comparative ease by incubating plates at temperatures in the range 40°–50°C (Travassos and Cury, 1971; see page 13).

B. Identification of Yeasts

Details of the criteria and procedures used in the identification and classification of yeasts are outside the scope of this brief article on methods for handling yeasts. These details can be found in the authoritative work edited by Lodder (1970). Briefly, four main classes of criteria are used in yeast classification, namely, morphological characteristics, cultural characteristics, sexual characteristics (particularly the ability to form spores), and physiological characteristics. A definitive taxonomic description of a newly isolated yeast is a protracted exercise and, to short-cut this procedure, several workers have devised abbreviated identification schemes for specific groups of yeasts such as those encountered in an industrial process. Beech *et al.* (1968) published details of two simplified schemes for identifying yeasts based on morphological and on biochemical characteristics. More recently, Barnett and Pankhurst (1974) devised a computer-made key for identifying yeasts, which can be used with Lodder's (1970) detailed key. It is based solely on the results of physiological tests, and features such as microscopic appearance of the vegetative cells and their capacity to produce ascospores are needed only for confirmation of identity and not in the key itself.

The highly specific serological technique for identifying species of microbes, exploited so successfully by the medical microbiologists, has recently been applied to yeasts. Campbell (1971a,b) pioneered this approach when he devised an identification scheme based on cell morphology and slide agglutination with six specific antisera. The scheme was used for rapid identification of groups of yeasts. The technique has now become widely accepted, particularly in the brewing industry for detecting contaminants in brewing yeasts (Campbell, 1972), mainly using antisera that have been conjugated with a fluorescent dye.

C.   Maintenance of Yeast Cultures

1.   LABORATORY MAINTENANCE METHODS

It is widely believed, and with some justification, that yeasts in general are fairly hardy microbes which can be maintained with little difficulty in the laboratory. The main reason for this supposition is that bulk quantities of pressed *S. cerevisiae* are marketed for baking purposes with little concern on the part of the producers about a loss of viability of the product. Most workers using a relatively small number of yeast strains satisfactorily maintain their species on slants of solidified malt extract or MYGP medium (see Section III,A) stored at 0°–4°C. These stock cultures are transferred at frequent intervals (about twice monthly). When the bottles containing slant cultures are filled with sterile paraffin oil, which minimizes drying out of the culture, the frequency of subculturing can be decreased (Ciblis, 1959). Alternatively, stock cultures can be kept in liquid MYGP medium at 0°–4°C, a technique that has been adopted by one of the larger yeast culture collections (Kirsop, 1954).

For yeasts that have special nutritional requirements, such as that of *P. ovale* for an unsaturated fatty acid (see Section II,A), the basal medium must be supplemented appropriately. Strains of *Brettanomyces* spp. produce excessive amounts of acetic acid, and solidified media for maintaining these yeasts should be supplemented with 2% calcium carbonate.

With repeated subculture it has been shown conclusively with many microorganisms that genetic variants are produced (Reusser, 1963). This is a cause for concern for many users of yeasts, particularly those such as brewers who exploit a variety of properties of a strain, a change in any one of which would be highly undesirable. As a result, several workers have examined alternatives to the use of slant cultures for maintaining strains in order to minimize the production of variants.

The labor incurred in repeated subculture can be decreased by storing yeast strains as freeze-dried or lyophilized cultures (Wynants, 1962), and many small and large collections now favor this maintenance method. Despite the economical advantages that accrue from using this technique, particularly for medium- or large-sized collections, serious doubt has been expressed about its suitability. The main disadvantage is the loss of viability often experienced when yeast cultures are freeze-dried. Atkin *et al.* (1949) reported losses of up to 99.8% when some yeasts were lyophilized. Not surprisingly, many attempts have been made to modify the lyophilization procedure in order to obtain a greater retention of viability. The main attention has been given to the composition of the medium in which the yeast cells are lyophilized, and compounds found to confer some protection on freeze-dried cells include glycerol, glucose, and sucrose with either peptone or glutamate. Nevertheless, difficulties are still experienced even when modified techniques are used, and the reservations widely held concerning the lyophilization technique were eloquently expressed by Beech and Davenport (1971) who stated that, "while lyophilisation offers a convenient method for storing large numbers of cultures, it is by no means the perfect method for storing yeasts with completely unchanged characteristics."

These worries have prompted other investigators to turn to the use of liquid-nitrogen refrigeration for maintaining yeast cultures. Wellman and Stewart (1973) reported on the storage of brewing strains of *S. cerevisiae* and *S. uvarum* with liquid-nitrogen refrigeration at $-196°C$. Strains varied in their sensitivity to freezing, and the degree of survival depended on the composition of the medium in which the yeasts were grown, the composition of the menstruum in which they were frozen, and the freezing conditions. Suspending cells in 10% (v/v) glycerol, cooling at 1°C per minute, warming rapidly, and plating on solidified MYGP medium (see Section III,A) in general produced the greatest percentage of viable cells with minimum alteration in metabolic characteristics. Liquid-nitrogen refrigeration seems to offer the greatest promise to workers interested in maintaining large collections of yeasts.

2. YEAST CULTURE COLLECTIONS

Several large collections of yeast cultures have been established in various countries of the world. As for collections of other microbes, these centers maintain a wide range of species, particularly of the type cultures of genera, and are able to supply catalogs and cultures on request. Locations of the main collections are given in Table II.

TABLE II

PRINCIPAL YEAST CULTURE COLLECTIONS IN THE WORLD

Country	Name and address of collection
Czechoslovakia	Czechoslovak Collection of Micro-organisms, J. E. Purkyne University, Brno
Great Britain	National Collection of Yeast Cultures, Brewing Industry Research Foundation, Lyttel Hall, Nutfield, Surrey, England
Netherlands	Centraalbureau voor Schimmelcultures, Baarn
United States	American Type Culture Collection, Rockville Maryland
	Yeast Genetics Stock Center, Donner Laboratory, University of California, Berkeley, California
U.S.S.R.	Department of Type Cultures, Institute of Microbiology, U.S.S.R. Academy of Sciences, Moscow

D. Assessment of Viability of Yeasts

1. DIRECT METHODS

The ability of yeasts to form single discrete colonies on solidified media makes direct assessment of yeast viability a comparatively easy procedure. Samples, after appropriate dilution, are plated on solidified malt extract or MYGP medium (see Section III,A), the plates are incubated for 24–48 hours at 25°–30°C, and the numbers of colonies on the plates are counted in the conventional manner (Postgate, 1969).

The principal drawback to this method for assessing yeast viability is that it gives only the number of viable cells in a sample and does not indicate the proportion of cells in the sample that are viable, i.e., the viability of the population. This assessment can conveniently be made with yeasts by using the slide culture technique, in which a portion of the population is incubated on a small microscopically visible slide culture, and the ability or lack of ability of the cells on the surface of this culture to form microcolonies is assessed by microscopic observation after a suitable period of incubation. Details of this technique as applied to microbes in general have been provided by Postgate (1967, 1969). The technique as described by Postgate (1969) involves placing molten medium in a small brass annulus (a section through a tube measuring about 2 cm in diameter and about 2 mm thick) on a microscope slide and spreading the drop of sample containing yeast cells on the surface of the agar medium. In our laboratory the technique has proved extremely successful in assessing the viability of a wide range of yeasts, and we recommend the use of MYGP medium solidified with 2% (w/v) agar. It is, however, absolutely essential to sterilize the molten medium by filtration through a membrane filter (1.2-μm pore size). Some of the constituents of the MYGP medium, particularly yeast extract, contain dead yeast cells which must be removed by filtration if they are not to interfere with the counting of nonviable cells on the slide culture. British brewers have advocated the use of a slide culture technique for assessing the viability of yeasts in the brewery, using a malt extract medium solidified with gelatin (Howard, 1971).

2. INDIRECT METHODS

Yeast viability has for many years been assessed by the ability of viable cells, in contrast to nonviable cells, to take up the dye methylene blue and to reduce it to the colorless leukomethylene blue. Nonviable cells, in contrast, are unable to reduce the dye and therefore stain blue (Pierce, 1970). Despite continued criticism over the years, the methylene blue method for measuring yeast viability continues to be used by discerning zymologists.

Several other dyes have been investigated for use in assessing yeast viability. Unfortunately, the physiological bases of these applications have not been explored, with the result that the use of these dyes has not been widely adopted. Among the more promising dyes are acridine orange (Strugger, 1948) and primuline (Graham, 1970), but neither is yet seen as a competitor of methylene blue. Zymologists would do well to examine more fully the use of dyes as indicators of yeast viability.

It has already been stressed that the main drawback in using direct methods for assessing yeast viability is the need to incubate plates or slide cultures for periods of 24–48 hours. Some industrial users of yeasts, notably brewers, have recently become interested in indirect methods for detecting the presence, in very low concentrations (as low as 10 to 100 cells/ml), of viable yeast cells in beers and related products. These methods aim to detect the presence of viable yeast cells in putatively sterile beer, for example, in as little as 5–10 hours. Several different types of method have been explored. The more successful of these are based on changes in such properties as pH value (Mitz, 1969) and conductance (Cady, 1973; Ur and Brown, 1973) that take place during limited multiplication of viable cells; another technique widely advocated involves the detection of radioactively labeled carbon dioxide as a result of respiration of labeled glucose or other carbon source (Mitz, 1969). Harrison and his colleagues (1974) have published a valuable review of indirect methods for assessing viability as applied to the detection of sparse populations of yeast cells and of brewery spoilage bacteria. They conclude that, for simple routine application, methods involving the detection of changes in pH value offer the greatest scope.

III.   Growth of Yeasts

A.   Composition of Growth Media

1.   GENERAL GROWTH MEDIA

The recommended general media for growing yeasts (van der Walt, 1970) are malt extract and Wickerham's (1951) MYGP medium. Either medium, liquid or solidified, can be used to study the morphological and cultural characteristics of yeast species.

Malt extract medium (Lodder and Kreger-van Rij, 1952) can easily be made from unhopped malt wort purchased from a brewery. The wort is adjusted to pH 5.4 and diluted to 10° Balling [the tables constructed by van Balling in 1843 relate the concentration of sucrose in a solution to the specific gravity of the solution at 17.5°C; they can with some reservations be applied to brewer's malt wort (Hough *et al.*, 1971)] . If unhopped brewer's malt wort cannot be purchased, malt extract medium can be made by using commercial malt extract as a 10% (w/v) solution with the pH value adjusted to 5.4. Few yeast workers nowadays make their own malt wort by mashing malted barley, but details of this procedure can be found in the article by van der Walt (1970). Malt extract medium is usually sterilized by autoclaving at 10 lb/in^2 for 10–15 minutes. Solidified malt extract medium is made by including 2% (w/v) agar in the medium.

Increasing preference is now given to the use of Wickerham's (1951) MYGP medium over malt extract as a general medium for yeast growth. MYGP medium has the following composition (percent, w/v): malt extract, 0.3; yeast extract, 0.3; glucose, 1.0; peptone, 0.5. The pH value of this medium is about 5.5; it does not need to be adjusted. Again, it can be solidified by incorporating 2% (w/v) agar.

The relatively nonfastidious nutritional requirements of the majority of yeast species makes it possible to growth these organisms in defined media with a fairly simple composition. In our laboratory we have for several years used a modification of a medium first described by Rose and Nickerson (1956). The modified medium, which differs from that originally described in that a mixture of vitamins is replaced by yeast extract, has the following composition (percent w/v): glucose, 2.0; $(NH_4)_2SO_4$, 0.3; KH_2PO_4, 0.3; yeast extract, 0.01; $MgSO_4 \cdot 7H_2O$, 0.025; $CaCl_2 \cdot 6H_2O$, 0.025; pH, 4.5.

More complex defined media are required for assessing accurately the nutritional requirements of yeasts. Fortunately, some of these are available commercially (marketed by Difco Laboratories, Detroit, Michigan). Table

TABLE III

COMPOSITION OF CHEMICALLY DEFINED MEDIA USED FOR GROWTH OF YEASTS

Ingredient (per liter of water)	Yeast nitrogen base	Yeast carbon base	Vitamin-free yeast base
Nitrogen sources			
Ammonium sulfate	5 gm	None	5 gm
Asparagine	None	None	None
Carbon source			
Glucose	None	10 gm	10 gm
Amino acids			
L-Histidine monohydrochloride	10 mg	1 mg	10 mg
DL-Methionine	20 mg	2 mg	20 mg
DL-Tryptophan	20 mg	2 mg	20 mg
Vitamins			
Biotin	20 μg	20 μg	None
Calcium pantothenate	2,000 μg	2,000 μg	None
Folic acid	2 μg	2 μg	None
Inositol	10,000 μg	10,000 μg	None
Nicotinic acid	400 μg	400 μg	None
p-Aminobenzoic acid	200 μg	200 μg	None
Pyridoxine hydrochloride	400 μg	400 μg	None
Riboflavin	200 μg	200 μg	None
Thiamine hydrochloride	400 μg	400 μg	None
Compounds supplying trace elements			
Boric acid	500 μg	500 μg	500 μg
Copper sulfate	40 μg	40 μg	40 μg
Potassium iodide	100 μg	100 μg	100 μg
Ferric chloride	200 μg	200 μg	200 μg
Manganese sulfate	400 μg	400 μg	400 μg
Sodium molybdate	200 μg	200 μg	200 μg
Zinc sulfate	400 μg	400 μg	400 μg
Salts			
Potassium phosphate, monobasic	0.85 gm	0.85 gm	0.85 gm
Potassium phosphate, dibasic	0.15 gm	0.15 gm	0.15 gm
Magnesium sulfate	0.5 gm	0.5 gm	0.5 gm
Sodium chloride	0.1 gm	0.1 gm	0.1 gm
Calcium chloride	0.1 gm	0.1 gm	0.1 gm

III lists the compositions of three of these media, which are referred to as yeast nitrogen base, yeast carbon base, and yeast base for vitamin requirements. They should be supplemented with a carbon source, a nitrogen source, and a mixture of vitamins, respectively, in order to obtain nutritionally complete media.

2. SELECTIVE MEDIA

Reference has already been made to the use of selective media for the growth of yeasts (see Section II,A). Most of the selective media already described were designed to encourage the growth of certain genera of yeasts at the expense of others. An example is the use of media containing nitrate as the sole source of nitrogen to encourage the growth of nitrate-utilizing yeasts such as *Hansenula* spp. Other more selective media have been described, media used to encourage the growth of certain strains of yeasts at the expense of other closely related strains which may even be members of the same genus. Some of the best examples of these more refined selective media have come from laboratories, such as those in the brewing industry, where relatively small numbers of closely related yeast strains are cultured and where the presence of other nearly related strains constitutes contamination. Walters and Thistleton (1953) advocate the use of media containing lysine as the sole nitrogen source to distinguish between cultured brewing strains of *S. cerevisiae* and related strains of yeasts known to the brewer as "wild" yeasts. Another example of the use of selective media in the brewing industry is the use of media containing the trisaccharide raffinose as sole carbon source to distinguish between ale yeasts (strains of *S. cerevisiae*) and lager yeasts (*S. carlsbergensis*). A vast number of yeast selective media is in use in various laboratories, and it is not the aim of this brief article to list them in detail. Other examples of these media can be found in the article by van der Walt (1970).

B. Cultural Conditions

On the whole, yeasts as single-celled organisms are easily grown under conditions used to grow such easily cultivated microbes as *Escherichia coli*, i.e., the use of static tube cultures, shaken or unshaken flask cultures, and fermenters with various capacities. The growth of yeasts is of course influenced by various cultural conditions, the more important of which are discussed in this section.

1. AERATION

Species of yeasts differ greatly in their need for molecular oxygen for growth. Some yeasts, including *Candida* spp., are obligate aerobes, and their requirement for oxygen is often so great that, in shake cultures and fermenters, the yield of cells is limited by the oxygen transfer rate in the cultures. In fermenters there is a need, with these aerobic yeasts, for vigorous stirring and for a continuous supply of air to provide large crops of cells. Many other yeasts, however, are facultative aerobes, and can grow well in media containing low concentrations of dissolved oxygen. These yeasts include strains of *S. cerevisiae*. Under stringent anaerobic conditions, which cannot be defined accurately in terms of concentrations of dissolved oxygen or of the redox potential in the culture because of limitations in the measuring devices available, strains of *S. cerevisiae* and possibly of other yeasts, although this has yet to be reported, become auxotrophic for a sterol (5 mg per liter) and for an unsaturated fatty acid (30 mg per liter), two cell components which require molecular oxygen for their synthesis (Andreasen and Stier, 1954). These anaerobically induced requirements have been exploited in studying the role of sterols and unsaturated fatty-acyl residues in phospholipids in the functioning of various yeast membranes (Alterthum and Rose, 1973; Linnane *et al.*, 1972).

2. TEMPERATURE

A temperature in the range 20°–25°C is suitable for growth of the majority of yeasts. It is also probably true that the majority of yeasts cannot grow at temperatures much above 30°C, and temperatures of 30°C and above must be avoided if the temperature range for the growth of a particular species has not been accurately established. We have already cited examples of psychrophilic and thermophilic yeasts (see Section II,A), and these species clearly need to be cultivated at temperatures appropriately lower or higher than those normally employed. Most yeasts have a fairly broad range of temperatures in which they will grow but, not surprisingly, there are exceptions. Among the most exacting of yeasts as far as growth temperature is concerned is *Saccharomycopsis guttulata* which grows only in the range 35°–40°C. The text by Phaff *et al.* (1966) has some valuable comments on the growth temperature requirements of yeasts, and merits perusal by zymologists concerned with this aspect of growing yeasts.

Often the microbiologist is not able to examine in detail the effects of temperature on the growth of a yeast, because a sufficiently large number of incubators or temperature-controlled water baths is not available. Those faced with such a problem might turn to the article by Patching and Rose (1970) which describes a versatile apparatus with which microbes can be grown in stirred culture at any temperature in the range 0°–95°C.

3. pH VALUE

In general, yeasts are not adversely affected by environments with pH values in the range 3.0–8.0. Optimum growth of the majority of strains occurs in the range 4.5–6.5. There are no reports of acidophilic or alkalinophilic yeasts, although this may simply reflect a lack of interest in these types of organisms.

C.   Measurement of Growth

As for most microbes, the best method for assessing the size of a population is to measure the number of cells in a unit volume of the culture or suspension. Because yeasts are 5 to 10 times the size of the majority of bacteria, they can very easily be counted on a hemocytometer slide. Electronic cell-sizing devices, such as the Coulter Counter® and the Celloscope®, are now quite widely available, and are well suited to measuring the numbers of yeast cells in a culture or suspension (Kubitschek, 1969). The size of yeast cells is such that the signal generated as the result of passage of a cell through the orifice of an electronic cell-sizing apparatus is well above the noise level of the apparatus, which is not the case when bacteria are counted.

Physiologists often prefer to assess yeast growth by measuring the wet or dry weight in a unit volume of culture or suspension. This can be done by centrifuging a portion of culture or suspension in a tared centrifuge tube and drying the crop after washing with water, or by filtering a portion through a tared membrane filter, washing the cells, and weighing the filter with the cells (Howard, 1971). A great advantage of using yeasts is that the cell concentration in a culture or suspension can be estimated by measuring the absorbance or extinction of the culture or suspension and relating the reading to the dry weight of the cells by a calibration curve. Caution must be exercised here, however, because the relationship between absorbance or extinction and cell dry weight can alter during the growth of yeasts in batch culture.

IV.   Procedures Used in Handling Yeast Populations

A.   Harvesting and Centrifugation

One of the principal advantages of using yeasts in laboratory studies is the ease with which the cells can be harvested from cultures and suspensions. The single-celled nature of yeasts means that the sedimented crop obtained by centifuging a culture or suspension compacts tightly, thereby allowing the supernatant liquid to be removed with ease. The gravitational force needed to sediment yeasts in the centrifuge is small compared with that needed to remove bacteria from suspension; centrifuging at about 500 g for about 10 minutes sediments most yeasts.

B.   Filtration

Almost all yeasts can be very easily and rapidly removed from a culture or suspension by filtration through a membrane filter. Filters with a pore size of about 1.2 μm are customarily used.

C. Storage

Most yeasts can be stored, as packed cell crops, for quite long periods of time (up to about 24 hours) at about 0°C without appreciable loss of viability or of individual enzyme activity.

REFERENCES

Alterthum, F., and Rose, A. H. (1973). *J. Gen. Microbiol.* **77**, 371–382.
Andreasen, A. A., and Stier, T. J. B. (1954). *J. Cell. Comp. Physiol.* **43**, 271–281.
Atkin, L. N., Moses, W. A., and Gray, P. P. (1949). *J. Bacteriol.* **57**, 575–580.
Barnett, J. A., and Pankhurst, R. J. (1974). "A New Key to the Yeasts. A Key for Identifying Yeasts Based on Physiological Tests Only". North-Holland Publ., Amsterdam.
Beech, F. W., and Davenport, R. R. (1971). *In* "Methods in Microbiology", (J. R. Norris *et al.*, eds.), Vol. 4, pp. 153–182. Academic Press, New York.
Beech, F. W., Davenport, R. R., Goswell, R. W., and Burnett, J. K. (1968). *In* "Identification Methods for Microbiologists: Part B" (B. M. Gibbs and D. A. Shapton, eds.), pp. 151–175. Academic Press, New York.
Cady, P. (1973). *In* "Symposium on Rapid Methods and Automation in Microbiology" (C-G. Hedén and T. Illenyi, eds.), Paper A 17. Stockholm.
Campbell, I. (1971a). *J. Appl. Bacteriol.* **34**, 237–242.
Campbell, I. (1971b). *J. Gen. Microbiol.* **67**, 223–231.
Campbell, I. (1972). *J. Inst. Brew.* **78**, 225–229.
Ciblis, E. (1959). *Proc. Int. Dairy Congr., 15th, 1959* Vol. 3, pp. 1361–1363.
Curey, A., Suassura, E. N., and Travassos, L. R. (1960). *An. Microbiol.* **8**, 13–64.
Graham, R. K. (1970), *J. Inst. Brew.* **76**, 16–21.
Harrison, J. S., Webb, T. J. B., and Martin, P. A. (1974). *J. Inst. Brew.* **80**, 390–398.
Hough, J. S., Briggs, D. E., and Stevens, R. (1971). "Malting and Brewing Science," Chapman & Hall, London.
Howard, G. A. (1971). *J. Inst. Brew.* **77**, 181–226.
Kirsop, B. H. (1954). *J. Inst. Brew.* **60**, 210–213.
Kubitschek, H. E. (1969). *In* "Methods in Microbiology" (J. R. Norris and D. W. Ribbons, eds.), Vol. 1, pp. 593–610. Academic Press, New York.
Linnane, A. W., Haslam, J. M., Lukins, H. B., and Nagley, P. (1972). *Annu. Rev. Microbiol.* **26**, 163–198.
Lodder, J., ed. (1970). "The Yeasts, a Taxonomic Study." North-Holland Publ., Amsterdam.
Lodder, J., and Kreger-van Rij, N. J. W. (1952). "The Yeasts, a Taxonomic Study." North-Holland Publ., Amsterdam.
Mitz, M. A. (1969). *Ann. N.Y. Acad. Sci.* **158**, 651–660.
Patching, J. W., and Rose, A. H. (1970). *In* "Methods in Microbiology" (J. R. Norris and D. W. Ribbons, eds.), Vol. 2, pp. 23–38. Academic Press, New York.
Phaff, J. J., Miller, M. W., and Mrak, E. M. (1966). "The Life of Yeasts." Harvard Univ. Press, Cambridge, Massachusetts.
Pierce, J. S. (1970). *J. Inst. Brew.* **76**, 442–443.
Postgate, J. R. (1967). *Advan. Microbial Physiol.* **1**, 1–23.
Postgate, J. R. (1969). *In* "Methods in Microbiology" (J. R. Norris and D. W. Ribbons, eds.), Vol. 1, pp. 611–628. Academic Press, New York.
Reusser, F. (1963). *Advan. Appl. Microbiol.* **5**, 189–215.
Richards, M., and Elliott, F. R. (1966). *Nature (London)*, **209**, 536–537.
Richle, R., and Scholer, H. J. (1961). *Pathol. Microbiol.* **24**, 783–790.

Rose, A. H., and Harrison, J. S., eds. (1969). "The Yeasts," Vol. 1. Academic Press, New York.

Rose, A. H., and Harrison, J. S., eds. (1970). "The Yeasts," Vol. 3. Academic Press, New York.

Rose, A. H., and Harrison, J. S., eds. (1971). "The Yeasts," Vol. 2. Academic Press, New York.

Rose, A. H., and Nickerson, W. J. (1956). *J. Bacteriol.* **72**, 324–328.

Shields, A. B., and Ajello, L. (1966). *Science* **151**, 208–209.

Shifrine, M., and Marr, A. G. (1963). *J. Gen. Microbiol.* **32**, 263–270.

Strugger, S. (1948). *Can. J. Res., Sect. C* **26**, 188–197.

Travassos, L. R. R. G., and Cury, A. (1971). *Annu. Rev. Microbiol.* **25**, 49–74.

Ur, A., and Brown, D. F. J. (1973). *In* "Symposium on Rapid Methods and Automation in Microbiology" (C-G. Hedén and T. Illenyi, eds.), Paper A 16. Stockholm.

van der Walt, J. P. (1961). *Antonie van Leeuwenhoek; J. Microbiol. Serol.* **27**, 332–339.

van der Walt, J. P. (1970). *In* "The Yeasts, a Taxonomic Study" (J. Lodder, ed.), pp. 34–113. North-Holland Publ., Amsterdam.

Walters, L. S., and Thistleton, M. R. (1953). *J. Inst. Brew.* **59**, 401–408.

Wellman, A. M., and Stewart, G. G. (1973). *Appl. Microbiol.* **26**, 577–583.

Wickerham, L. J. (1951). *U.S., Dep. Agr., Tech. Bull.* **1029**.

Wynants, J. (1962). *J. Inst. Brew.* **68**, 350–356.

Chapter 2

Inhibitors of Macromolecular Synthesis in Yeast

DANIEL SCHINDLER AND JULIAN DAVIES

Department of Biochemistry,
University of Wisconsin,
College of Agricultural and Life Sciences,
Madison, Wisconsin

I. Introduction

Studies of macromolecular metabolism in eukaryotic cells have been aided by the availability of inhibitors that specifically block nucleic acid or protein synthesis; most of the work to date has been done on cultured mammalian cells such as HeLa cells. Examples of inhibitors that have been used are actinomycin D, which suppresses rRNA synthesis at low concentrations (Perry, 1963; Darnell *et al.*, 1971), cordycepin, which prevents the formation of poly A during nuclear mRNA metabolism (Mendecki *et al.*, 1972), and camptothecin, a reversible inhibitor of nuclear RNA synthesis which allows continued mitochondrial RNA synthesis (Perlman *et al.*, 1973). Such inhibitors cannot be used in similar studies of yeasts such as *Saccharomyces cerevisiae*, since yeasts are impermeable to many drugs. However, recently, the modes of action of several antibiotics affecting yeast growth have been elucidated and found to affect protein or nucleic acid synthesis as their primary mechanism of action. As the collection of inhibitors is further investigated and expanded, it is presumed that some will be found that block specific processes and yield useful information on the workings of the cell. These inhibitors can also be used to study the interrelationships among the different macromolecular synthetic processes.

Since knowledge of the genetics of *S. cerevisiae* is well advanced, inhibitor studies in yeasts should aid in the analyses of macromolecular synthesis and its regulation in all eukaryotes. In addition, inhibitors can aid in identifying defects in temperature-sensitive mutants.

This chapter is divided into two parts. The first catalogs the known inhibitors of macromolecular functions that inhibit the growth of yeast and discusses the inhibitors that have been used in physiological studies and their specificity in *in vitro* systems. It is possible that other inhibitors which have not been extensively used in physiological studies will prove useful in the future. The second part considers the interrelationships between DNA, RNA, and protein synthesis, and examines the effects of inhibitors on such interdependent processes in yeasts.

Several excellent reviews are available to provide detailed discussions of the mechanisms of many of the inhibitors mentioned here (Pestka, 1971; Goldberg and Friedman, 1971; Gale *et al.*, 1972). A recent review on the biochemical genetics of yeasts is a useful source for our current understanding of the details of macromolecular synthesis in yeasts (Hartwell, 1970).

II. Inhibitors

A. DNA Synthesis Inhibitors

The DNA content of haploid *S. cerevisiae* is 9×10^9 daltons arranged in at least 17 chromosomes (Hartwell, 1970). Five genes have been identified that are required for DNA synthesis; three of these are necessary for initiation or steps prior to initiation of DNA synthesis, and the other two are necessary for the continuation of DNA synthesis (Hartwell, 1973). The chromosomes are replicated in the early part (0.1 to 0.35) of the cell cycle. DNA replication on most, if not all, of the chromosomes is initiated internally, and each chromosome contains more than one replication unit. The size of the replication unit in yeasts is about the same as in higher eukaryotes (Newlon *et al.*, 1974).

The study of DNA synthesis in yeasts is complicated by the fact that specific labeling of the DNA is not possible. Thymine cannot be used as a precursor, since yeasts lack thymidine kinase, the enzyme that phosphorylates thymidine to 5'-thymidine monophosphate (TMP) (Grivell and Jackson, 1968). A mutant strain of *S. cerevisiae* has been isolated which incorporates TMP into DNA (Jannsen and Lochmann, 1970; Brendel and Haynes, 1973). Other mutants that can take up larger amounts of TMP have been isolated using a selection technique (Jannsen *et al.*, 1973; Wickner, 1974) involving inhibition of growth with sulfanilamide and aminopterin. If endogenous TMP production is not blocked in this way, these mutants do not use exogenous TMP. Another difficulty in the labeling of DNA is the large excess of RNA over DNA existing in yeasts. If the nucleic acid of the cell is labeled with a precursor such as phosphorus or adenine, it is difficult to eliminate radioactive RNA contamination of the DNA. Attempts have been made to permeabilize yeast cells to nucleotides such as TMP using toluene and detergent treatment (Hereford and Hartwell, 1971). Toluene-treated cells were inactive in DNA synthesis, and DNA synthesis in Brij-58-treated cells occurred mostly in the mitochondria (Banks, 1973). For these reasons DNA synthesis in yeasts is not as well studied as in other organisms, although there is no lack of cell division mutations from the work of Hartwell (1974).

Hydroxyurea is probably the most specific inhibitor of yeast DNA synthesis (Slater, 1973, 1974). It inhibits the enzyme ribonucleotide reductase (Elford, 1968; Krakoff *et al.*, 1968; Moore, 1969), thus preventing the formation of deoxynucleotide substrates; it has no direct effect on template function or on the polymerization reaction. At a concentration of 0.075 *M* hydroxyurea greatly depresses DNA synthesis, while allowing RNA and protein synthesis to continue almost at control rates. This inhibition is reversible so that, on removal of the inhibitor, DNA synthesis resumes without a lag.

Triaziquone (Trenimon) is a potent alkylating agent which has been used to inhibit DNA synthesis in yeasts (Jaenicke *et al.*, 1970; Hartwell *et al.*, 1974). Triaziquone alkylates DNA *in vitro*, causing the formation of apurinic acid and single-strand breaks leading to loss of ability to serve as a template for *Escherichia coli* RNA polymerase *in vitro*. In addition, it generates crosslinks in the DNA which prevent unwinding of the strands, and this may be the cause of the inhibition of DNA synthesis (Klamerth and Kopun, 1971). Triaziquone was shown to inhibit DNA synthesis preferentially in yeasts at a concentration of 10 μg/ml.

Porfiromycin and *mitomycin C*, two other alkylating agents (Szybalski and Iyer, 1967), inhibit yeast growth without blocking any specific stage of the cell cycle. Presumably, these compounds also affect other processes in the cell (Jaenicke *et al.*, 1970). Mitomycin C (400 μg/ml) is known to inhibit DNA synthesis without having any effect on respiratory adaptation (Bartley and Tustanoff, 1966).

Saccharomyces cerevisiae has been found to be resistant to many folic acid antagonists (Nickerson and Webb, 1956). This resistance may not be due to impermeability, since aminopterin can inhibit the activity of intracellular dihydrofolate reductase (Scholz and Jaenicke, 1968). Endogenous synthesis of folic acid would then allow resynthesis of the depleted tetrahydrofolate. Sulfanilamides can prevent this resynthesis, and in this way TMP-permeable mutants were obtained as mentioned above. It is possible that such a method can be used to starve yeasts for thymine; this may provide a way of blocking DNA synthesis in a specific manner.

Streptonigrin is an antitumor agent which selectively inhibits DNA synthesis in bacteria and mammalian cells (Mizuno, 1965; White and White, 1968) by causing single-strand breaks in DNA (Szybalski and Iyer, 1967). It is lethal and thus nonreversible. Streptonigrin inhibits the growth of *Candida albicans* (25 μg/ml) and *S. cerevisiae* (50 μg/ml) (Bhuyan, 1967), but to our knowledge has not been used in studies of DNA synthesis in yeasts.

Phenylethanol, an inhibitor of the initiation of DNA synthesis in *E. coli* (Lark and Lark, 1966), inhibits yeast growth and respirator adaptation at quite high concentrations (2–3 mg/ml) (Bartley and Tustanoff, 1966). In yeasts phenylethanol inhibits DNA synthesis, but RNA and protein synthesis are also affected and the drug is suspected of having more than one mechanism of action, probably an additional effect on the membrane (Weber *et al.*, 1970; Burns, 1968). This drug also inhibits RNA and protein synthesis in *E. coli* and in mammalian cells (Rosenkrantz *et al.*, 1967; Plagemann, 1968). The inhibitory effects of phenylethanol are partially reversed by the growth of cells in rich medium (Weber *et al.*, 1970).

Nalidixic acid, an inhibitor of DNA synthesis in bacteria, was found to inhibit the growth of a petite negative strain of *Kluyveromyces lactis* grown on glycerol or glucose (Luha *et al.*, 1971). However, in *S. cerevisiae* it has only a transient effect on RNA and DNA synthesis; both are inhibited for several hours, but then growth resumes normally. The antibiotic is not inactivated, nor do the yeasts become petites. Therefore mitochondrial DNA synthesis is not being specifically inhibited. This transient inhibition is not understood (Weber *et al.*, 1970; Michels *et al.*, 1973). Cytosine arabinoside, another DNA synthesis inhibitor, has no effect on yeast growth (Slater, 1973; Jaenicke *et al.*, 1970).

B. RNA Synthesis Inhibitors

Saccharonyces cerevisiae contains about 5.0×10^{11} daltons of RNA (Hartwell, 1970), 85% of which is rRNA. Yeasts contain three major nuclear RNA polymerases which have been numbered according to the order of their elution from anionic ion-exchange columns, as with the higher euka-ryotes (Ponta et al., 1971, 1972; Brogt and Planta, 1972; Adman et al., 1972). Their properties resemble those of the corresponding mammalian RNA polymerases (Jacob, 1973). Yeast polymerases differ in their salt dependence and template specificity, but all three are resistant to rifampicin (Winters-berger and Wintersberger, 1970; Dezelee et al., 1970). Polymerases I and III are resistant to α-amanitin, while polymerase II is sensitive to this toxin.

Actinomycin D is the most widely used inhibitor of RNA synthesis; it acts through its preferential binding to GC regions in DNA (Wells and Larson, 1970). As mentioned previously, it can be used to inhibit the production of rRNA in HeLa cells (Perry, 1963; Darnell et al., 1971). Actinomycin D can be used at relatively high levels (80 μg/ml) to block RNA synthesis in *Saccharomyces carlsbergensis* spheroplasts (van Dam et al., 1965), but it has been reported to inhibit the growth of *S. cerevisiae* cells at about 20 μg/ml, and *S. carlsbergensis* cells at about 25 μg/ml. Inhibition of glutamate dehydro-genase on derepression required higher concentrations (35 μg/ml) and prolonged incubation (Westphal and Holzer, 1964; Holzer and Hierholzer, 1963). Actinomycin D was shown to inhibit the induction of α-glucosidase and allantoinase (van Wijk, 1968; van de Poll, 1970). The kinetics of inhibi-tion were consistent with the notion that actinomycin D enters the cells slowly (van Dam et al., 1965). The degree of inhibition varied with the batch of spheroplasts used; pulse-labeled RNA was inhibited by 50–60%. In yeasts the synthesis of rapidly labeled AU-rich polysomal RNA was inhibited pre-ferentially compared to other RNA species; this suggests that actinomycin D may preferentially inhibit mRNA synthesis in yeasts (Hartlief and Konigs-berger, 1968).

Daunomycin (daunorubicin) is a member of the anthracycline group of antibiotics, which also includes nogalamycin and ruticulomycin. It inhibits elongation of RNA chains, probably by intercalating into the DNA template and blocking RNA synthesis similarly to actinomycin D (Sentenac et al., 1968). Daunomycin inhibits the growth of *S. carlsbergensis* at 6 μg/ml, but *C. albicans* is resistant to over 100 μg/ml (Sanfilippo and Mazzolini, 1964). It has also been used to inhibit total RNA synthesis in *S. cerevisiae*, but high concentrations (313 μg/ml) were needed (Tonnesen and Friesen, 1973). This is one example of the wide differences in antibiotic sensitivity among various strains of yeasts. If a judicious choice of strain is made, low concen-trations of inhibitor may be used to give results which may be less question-able in terms of secondary inhibitory effects. Daunomycin has been shown to have a preferential effect on mitochondrial RNA synthesis in *S. cerevisiae* (Rhodes and Wilkie, 1971).

Ethidium bromide, a phenanthridine drug, and *acriflavine*, an acridine, have the same specificity and inhibit mitochondrial RNA synthesis at concentrations lower than those required to inhibit nuclear RNA synthesis (Fukuhara and Kujawa, 1970). Ethidium bromide at 19 μg/ml has been used to inhibit total RNA synthesis in *S. cerevisiae* (Tonnesen and Friesen, 1973). As further evidence of their mitochondrial specificity, these drugs are known to cause a reversible respiratory deficiency in petite negative mutants (Roodyn and Wilkie, 1968). An interesting explanation for this specificity has been proposed (Gale *et al.*, 1972); these drugs bind readily to super-coiled DNA molecules by intercalation *in vitro* (Bauer and Vinograd, 1970) and, since yeast mitochondrial DNA can be isolated as covalently closed circles (Hollenberg *et al.*, 1970; Avers *et al.*, 1968), this DNA may be super-coiled *in vivo* and thus allow preferential binding of the drug and inhibition of nucleic acid synthesis. This specificity has proved valuable in studying the interrelationships of nuclear RNA synthesis and mitochondrial function in the absence of mitochondrial RNA synthesis (Ashwell and Work, 1970).

Lomofungin, a phenazine antibiotic, is a potent inhibitor of yeast growth (Gottlieb and Nicolas, 1969). It has been shown to inhibit RNA synthesis at low concentrations in spheroplasts (Cannon *et al.*, 1973; Kuo *et al.*, 1973), and stops the synthesis of all high-molecular-weight RNA; however, 4–5 S RNA is particularly resistant to inhibition (Fraser *et al.*, 1973). It also blocks the maturation of precursor rRNA. Lomofungin appears to interfere with the elongation of RNA chains during transcription. (Cano *et al.*, 1973). This drug also inhibits DNA synthesis. The induction of invertase is prevented by lomofungin, indicating that it blocks mRNA synthesis; it does not inhibit protein synthesis *in vitro*. Lomofungin has been shown to inhibit RNA synthesis because of its ability to bind divalent cations, possibly the zinc in RNA polymerase (Fraser and Creanor, 1974; Pavletich *et al.*, 1974). *8-Hydroxyquinoline*, which is also a chelating agent, has also been found to specifically inhibit nucleic acid synthesis in yeast in a similar manner to lomofungin. Because of hydroxyquinoline's ease of availability and solubility, this may become the drug of choice in inhibiting RNA synthesis in yeast (Fraser and Creanor, 1974).

Thiolutin is a member of the pyrrothin class of antibiotics, and is a potent but reversible inhibitor of yeast growth. It inhibits RNA synthesis and rRNA precursor processing in whole cells and spheroplasts, with subsequent inhibition of protein synthesis (Jimenez *et al.*, 1973; Tipper, 1973). Thiolutin has little or no effect on protein synthesis *in vitro* and in spheroplasts causes decay of polysomes, suggesting that the synthesis of mRNA is inhibited. Studies with RNA polymerases *in vitro* showed that all three enzymes were inhibited; the addition of excess template DNA protected the enzymes from inhibition, which was interpreted to show that thiolutin can only affect free RNA polymerase and thus may be a specific inhibitor of the initiation of RNA synthesis *in vivo*.

Cordycepin (3'-deoxyadenosine) is an adenosine analog which, when incorporated into a polynucleotide chain, prevents elongation of the chain since it lacks a 3'-hydroxy group (Kaczka *et al.*, 1964); it has been shown to have differential effects on RNA metabolism in eukaryotes. Cordycepin prevents rRNA synthesis and mRNA synthesis, but not heterogeneous nuclear RNA synthesis in HeLa cells (Penman *et al.*, 1970). It blocks the addition of poly A to heterogeneous RNA, and in this way prevents processing to mRNA. Unfortunately, yeasts are normally impermeable to this drug. However, an adenosine-requiring mutant strain has recently been shown to be sensitive, and studies of the effect of cordycepin on yeast RNA metabolism will be possible (Anderson and Roth, 1974).

The effects of a large number of base analogs on nucleic acid metabolism in yeasts have been studied. *5-Fluorouracil* inhibits the synthesis of RNA, but not DNA or protein synthesis (Mayo *et al.*, 1968; Kempner 1961; de Kloet and Strijkert, 1966); 80% inhibition of RNA synthesis occurs at a concentration of 20 μg/ml. In the absence of exogenous uracil, 5-fluorouracil is incorporated into RNA, replacing a large fraction of the uracil. In the presence of 5-fluorouracil, a high-molecular-weight AU-rich RNA is formed, which probably consists of precursor rRNA which cannot be processed (Mayo and de Kloet, 1972). Induction of α-glucosidase and β-galactosidase was not affected by 5-fluorouracil, indicating that functional mRNA is made (de Kloet, 1968). Consistent with this is the finding that 5-fluorouracil inhibits ribosome formation much more strongly than it inhibits overall protein synthesis (Kempner and Miller, 1963; de Kloet, 1968).

6-Azauracil also inhibits RNA synthesis in yeast but, unlike 5-fluorouracil, does not cause the accumulation of RNA of greater molecular weight than mature rRNA (de Kloet, 1968). Reports on the effects of 8-azaguanine are inconclusive, with one vote for inhibition of yeast growth (Pickering and Woods, 1972), one vote for lack of inhibition (Harris and MacWilliam, 1965), and one vote for irreproducible results (Miller and Kempner, 1963). Another base analog, 6-methylpurine, inhibits the conversion of guanine to adenine in *Candida utilis*; it inhibits the synthesis of all classes of RNA, even the synthesis of tRNA (Miller and Kempner, 1963). *Saccharomyces cerevisiae* is sensitive to the adenine analog 4-aminopyrazolo-3,4d-pyrimidine (Pickering and Woods, 1972), and a strain of *S. cerevisiae* has been isolated that is sensitive to 2,6-diaminopurine (Lomax and Woods, 1969). However, neither of these inhibitors has been investigated for use in the study of nucleic acid metabolism. Several other base analogs have been tested and found ineffective (Lomax and Woods, 1969; Pickering and Woods, 1972; de Kloet and Strijkert, 1966; Kempner and Miller, 1963).

Amphotericin B is a polyene antibiotic which affects cell membranes containing sterols; the sensitive cells leak ions and macromolecules, leading to cell death (Lampen, 1969). However, at sublethal concentrations it acts synergistically with other drugs, presumably by altering the permeability of the membrane to allow compounds to enter the cell. This method has been used to permeabilize yeast to several drugs to which they are usually resistant (Medoff *et al.*, 1972). Two inhibitors of RNA synthesis that have been studied in such permeabilized cells are *rifampicin* and *5-fluorocytosine.* Rifampicin (rifampin) is a rifamycin derivative which inhibits initiation of RNA synthesis in bacteria; it has no effect on the activity of the RNA polymerases isolated from the nuclei of eukaryotes, but may inhibit the RNA polymerases of mitochondria. When yeast cells permeabilized with amphotericin B are treated with rifampicin, either the synthesis of all classes of RNA is inhibited (Battaner and Kumar, 1974), or the synthesis of rRNA is specifically inhibited (Medoff *et al.*, 1972). Inhibition of RNA synthesis by rifampicin has also been observed without amphotericin-B addition, using an osmotically sensitive mutant permeable to refampicin (Venkov *et al.*, 1974). The *in vivo* results are not consistent with the results of studies with isolated RNA polymerases *in vitro*, where rifampicin had no effect on nuclear enzymes. However, it is possible that rifampicin may affect a factor(s) present in whole yeast but not in purified polymerase preparations.

Three other RNA synthesis inhibitors are of potential interest. *Netropsin* is a basic polypeptide antibiotic which is a potent inhibitor of yeast growth (P. Grant, personal communication). It also inhibits yeast RNA polymerase *in vitro* (D. Tipper, personal communication). It inhibits by binding to the minor groove of DNA, not by intercalation, as actinomycin D does. Guanine prevents its binding, giving netropsin a specificity for AT-rich DNA or a specificity complementary to actinomycin D (Wartell *et al.*, 1974). *Chetomin* is one of the epipolythiadioxopiperazines. It inhibits yeast RNA polymerases, but yeasts are impermeable to it. Several strains of yeasts sensitive to this drug have been selected (D. Tipper and B. Littlewood, personal communication). *Gliotoxin*, another member of this class of antibiotics, has been shown to inhibit nucleic acid synthesis in *S. carlsbergensis* (Kerridge, 1958). It does not affect *in vitro* protein synthesis in extracts from many yeasts (Jimenez *et al.*, 1972).

C.   Protein Synthesis Inhibitors

Yeast ribosomes are typical of eukaryotic systems. Each 80 S ribosome is made up of a 60 S subunit and a 40 S subunit with a combined molecular weight of about 3.6×10^6 daltons; about 53% of the ribosome is made up of RNA (Mazelis and Petermann, 1973). The rRNA is present as four distinct species, 25, 18, 5.8, and 5 S. The protein component is a collection of about

80 proteins (Warner, 1971). Other translation components, such as the initiation and termination factors in yeasts, have not been well studied, but much work has been done on the elongation factors, which are similar to those of other eukaryotes (Richter, 1971; Heredia *et al.*, 1971).

Cycloheximide is the most commonly used inhibitor of protein synthesis in yeasts and other eukaryotes; it has no effect on prokaryotic protein-synthesizing systems. It is one of the glutarimide antibiotics, along with streptovitacin A and streptimidone. Yeast species have varying sensitivities to cycloheximide, *Saccharomyces pastorianus* being sensitive to low concentrations (less than 1 μg/ml) and *Kluyveromyces fragilis* being resistant to 1 mg/ml (Westcott and Sisler, 1964); in general, strains of *Saccharomyces* are sensitive to cycloheximide, but strains now known as *Kluyveromyces* (*K. lactis*, *K. fragilis*, etc.) are resistant (Lodder, 1971; van der Walt, 1965). Resistance in these strains is due to the presence of resistant ribosomes (Siegel and Sisler, 1964). In addition, a single recessive mutation in a strain of *S. cerevisiae* leading to resistance to cycloheximide has been shown to be due to an alteration in the 60 S subunit of the ribosome (Jimenez *et al.*, 1972). Low-level-resistance mutations at other loci which lead to cycloheximide resistance have been isolated, but the nature of the alterations is not known (Wilkie and Lee, 1965).

Cycloheximide affects a step in the elongation of peptide chains, possibly the translocation of nascent peptidyl-tRNA from the A site to the P site, by interfering with the action of elongation factor II (Baliga *et al.*, 1969; Baliga and Munro, 1971). In addition, cycloheximide may interfere with the process of protein synthesis initiation (Lin *et al.*, 1966; Munro *et al.*, 1968). It affects other processes *in vivo*, but this is probably secondary to its effects on protein synthesis. Such effects are considered in Section III,A. There have been several reports that cycloheximide does (Timberlake *et al.*, 1972a,b; Horgen and Griffin, 1971) and does not inhibit eukaryotic RNA polymerases *in vitro*. However, it does not inhibit any of the yeast RNA polymerases *in vitro* or RNA synthetic activity in isolated nuclei from yeasts (Ponta *et al.*, 1972; Gross and Pogo, 1974). The fact that a single recessive mutation affecting ribosome structure leads to resistance is a strong argument that the primary effect of the drug is on translation.

The *phenanthrene alkaloids* cryptopleurine, tylophorine, and tylocrebrine inhibit elongation of protein synthesis in yeasts and other eukaryotes (Donaldson *et al.*, 1968; Huang and Grollman, 1972). Yeast mutants that are resistant to this class of compounds have an alteration in the small subunit of the ribosome (Skogerson *et al.*, 1973; Grant *et al.*, 1974). Cytoplasmic and mitochondrial protein syntheses are differentially inhibited (Haslam *et al.*, 1968), cytoplasmic ribosomes being more sensitive to inhibition, particularly at low drug concentrations.

The *12,13-epoxytrichothecenes* are a group of fungal toxins of which trichodermin, trichothecin, and verrucarin A are typical examples. They are potent inhibitors of yeast growth and protein synthesis *in vivo* and *in vitro* (Stafford and McLaughlin, 1973; Cundliffe *et al.*, 1974). This class of toxins blocks the peptidyl transferase reaction *in vitro* (Carrasco *et al.*, 1973); however, they have somewhat different effects on protein synthesis *in vivo*. Verrucarin A, diacetoxyscirpenol, and T-2 toxin cause runoff of polysomes, indicating an effect on the initiation of protein synthesis, while trichodermin and trichothecin stabilize polysomes, suggesting a block in elongation or termination (Cundliffe *et al.*, 1974). Trichodermin appears to inhibit termination of protein synthesis preferentially, since polysomes can reform in the presence of trichodermin, but not in the presence of cycloheximide, after the release of inhibition of initiation of protein synthesis in a temperature-sensitive mutant of *S. cerevisiae* (Stafford and McLaughlin, 1973; Wei *et al.*, 1974). Mutants of *S. cerevisiae* resistant to trichodermin have an altered 60 S ribosomal subunit; this provides further evidence that trichodermin specifically inhibits protein synthesis (Schindler *et al.*, 1974).

Anisomycin inhibits elongation of protein synthesis in all eukaryotic systems that have been tested; it has been used to inhibit protein synthesis in yeasts (Foury and Goffeau, 1973; Slater, 1974). Anisomycin is an inhibitor of the peptidyl transferase reaction *in vitro* and stabilizes polysomes *in vivo* (Vazquez *et al.*, 1969; Grollman, 1967).

2-(4-Methyl-2,6-dinitroanilino)-N-methylpropionamide (MDMP) is a herbicide which blocks initiation of protein synthesis in eukaryotes by preventing the binding of the 60 S subunit to the initiation complex containing the 40 S subunit, mRNA, and initiator tRNA (Weeks and Baxter, 1972). MDMP inhibits yeast growth (Baxter *et al.*, 1973) and causes the disaggregation of yeast polysomes; this disaggregation can be prevented by cycloheximide, which is consistent with the notion that the drug blocks initiation of protein synthesis (D. Schindler and J. Davies, unpublished observation).

Pederine, the toxic principle from blister beetles, is a potent and specific inhibitor of eukaryotes and inhibits the growth of yeasts at low concentrations. Protein synthesis *in vitro* with endogenous mRNA and polyuridylic acid–dependent systems was inhibited (Tiboni *et al.*, 1968). It does not affect aminoacyl-tRNA synthetase activity or phenylalanine–tRNA binding in extracts from yeast. In a heterologous protein-synthesizing system using ribosomes and supernatant fractions from yeasts and insects, resistance was shown to be a property of the ribosomes.

All the inhibitors described above specifically inhibit eukaryotic protein synthesis without having an effect on prokaryotes. As has been shown for cycloheximide (Lamb *et al.*, 1968), such compounds should inhibit cytoplasmic protein synthesis without affecting mitochondrial protein synthesis, and thus be useful in studies of the autonomy and biogenesis of mitochondria (Ashwell and Work, 1970).

Puromycin is an aminonucleoside-containing antibiotic resembling the 3'-terminus of a tyrosinyl-tRNA that accepts and releases nascent polypeptide chains from the ribosome; it inhibits both eukaryotic and prokaryotic protein synthesis. Limited uptake of puromycin has been detected in yeasts, but high levels are needed to inhibit protein synthesis, since the drug is metabolized to inactive derivatives (Melcher, 1971). Nonetheless, puromycin can be used to inhibit protein synthesis in yeast spheroplasts (van Wijk, 1968; van de Poll, 1970; Westphal *et al.*, 1966).

The *4'-aminohexosylcytosinyl* antibiotics, which include blasticidin S, gougerotin, and amicetin, inhibit protein synthesis by blocking a step in the peptidyl transferase reaction (Cerna *et al.*, 1973). Although *S. cerevisiae* has been reported to be resistant to gougerotin and blasticidin S (Kanzaki *et al.*, 1962; Takeuchi *et al.*, 1958), protein synthesis on ribosomes from *S. cerevisiae* is sensitive to gougerotin and blasticidin S *in vitro* (Battaner and Vazquez, 1971). The latter drug inhibits protein synthesis in a strain of *S. cerevisiae* (*S. sake*) (Hashimoto *et al.*, 1963), and it is our general experience that most strains of yeasts are sensitive to blasticidin S. *Saccharomyces cerevisiae* is resistant to amicetin both *in vitro* and *in vivo* (Battaner and Vazquez, 1971).

Yeasts are resistant to other inhibitors of protein synthesis such as chartreusin, pactamycin, actinobolin, borrelidin, hikizimycin, fusidic acid (Jimenez *et al.*, 1972), and tenuazonic acid (Carrasco and Vazquez, 1973). Sparsomycin inhibits the peptidyl transferase reaction *in vitro*, but only partially inhibits the growth of *C. albicans* at 100 μg/ml (Owen *et al.*, 1962). Streptomycin and neomycin are antibacterial antibiotics which usually have no effect on cytoplasmic ribosomes of eukaryotes such as yeasts, but several strains of yeasts have been isolated that are sensitive to these drugs (Bayliss and Ingraham, 1974; B. Littlewood, personal communication).

III. Effects of Inhibitors on Yeast Metabolism

A. Effects of Protein Synthesis Inhibitors on Yeast Metabolism

Yeast RNA and DNA synthesis stops when protein synthesis is inhibited in a variety of ways. Inhibition of yeast protein synthesis with cycloheximide (Roth and Dampier, 1972; Gross and Pogo, 1974; Kerridge, 1958; de Kloet, 1965), trichodermin, T-2 toxin, verrucarin A (Cundliffe *et al.*, 1974), or amino acid starvation (Roth and Dampier, 1972; de Kloet, 1966;

Gross and Pogo, 1974; Ycas and Brawerman, 1957) causes a reduction in the rate of RNA synthesis. In contrast, cycloheximide stimulates RNA synthesis under conditions of amino acid starvation, release from glucose repression, or sporulation (de Kloet, 1966; Sogin et al., 1972; Foury and Goffeau, 1973; Gross and Pogo, 1974). This indicates that two different mechanisms might be involved in the cessation of RNA synthesis subsequent to the inhibition of protein synthesis: inhibitors on the one hand, or amino acid starvation on the other (de Kloet, 1965; Gross and Pogo, 1974). In addition, anisomycin and trichodermin stimulate RNA synthesis in a superrepressed mutant of Schizosaccharomyces pombe, thus the phenomenon is not specific for any one inhibitor of protein synthesis (Foury and Goffeau, 1973). In other experiments amino acid starvation was found to inhibit the synthesis of poly-A-containing RNA (mRNA), as well as rRNA (Gross and Pogo, 1974).

A variety of protein synthesis inhibitors, including cycloheximide, anisomycin, and cryptopleurine (de Kloet, 1965; Taber and Vincent, 1969; Udem and Warner, 1972; Mayo et al., 1968; Jimenez et al., 1973), have been shown to inhibit the processing of the large rRNA precursor species in both budding and fission yeasts; this effect also occurs during sporulation (Sogin et al., 1972). On removal of the inhibitor, the precursor is cleaved into mature ribosomal products in the normal manner. This coordinate effect may indicate that rRNA synthesis is coupled to ribosomal protein synthesis; e.g., binding of one or more ribosomal proteins may be required for correct processing of precursor RNA. This notion is consistent with the finding that, in yeasts, synthesis of rRNA and ribosomal protein is continuous (Shulman et al., 1973, and references therein), whereas many enzymes are made at specific times in the cell cycle (e.g., Tauro et al., 1968).

As mentioned earlier, cycloheximide does not inhibit the activity of isolated RNA polymerases from yeasts, nor RNA synthesis in isolated nuclei; however, nuclei from cycloheximide-treated or amino acid–starved yeasts have been shown to have reduced RNA synthetic activity when tested in vitro. The amount and relative proportion of the different RNA polymerases were unchanged in these nuclei, but both α-amanitin-resistant and α-amanitin-sensitive activities were lower in nuclei isolated from cycloheximide or amino acid–starved cells. The nucleolar enzyme is believed to be responsible for the synthesis of rRNA (Jacob, 1973), although this has not been confirmed in yeasts by the in vitro synthesis of rRNA by isolated RNA polymerase I (Hollenberg, 1973). Thus cycloheximide has a marked effect on rRNA synthesis in vivo, but this effect is not observed in vitro. Stimulation of RNA synthesis in vivo by cycloheximide in amino acid–starved cells cannot be demonstrated in isolated nuclei from treated yeast cells; but it is possible that the stimulation may simply represent an increase in the intracellular pool of RNA precursors. To explain these findings, Foury and Goffeau have suggested that cycloheximide stimulates the production of a positive regulator of RNA synthesis during derepression of a superrepressed strain of yeast (Foury and Goffeau, 1973). However, Gross and Pogo proposed that two controls exist, one a negative control similar to the stringent control in bacteria and the other a positive control, in which a protein with a high turnover is necessary for the binding of polymerase to DNA (Gross and Pogo, 1974).

Cycloheximide prevents the inactivation of several enzymes following repression by glucose, and this may be relevant to the finding that cycloheximide stimulates RNA synthesis in derepressed cells. When glucose is added to cells growing under derepressed conditions, several enzymatic activities decrease faster than can be explained by dilution on cell multiplication; in fact, the enzymes are known to be inactivated by proteolysis in some cases (Schött and Holzer, 1974). Addition of cycloheximide prevented the inactivation of malate dehydrogenase (Ferguson et al., 1967), ornithine transcarbamylase (Béchet and Wiame, 1965), and the maltose uptake system (Görts, 1969). Preincubation of the cells with cycloheximide was necessary to prevent the inactivation of fructose-1,6-diphosphatase (Gancedo, 1971). In addition, recovery of activity of fructose-1,6-diphosphatase and the maltose uptake system in S. cerevisiae was prevented by cycloheximide, and the reappearance of malate dehydrogenase was prevented by amino acid starvation, suggesting that de novo enzyme synthesis was necessary. Amino acid starvation, or inhibition of protein synthesis with amino acid analogs, did not prevent enzyme inactivation (Ferguson et al., 1967; Béchet and Wiame, 1965; Duntze et al., 1968). Thus cycloheximide seems to interfere with protein degradation, but it does not affect this process in growing mammalian cells (Feldman and Yagil, 1968). However, protein degradation in untransformed mouse fibroblasts induced by serum deprivation is inhibited by this drug (Hershko et al., 1971). This effect results from a reduction in the cyclic-AMP levels in starved fibroblasts.

Tomkins and his colleagues have described the response of animal cells to the withdrawal of growth-promoting substances as "negative pleiotypic responses" (Hershko et al., 1971). Protein and RNA synthesis is inhibited, and protein degradation increases. The response is blocked by cycloheximide which prevents protein degradation and stimulates the uptake of most nucleic acid precursors; cyclic AMP has been implicated as the mediator of this response (Kram et al., 1973). In yeasts amino acid starvation, derepression, and sporulation are in many ways similar to this; levels of cyclic AMP have been shown to increase during derepression (van Wijk and Konijn, 1971; Sy and Richter, 1972). Inhibition of all classes of RNA synthesis presumably occurs in yeasts under these conditions and, by analogy with the negative pleiotypic response in animal cells, cycloheximide relieves part of this inhibition by reversing the pleiotypic response. However, since ribosomal proteins cannot be made and are probably necessary for rRNA precursor processing, maturation of rRNA is presumably blocked. Accumulation of ribosomal precursor RNA may prevent the synthesis of rRNA by an unknown type of feedback mechanism (Willems et al., 1969). In such a model cyclic AMP would be expected to inhibit and not stimulate RNA synthesis, as has been reported (Foury and Goffeau, 1973). However, phosphorylated compounds are not normally transported into yeasts, and anomalous results may result from the effects of cyclic AMP on membrane processes; this is believed to be the case in animal cells (Kram et al., 1973).

We must also consider the effect of cycloheximide on the degradation of RNA. In a mutant that is temperature-sensitive for RNA synthesis, polysomes decay with a half-life of 20 minutes following cessation of RNA synthesis at the nonpermissive temperature. This time is consistent with the rate of appearance of newly synthesized mRNA in polysomes, and probably indicates the rate of decay of the mRNA rather than an effect on the initiation of protein synthesis as seen in mammalian cells (Goldstein and Penman, 1973). Cycloheximide does not prevent the breakdown of polysomes when RNA synthesis is inhibited, indicating that it does not lengthen mRNA half-life (Hartwell et al., 1970). This is in contrast to the results of a study on the decay of a class of RNA in S. cerevisiae which has the characteristics of mRNA (half-life of 15–20 minutes); cycloheximide and glucose starvation inhibited the degradation of this RNA (Johnson, 1970). When initiation of protein synthesis was slowed by a temperature-sensitive defect and elongation of protein chains inhibited by cycloheximide, the polysomes remained stable; this indicates that cycloheximide does not affect the synthesis of new mRNA. As further evidence that cycloheximide does not inhibit mRNA synthesis, newly synthesized RNA (which is likely to be mRNA, since it does not cosediment with the ribosomal subunits) enters the polysomes in the presence of cycloheximide. As suggested by Hartwell et al., cycloheximide may be a useful (if indirect) inhibitor of rRNA synthesis in studies of RNA metabolism (Hartwell et al., 1970). It would be interesting to see if the mRNA for certain enzymes can be induced in the presence of cycloheximide and expressed when the drug is removed.

Inhibition of protein synthesis with cycloheximide or by amino acid starvation inhibits nuclear DNA replication, without having an effect on mitochondrial DNA synthesis (Grossman et al., 1969). By using synchronized cultures it was shown that cycloheximide blocks the initiation of DNA synthesis, but not its elongation (Hereford and Hartwell, 1973; Williamson, 1973); this result is independent of the method of synchronization used. In asynchronous cultures the small increase in DNA content that occurs following protein synthesis inhibition is consistent with the notion that cells in the process of DNA synthesis complete one round of synthesis but cannot reinitiate. The protein(s) necessary for initiation was stable for 3.5 hours, and concurrent protein synthesis was not necessary for DNA replication (Slater, 1974). Synthesis of the proteins necessary for a complete round of replication is completed shortly before the round begins (Slater, 1974; Hereford and Hartwell, 1974). Nuclear division, cell division, and the start of new budding cycles are blocked when DNA synthesis is inhibited, but bud emergence and nuclear migration are unaffected (Hartwell et al., 1974). Protein synthesis is required for the recovery of budding and cell division, although the synthesis of DNA itself does not require protein synthesis (Slater, 1973).

Other effects of cycloheximide on yeast physiology are probably related to the inhibition of protein synthesis. This antibiotic blocks mannan synthesis by preventing the formation of the polypeptide acceptor of GDP-mannose, and the latter accumulates in the cell (Sentandreu and Lampen, 1970; Sentandreu et al., 1972). The mechanism of this inhibition is not known; transfer of mannose to the peptide was not inhibited by cycloheximide in vitro, nor did the drug inhibit mannan synthesis in K. fragilis, a species that has ribosomes resistant to cycloheximide. Incorporation of threonine into the cell wall was completely inhibited (Sentandreu and Northcote, 1969). Cycloheximide does not prevent the uptake of amino acids into the soluble fraction of the cell (Kotyk et al., 1971, Greasham and Moat, 1973), except indirectly by feedback inhibition of transport (Grenson et al., 1968). Potassium and phosphate ion uptake were reported to be inhibited substantially by high concentrations of cycloheximide, or by prolonged incubation in its presence (Reilly et al., 1970), however, another study found no effect of cycloheximide on the potassium ion content or on glucose oxidation, although the rate of endogenous metabolism increased (Kotyk et al., 1971).

B.   Effects of RNA Synthesis Inhibitors on Yeast Metabolism

Inhibition of yeast RNA synthesis with thiolutin (Jimenez et al., 1973), lomofungin (Cannon et al., 1973), daunomycin, or ethidium bromide (Tonnesen and Friesen, 1973) leads to the disaggregation of polyribosomes with a half-life of about 20 minutes. In mammalian cells disaggregation of polyribosomes subsequent to the inhibition of RNA synthesis is the result of a slowing of the initiation of protein synthesis (Goldstein and Penman, 1973). However, as mentioned previously, this does not seem to be the case with yeasts. As in higher eukaryotes, the mRNA in yeasts contains poly A;

the poly-A sequence in yeasts is about 50 residues, but is four times this in higher organisms (McLaughlin *et al.*, 1973; Reed and Wintersberger, 1973). Although bulk mRNA in yeasts seems to have a half-life of about 20 minutes, there is evidence that the mRNAs for certain induced enzymes are more stable. The synthesis of glutamate dehydrogenase continues for some time after the synthesis of RNA is inhibited (Westphal *et al.*, 1966); this is also the case for α-glucosidase (van Wijk, 1968) following the inhibition of RNA synthesis by actinomycin D. The fact that cycloheximide blocks this increase indicates that protein synthesis is necessary for enzyme activity, and that posttranslational control is not involved. Lomofungin, which has been shown to inhibit the synthesis of all classes of higher-molecular-weight RNA, allows continued synthesis of invertase and alkaline phosphatase for 40 minutes after complete inhibition of RNA synthesis (Cano *et al.*, 1973). This tends to rule out the possibility that mRNA is still made; since lomofungin has been shown to cause the disaggregation of polysomes with a half-life of 20 minutes, it probably does not lengthen the half-life of the total mRNA.

If mRNAs for induced enzymes are in fact more stable than those for constitutive enzymes, substantial purification of certain mRNA species may be achieved by inhibiting RNA synthesis and isolating the poly-A-containing RNA after the bulk of the RNA has decayed. It may be that RNA synthesis is necessary for the deinduction of the enzymes. Actinomycin D slightly increases derepressed α-glucosidase synthesis in yeasts (van Wijk, 1968), which resembles the superinduction phenomenon seen in other systems (Tomkins *et al.*, 1972).

During the induction of α-glucosidase (van Wijk, 1968; Cano *et al.*, 1973), inhibitors of RNA synthesis are effective only during the early phase of induction, presumably during the time when the mRNA is synthesized. Actinomycin D does not inhibit the derepression of α-glucosidase in the presence of low concentrations of glucose, or of allantoinase in low concentrations of nitrogen, although cycloheximide blocks the increase in the activities of both these enzymes under these conditions (van Wijk, 1968; van de Poll *et al.*, 1968). Glucose causes repression of α-glucosidase or allantoinase activity even in the presence of actinomycin D; this effect was shown not to be the result of mRNA degradation (van Wijk *et al.*, 1969). This suggests that new mRNA synthesis is not necessary for all cases of repression or derepression in yeasts, and that translational controls exist. This type of control may represent a means of regulating the synthesis of proteins on long-lived messages. However, lomofungin prevents the increase in activity of α-glucosidase when synthesis of this enzyme is derepressed in yeast (Cano *et al.*, 1973).

In mammalian cells, α-amanitin does not inhibit rRNA synthesis, although it inhibits the isolated, purified nucleolar enzyme *in vitro* (Jacob, 1973; Jacob *et al.*, 1970; Tata *et al.*, 1972; Hadjiolov *et al.*, 1974). This may indicate that the synthesis of extranucleolar RNA is necessary for rRNA synthesis and/or processing *in vivo*. Alternatively, a factor(s) may be purified away from the polymerase *in vitro*, which protects the enzyme from inhibition by α-amanitin. This may explain why thiolutin and lomofungin, which appear to be specific inhibitors of RNA synthesis in yeasts, also block precursor rRNA processing (Fraser *et al.*, 1973; Jimenez *et al.*, 1973). For example, ribosomal proteins may be made from a class of mRNAs with a rapid turnover (Petersen and McLaughlin, 1974); inhibition of RNA synthesis would thus prevent the synthesis of certain ribosomal proteins necessary for the maturation of precursor RNA. As precursor ribosomal particles accumulate or ribosomal proteins are used up, rRNA synthesis would be inhibited, possibly at the initiation step.

IV. Conclusion

Yeasts, because of easy genetic and biochemical manipulation, are ideal, simple models for the study of eukaryotic macromolecules, and their synthesis and regulation. As described above, many inhibitors are available as tools for these studies which when combined with the use of well-characterized mutants will enhance our knowledge of the workings of eukaryotic cells.

ACKNOWLEDGMENTS

This work was supported by the College of Agricultural and Life Sciences, University of Wisconsin-Madison, The Upjohn Company, and the National Institutes of Health.

REFERENCES

Adman, R., Schultz, L. D., and Hall, B. D. (1972). *Proc. Nat. Acad. Sci. U.S.* **69**, 1702–1706.
Anderson, J. M., and Roth, R. M. (1974). *Biochim. Biophys. Acta* **335**, 285–289.
Ashwell, M., and Work, T. S. (1970). *Annu. Rev. Biochem.* **39**, 251–290.
Avers, C. J., Billheimer, F. E., Hoffmann, H. P., and Pauli, R. M. (1968). *Proc. Nat. Acad. Sci. U.S.* **61**, 90–97.
Baliga, B. S., and Munro, H. N. (1971). *Nature (London), New Biol.* **233**, 257–258.
Baliga, B. S., Pronczuk, A. W., and Munro, H. N. (1969). *J. Biol. Chem.* **244**, 4480–4489.
Banks, G. R. (1973). *Nature (London), New Biol.* **245**, 196–199.

Bartley, W., and Tustanoff, E. R. (1966). *Biochem. J.* **99**, 599–603.

Battaner, E., and Kumar, B. V. (1974). *Antimicrob. Ag. Chemother.* **5**, 371–376.

Battaner, E., and Vazquez, D. (1971). *Biochem. Biophys. Acta* **254**, 316–330.

Bauer, W., and Vinograd, J. (1970). *J. Mol. Biol.* **47**, 419–435.

Baxter, R., Knell, V. C., Somerville, H. J., Swain, H. M., and Weeks, D. P. (1973). *Nature (London), New Biol.* **243**, 139–142.

Bayliss, F. T., and Ingraham, J. L. (1974). *J. Bacteriol.* **118**, 319–328.

Béchet, J., and Wiame, J. M. (1965). *Biochem. Biophys. Res. Commun.* **21**, 226–234.

Bhuyan, B. K. (1967). *In* "Antibiotics" (D. Gottlieb and P. D. Shaw, eds.), Vol. 1, p. 175. Springer-Verlag, Berlin and New York.

Brendel, M., and Haynes, R. H. (1973). *Mol. Gen. Genet.* **126**, 337–348.

Brogt, T. M., and Planta, R. J. (1972). *FEBS (Fed. Eur. Biochem. Soc.) Lett.* **20**, 47–52.

Burns, V. W. (with technical assistance of D. Wong). (1968). *J. Cell. Physiol.* **72**, 97–108.

Cannon, M., Davies, J. E., and Jimenez, A. (1973). *FEBS (Fed. Eur. Biochem. Soc.) Lett.* **32**, 277–280.

Cano, F. R., Kuo, S.-C., and Lampen, J. O. (1973). *Antimicrob. Ag. Chemother.* **3**, 723–728.

Carrasco, L., and Vazquez, D. (1973). *Biochim. Biophys. Acta* **319**, 209–215.

Carrasco, L., Barbacid, M., and Vazquez, D. (1973). *Biochim. Biophys. Acta* **312**, 368–376.

Cerna, J., Rychlik, I., and Lichtenthaler, F. W. (1973). *FEBS (Fed. Eur. Biochem. Soc.) Lett.* **30**, 147–150.

Cundliffe, E., Cannon, M., and Davies, J. E. (1974). *Proc. Nat. Acad. Sci. U.S.* **71**, 30–34.

Darnell, J. E., Phillipson, L., Wall, R., and Adesnik, M. (1971). *Science* **174**, 507–510.

de Kloet, S. R. (1965). *Biochem. Biophys. Res. Commun.* **19**, 582–586.

de Kloet, S. R. (1966). *Biochem. J.* **99**, 566–581.

de Kloet, S. R. (1968). *Biochem. J.* **106**, 167–178.

de Kloet, S. R., and Strijkert, P. J. (1966). *Biochem. Biophys. Res. Commun.* **23**, 49–55.

Dezelee, S., Sentenac, A., and Fromageot, P. (1970). *FEBS (Fed. Eur. Biochem. Soc.) Lett.* **7**, 220–222.

Donaldson, G. R., Atkinson, M. R., and Murray, A. E. (1968). *Biochem. Biophys. Res. Commun.* **31**, 104–109.

Duntze, W., Neumann, D., and Holzer, H. (1968). *Eur. J. Biochem.* **3**, 326–331.

Elford, H. L. (1968). *Biochem. Biophys. Res. Commun.* **33**, 129–135.

Feldman, M., and Yagil, G. (1968). *Biochem. Biophys. Res. Commun.* **37**, 198–203.

Ferguson, J. J., Boll, M., and Holzer, H. (1967). *Eur. J. Biochem.* **1**, 21–25.

Foury, F., and Goffeau, A. (1973). *Nature (London), New Biol.* **245**, 44–47.

Fraser, R. S. S., Creanor, J., and Mitchison, J. M. (1973). *Nature (London)* **244**, 222–224.

Fraser, R. S. S., and Creanor, J. (1974). *Eur. J. Biochem.* **46**, 67–73.

Fukuhara, H. and Kujawa, C. (1970). *Biochem. Biophys. Res. Commun.* **41**, 1002–1008.

Gale, E. F., Cundliffe, E., Reynolds, R., Richmond, M. H., and Waring, M. J. (1972). "The Molecular Basis of Antibiotic Action." Wiley, New York.

Gancedo, C. (1971). *J. Bacteriol.* **107**, 401–405.

Goldberg, I. H., and Friedman, P. A. (1971). *Annu. Rev. Biochem.* **40**, 775–810.

Goldstein, E. S., and Penman, S. (1973). *J. Mol. Biol.* **80**, 243.

Görts, C. P. M. (1969). *Biochim. Biophys. Acta* **184**, 299–305.

Gottlieb, D., and Nicolas, G. (1969). *Appl. Microbiol.* **18**, 35–40.

Grant, P. Sanchez, L., and Jimenez, A. (1974). *J. Bacteriol.* **120**, 1308–1314.

Greasham, R. L., and Moat, A. G. (1973). *J. Bacteriol.* **115**, 975–981.

Grenson, M., Crabeel, M., Wiame, J. M., and Béchet, J. (1968). *Biochem. Biophys. Res. Commun.* **30**, 414–419.

Grivell, A. R., and Jackson, J. F. (1968). *J. Gen. Microbiol.* **54**, 307–317.

Grollman, A. P. (with technical assistance of M. Walsh). (1967). *J. Biol. Chem.* **242**, 3226–3233.

Gross, K. J., and Pogo, A. O. (1974). *J. Biol. Chem.* **249**, 568–576.

Grossman, L. I., Goldring, E. S., and Marmur, J. (1969). *J. Mol. Biol.* **46**, 367–376.

Hadjiolov, A. A., Dabeva, M. D., and MacKedonski, V. V. (1974). *Biochem. J.* **138**, 321–334.

Harris, G., and MacWilliam, I. C. (1965). *Biochim. Biophys. Acta* **95**, 205–208.

Hartlief, R., and Konigsberger, V. V. (1968). *Biochim. Biophys. Acta* **166**, 512–531.

Hartwell, L. H. (1970). *Annu. Rev. Genet.* **4**, 373–396.

Hartwell, L. H. (1973). *J. Bacteriol.* **115**, 966–974.

Hartwell, L. H. (1974). *Bacteriol. Rev.* **38**, 164–198.

Hartwell, L. H., Hutchinson, H. T., Holland, T. M., and McLaughlin, C. S. (1970). *Mol. Gen. Genet.* **106**, 347–361.

Hartwell, L. H., Culotti, J., Pringle, J. R., and Reid, B. J. (1974). *Science* **183**, 46–51.

Hashimoto, K., Katagiri, M., and Misato, T. (1963). *J. Agr. Chem. Soc. Jap.* **37**, 245–250.

Haslam, J. M., Davey, P. J., and Linnane, A. W. (1968). *Biochem. Biophys. Res. Commun.* **33**, 368–378.

Heredia, C. F., Torǎno, A., Ayuso, M. S., Sandoval, A., and San José, C. (1971). *In* "Gene Expression and its Regulation" (F. T. Kenney *et al.*, eds.), pp. 429–443. Plenum, New York.

Hereford, L. M., and Hartwell, L. H. (1971). *Nature (London), New Biol.* **234**, 171–172.

Hereford, L. M., and Hartwell, L. H. (1973). *Nature (London), New Biol.* **244**, 129–131.

Hereford, L. M., and Hartwell, L. H. (1974). *J. Mol. Biol.* **84**, 445–461.

Hershko, A., Mamont, P., Shields, R., and Tomkins, G. (1971). *Nature (London), New Biol.* **232**, 206–211.

Hollenberg, C. P. (1973). *Biochemistry* **12**, 5320–5325.

Hollenberg, C. P., Borst, P., and van Bruggen, E. F. J. (1970). *Biochim. Biophys. Acta* **209**, 1–15.

Holzer, H., and Hierholzer, G. (1963). *Biochim. Biophys. Acta* **89**, 42–46.

Horgen, P. A., and Griffin, O. H. (1971). *Proc. Nat. Acad. Sci. U.S.* **68**, 338–341.

Huang, M. T., and Grollman, A. P. (1972). *Mol. Pharmacol.* **8**, 538–550.

Jacob, S. T. (1973). *Progr. Nucl. Acid Res. Mol. Biol.* **13**, 93–126.

Jacob, S. T., Muecke, W., Sajdal, E. M., and Munro, H. N. (1970). *Biochem. Biophys. Res. Commun.* **40**, 334–342.

Jaenicke, L., Scholz, K., and Donike, M. (1970). *Eur. J. Biochem.* **13**, 137–141.

Jannsen, S., and Lochmann, E. R. (1970). *FEBS (Fed. Eur. Biochem. Soc.) Lett.* **8**, 113–115.

Jannsen, S., Witte, I., and Megnet, R. (1973). *Biochim. Biophys. Acta* **299**, 681–685.

Jimenez, A., Littlewood, B., and Davies, J. (1972). *In* "Molecular Mechanisms of Antibiotic Action of Protein Biosynthesis and Membranes" (E. Muñoz, F. Garcia-Ferrândiz, and D. Vasquez, eds.), pp. 292–306. Elsevier, Amsterdam.

Jimenez, A., Tipper, D. J., and Davies, J. E. (1973). *Antimicrob. Ag. Chemother.* **3**, 729–738.

Johnson, R. (1970). *Biochem. J.* **119**, 699–706.

Kaczka, E. A., Trenner, N. R., Arison, B., Walker, R. W., and Folkers, K. (1964). *Biochem. Biophys. Res. Commun.* **14**, 456–457.

Kanzaki, T., Higashide, E., Yamamoto, H., Shibata, M., Nakazawa, K., Iwasaki, H., Takewaka, T., and Miyake, A. (1962). *J. Antibiot.* **15**, 93–97.

Kempner, E. S. (1961). *Biochim. Biophys. Acta* **53**, 111–122.

Kempner, E. S., and Miller, J. H. (1963). *Biochim. Biophys. Acta* **76**, 341–346.

Kerridge, D. (1958). *J. Gen. Microbiol.* **19**, 497–506.

Klamerth, O. L., and Kopun, M. (1971). *Eur. J. Biochem.* **21**, 199–203.

Kotyk, A., Ponec, M., and Rikovo, L. (1971). *Folia Microbiol. (Prague)* **16**, 432–444.

Krakoff, I. H., Brown, N. C., and Reichard, P. (1968). *Cancer Res.* **28**, 1559–1565.

Kram, R., Mamont, P., and Tomkins, G. M. (1973). *Proc. Nat. Acad. Sci. U.S.* **70**, 1432–1436.
Kuo, S.-C., Cano, F. R., and Lampen, J. O. (1973). *Antimicrob. Ag. Chemother.* **6**, 716–722.
Lamb, A. J., Clark-Walker, G. D., and Linnane, A. W. (1968). *Biochim. Biophys. Acta* **161**, 415–427.
Lampen, J. O. (1969). *Amer. J. Clin. Pathol.* **52**, 138–146.
Lark, K. G., and Lark, C. (1966). *J. Mol. Biol.* **20**, 9–19.
Lin, S.-Y., Mosteller, R. D., and Hardesty, B. (1966). *J. Mol. Biol.* **21**, 51–69.
Lodder, J., ed. (1971). "The Yeasts." North-Holland Publ., Amsterdam.
Lomax, C. A., and Woods, R. A. (1969). *J. Bacteriol.* **100**, 817–822.
Luha, A., Sarcoe, L. E., and Whittaker, P. A. (1971). *Biochem. Biophys. Res. Commun.* **44**, 396–402.
McLaughlin, C. S., Warner, J. R., Edmonds, M., Nakazato, H., and Vaughan, M. H. (1973). *J. Biol. Chem.* **248**, 1466–1471.
Mayo, V. S., and de Kloet, S. R. (1972). *Arch. Biochem. Biophys.* **153**, 508–514.
Mayo, V. S., Andrean, B. A. G., and de Kloet, S. R. (1968). *Biochim. Biophys. Acta* **169**, 297–305.
Mazelis, A. G., and Petermann, M. L. (1973). *Biochim. Biophys. Acta* **312**, 111–121.
Medoff, G., Kobayashi, G. S., Kwan, C. N., Schlessinger, D., and Venkov, P. (1972). *Proc. Nat. Acad. Sci. U.S.* **69**, 196–199.
Melcher, U. (1971). *Biochim. Biophys. Acta* **246**, 216–224.
Mendecki, J., Lee, Y., and Brawerman, G. (1972). *Biochemistry* **11**, 792–798.
Michels, C. A., Blamire, J., Goldfinger, B., and Marmur, J. (1973). *Antimicrob. Ag. Chemother.* **3**, 562–567.
Miller, J. H., and Kempner, E. S. (1963). *Biochim. Biophys. Acta* **76**, 333–340.
Mizuno, N. S. (1965). *Biochim. Biophys. Acta* **108**, 394–403.
Moore, E. C. (1969). *Cancer Res.* **29**, 291–295.
Munro, H. N., Baliga, B. S., and Pronczak, A. W. (1968). *Nature (London)* **219**, 944–946.
Newlon, C. S., Petes, T. D., Hereford, L. M., and Fangman, W. L. (1974). *Nature (London)* **247**, 32–35.
Nickerson, W. J., and Webb, M. (1956). *J. Bacteriol.* **71**, 129–139.
Owen, S. P., Dietz, A., and Camrener, G. W. (1962). *Antimicrob. Ag. Chemother.* pp. 772–779.
Pavletich, K., Kuo, S.-C., and Lampen, J. O. (1974). *Biochem. Biophys. Res. Commun.* **60**, 942–950.
Penman, S., Fan, H., Perlman, S., Rosbash, M., Weinberg, R., and Zylber, E. (1970). *Cold Spring Harbor Symp. Quant. Biol.* **35**, 561–575.
Perlman, S., Abelson, H. T., and Penman, S. (1973). *Proc. Nat. Acad. Sci. U.S.* **70**, 350–353.
Perry, R. P. (1963). *Exp. Cell Res.* **29**, 400–406.
Pestka, S. (1971). *Annu. Rev. Microbiol.* **25**, 487–562.
Petersen, N. S., and McLaughlin, C. S. (1974). *Mol. Gen. Genet.* **129**, 189–200.
Pickering, W. R., and Woods, R. A. (1972). *Biochim. Biophys. Acta* **264**, 45–58.
Plagemann, P. G. W. (1968). *Biochim. Biophys. Acta* **155**, 202–218.
Ponta, H., Ponta, U., and Wintersberger, E. (1971). *FEBS (Fed. Eur. Biochem. Soc.) Lett.* **18**, 204–208.
Ponta, H., Ponta, U., and Wintersberger, E. (1972). *Eur. J. Biochem.* **29**, 110–118.
Reed, J., and Winterberger, E. (1973). *FEBS (Fed. Eur. Biochem. Soc.) Lett.* **32**, 213.
Reilly, C., Fuhrmann, G. F., and Rothstein, A. (1970). *Biochim. Biophys. Acta* **203**, 583–585.
Rhodes, P. M., and Wilkie, D. (1971). *Heredity* **26**, 347.
Richter, D. (1971). *In* "Methods in Enzymology" (K. Moldave and L. Grossman, eds.), Vol. 20, Part C, pp. 349–359. Academic Press, New York.

Roodyn, D. B., and Wilkie, D. (1968). "Biogenesis of Mitochondria." Methuen, London.
Rosenkranz, H. S., Mednis, A., Marko, P., and Rose, H. M. (1967). *Biochim. Biophys. Acta* **149**, 513–518.
Roth, R. M., and Dampier, C. (1972). *J. Bacteriol.* **109**, 773–779.
Sanfilippo, A., and Mazzolini, R. (1964). *G. Microbiol.* **12**, 83–92.
Schindler, D., Grant, P., and Davies, J. E. (1974). *Nature (London)* **248**, 535–536.
Scholz, K., and Jaenicke, L. (1968). *Eur. J. Biochem.* **4**, 448–457.
Schött, E. H., and Holzer, H. (1974). *Eur. J. Biochem.* **42**, 61–66.
Sentandreu, R., and Lampen, J. O. (1970). *FEBS (Fed. Eur. Biochem. Soc.) Lett.* **11**, 95–99.
Sentandreu, R., and Northcote, D. H. (1969). *Biochem. J.* **115**, 231–239.
Sentandreu, R., Elorza, M. V., and Lampen, J. O. (1972). *In* "Molecular Mechanisms of Antibiotic Action on Protein Biosynthesis and Membranes" (E. Muñoz, F. Garcia-Ferrândiz, and D. Vazquez, eds.), pp. 438–454. Elsevier, Amsterdam.
Sentenac, A., Simon, E. J., and Fromagenot, P. (1968). *Biochim. Biophys. Acta* **161**, 299–308.
Shulman, R. W., Hartwell, L. H., and Warner, J. R. (1973). *J. Mol. Biol.* **73**, 513–525.
Siegel, M. R., and Sisler, H. D. (1964). *Biochim. Biophys. Acta* **87**, 83–89.
Skogerson, L., McLaughlin, C., and Wakatama, E. (1973). *J. Bacteriol.* **116**, 818–822.
Slater, M. L. (1973). *J. Bacteriol.* **113**, 263–270.
Slater, M. L. (1974). *Nature* **247**, 275–276.
Sogin, S. J., Haber, J. E., and Halvorson, H. O. (1972). *J. Bacteriol.* **112**, 806–814.
Stafford, M. E., and McLaughlin, C. (1973). *J. Cell. Physiol.* **82**, 121–128.
Sy, J., and Richter, D. (1972). *Biochemistry* **11**, 2788–2791.
Szybalski, W., and Iyer, V. N. (1967). *In* "Antibiotics" (D. Gottlieb, D. and P. D. Shaw, eds.), Vol. 1, pp. 211–245. Springer-Verlag, Berlin and New York.
Taber, R. L., and Vincent, W. S. (1969). *Biochem. Biophys. Res. Commun.* **34**, 488–494.
Takeuchi, S., Hikayama, K., Veda, K., Sakai, H., and Yonehara, H. (1958). *J. Antibiot.* **11**, 1–5.
Tata, J. R., Hamilton, M. J., and Shields, D. (1972). *Nature (London), New Biol.* **238**, 161–164.
Tauro, P. Halvorson, H., and Epstein, R. L. (1968). *Proc. Nat. Acad. Sci. U.S.* **59**, 277–284.
Tiboni, O., Parisi, B., and Ciferri, O. (1968). *G. Bot. Ital.* **102**, 337–345.
Timberlake, W. E., McDowell, L., and Griffin, D. H. (1972a). *Biochem. Biophys. Res. Commun.* **46**, 942–947.
Timberlake, W. E., Hagen, G., and Griffin, D. H. (1972b). *Biochem. Biophys. Res. Commun.* **48**, 823–827.
Tipper, D. J. (1973). *J. Bacteriol.* **116**, 245–256.
Tomkins, G. M., Levinson, B. B., Baxter, J. D., and Dethlefsen, L. (1972). *Nature (London), New Biol.* **239**, 9.
Tonnesen, T., and Friesen, J. D. (1973). *J. Bacteriol.* **115**, 889–896.
Udem, S. A., and Warner, J. R. (1972). *J. Mol. Biol.* **65**, 227–242.
van Dam, G. J. W., Bloemers, H. P. J., Hartlief, R., van de Meene, J. G. C., van de Poll, K. W., van der Saag, P. T. M., and Konigsberger, V. V. (1965). *Proc. Kon. Ned. Akad. Wetensch., Ser. B* **68**, 281–230.
van de Poll, K. W. (1970). *Proc. Kon. Ned. Akad. Wetensch., Ser. B* **73**, 10–21.
van de Poll, K. W., Verweg, A. A. G., and Konigsberger, V. V. (1968). *Proc. Kon. Ned. Akad. Wetensch., Ser. B* **71**, 344–358.
van der Walt, J. P. (1965). *Antonie von Leeuwenhoek; J. Microbiol. Serol.* **31**, 341–348.
van Wijk, R. (1968). *Proc. Kon. Ned. Akad. Wetensch., Ser. C* **71**, 302–313.
van Wijk, R., and Konijn, T. M. (1971). *FEBS (Fed. Eur. Biochem. Soc.) Lett.* **13**, 184–186.
van Wijk, R., Ourvehand, J., Van den Bos, T., and Konigsberger, V. V. (1969). *Biochim. Biophys. Acta* **186**, 178–191.

Vazquez, D., Battaner, E., Neth, R., Heller, G., and Monro, R. E. (1969). *Cold Spring Harbor Symp. Quant. Biol.* **34**, 369–375.

Venkov, P., Battaner, E., Hadjalov, A., and Schlessinger, D. (1974). *Biochem. Biophys. Res. Commun.* **56**, 599–604.

Warner, J. R. (1971). *J. Biol. Chem.* **246**, 447–454.

Wartell, R. M., Larson, J. E., and Wells, R. D. (1974). *J. Biol. Chem.* **249**, 6719–6731.

Weber, C. T., Kudrna, R. D., and Parks, L. W. (1970). *J. Bacteriol.* **102**, 636–641.

Weeks, D. P., and Baxter, R. (1972). *Biochemistry* **11**, 3060–3064.

Wei, C.-M., Hansen, B., Vaughan, M. H., and McLaughlin, C. W. (1974). *Proc. Nat. Acad. Sci. U.S.* **71**, 713–717.

Wells, R. D., and Larson, J. E. (1970). *J. Mol. Biol.* **49**, 319–342.

Westcott, E. W., and Sisler, H. D. (1964). *Phytopathology* **54**, 1261–1264.

Westphal, H., and Holzer, H. (1964). *Biochim. Biophys. Acta* **89**, 42–46.

Westphal, H., Oeser, A., and Holzer, H. (1966). *Biochem. Z.* **346**, 252–263.

White, H. L., and White, J. R. (1968). *Mol. Pharmacol.* **4**, 549–565.

Wickner, R. B. (1974). *J. Bacteriol.* **117**, 252–260.

Wilkie, D., and Lee, B. K. (1965). *Genet. Res.* **6**, 130–138.

Willems, M., Penman, M., and Penman, S. (1969). *J. Cell Biol.* **41**, 177–187.

Williamson, D. H. (1973). *Biochem. Biophys. Res. Commun.* **52**, 731–740.

Wintersberger, E., and Wintersberger, U. (1970). *FEBS (Fed. Eur. Biochem. Soc.) Lett.* **6**, 58–60.

Ycas, M., and Brawerman, G. (1957). *Arch. Biochem. Biophys.* **68**, 118–129.

Chapter 3

Isolation of Yeast DNA

D. R. CRYER, R. ECCLESHALL, AND J. MARMUR

Department of Biochemistry,
Albert Einstein College of Medicine,
New York, New York

I.   Introduction

The following procedure for the isolation of DNA from the yeast *Saccharomyces cerevisiae* is a modification of an earlier method for isolating DNA from bacteria (Marmur, 1961). This modification is necessary for several reasons. First, the yeast cell wall is a thick, rigid structure (Ballou, 1974) which must be removed by special means. Second, the usual yeast DNA preparations are contaminated with RNA, since DNA comprises only 1–2% of the total cellular nucleic acid, and also with polysaccharides which copurify with the DNA.

The following procedure yields a preparation of total yeast DNA including all three buoyant density species observed in neutral isopycnic cesium chloride gradients: nuclear, 1.699 gm/cm³; nuclear heavy satellite, 1.704 gm/cm³; and mitochondrial, 1.683 gm/cm³. These preparations have not been examined for the presence of the 2 μm circular DNA molecules of nuclear density described by Guerineau *et al.* (1971) and by Clark-Walker and Miklos (1974). If spheroplasting of the cells is adequate and lysis is complete, there are no preferential losses of any of the observed DNA species. Preparations free of DNA of mitochondrial density can be obtained from petite yeast strains that lack mitochondrial DNA (Goldring *et al.*, 1970). The various DNA species obtained from normal strains by this method may be separated and purified using cesium chloride gradients (Grossman *et al.*, 1969), cesium chloride–ethidium bromide gradients (Radloff *et al.*, 1967), cesium sulfate–mercury (Retel and Planta, 1972), or cesium sulfate–silver gradients (Bernardi *et al.*, 1972), and hydroxyapatite chromatography (Britten *et al.*, 1969; Bernardi, 1971).

In outline the procedure consists of the following general operations: (1) preparation of osmotically fragile spheroplasts by enzymatic digestion of the cell wall, (2) lysis of the spheroplasts and partial proteolysis of the lysate, (3) deproteinization and extraction of lipids from the lysate, (4) enzymatic digestion of RNA and elimination of polysaccharides by centrifugation or digestion, followed by separation of the DNA from the digestion products by selective isopropanol precipitation.

II.   Isolation Procedure

All operations are carried out at room temperature unless otherwise stated. Volumes are only approximate, generally minimal, and may be increased at least 2-fold in steps 1 to 9; for smaller quantities of cells, volumes should not be scaled down proportionately.

1. Ten grams (wet weight; laboratory-grown cells are weighed immediately on harvesting at 6000 g for 1 minute) of commercial bakers' yeast or laboratory-grown cells harvested in the late log phase of growth are washed by suspending in 30 ml of 0.05 *M* disodium ethylenediamine tetraacetate (EDTA) (pH 7.5) and collected by centrifugation at 6000 g for 1 minute. (It is important to have EDTA present during the early stages of the procedure to inhibit divalent ion-dependent yeast nucleases as well as nucleases that may be introduced with the cell wall lytic enzymes.)

2. The cell pellet is then suspended in 20 ml of deionized water; 2 ml of 0.5 *M* EDTA (pH 9.0) and 0.5 ml of β-mercaptoethanol are added, and the suspension is incubated at room temperature for 10–15 minutes. (Increasing the time of exposure to the thiol reagent generally increases the sensitivity of the cell wall to subsequent enzymatic digestion.)

3. The treated cells are collected by centrifugation at 6000 g for 1 minute, and the pellet is suspended in sorbitol–EDTA [1 *M* sorbitol, 0.1 *M* EDTA (pH 7.5)] at a ratio of 2 ml/gm of cells.

4. Zymolyase [zymolyase "crude," 5500 units/gm, obtained from the research laboratories of Kirin Brewery Company, Ltd., Takasaki, Gumma Prefecture, Japan; zymolyase is a mixture of β-1, 3-glucanases produced by certain *Arthrobacter* species (Kitamura and Yamamoto, 1972; Kaneko,

Kitamura, and Yamamoto, 1973); the crude preparation seems to contain little or no DNase activity] is dissolved in 2 ml of sorbitol–EDTA immediately before use and is added to the cell suspension to give a final concentration of 0.1–0.5 mg enzyme per gram of cells. (The amount of enzyme necessary to produce spheroplasts can vary with the yeast strain used and its phase of growth when harvested. Cells grown to the midlog phase are usually significantly easier to spheroplast than stationary-phase cells.) The cell suspension is then incubated with the enzyme at 37°C until the cells are converted to spheroplasts (usually 1–2 hours). If necessary, the cells can be incubated with the enzyme for several hours to form spheroplasts, and the DNA obtained will be comparable to that obtained when spheroplast formation is rapid. Incubation at room temperature will also yield spheroplasts, however, too much zymolyase or too long an exposure to the enzyme may result in lysis of the spheroplasts. If lysis occurs during incubation, the suspension is diluted twofold with saline-EDTA [0.15 M NaCl, 0.1 M EDTA (pH 8.0)] and the procedure is continued at step 7 with the elimination of steps 5 and 6. Conversion to spheroplasts is monitored at 20 to 30-minute intervals by diluting 1 drop of the cell suspension in 1 ml of water and adding 2 drops of 25% (w/v) sodium dodecyl sulfate (SDS); when shaken, a clear, viscous solution is obtained if spheroplast formation is good. An alternative to zymolyase for spheroplast formation is Glusulase (Endo Laboratories, Inc., Garden City, N.Y.), a crude enzyme preparation from the hepatopancreas of the snail *Helix pomatia*, which contains exoglucosidase and β-1, 3- and β-1, 6-glucanase activities (Anderson and Millbank, 1966). When Glusulase is used, steps 2 through 5 are replaced by the following. The washed yeast cells are first incubated in 20 ml of 0.1 M tris–sulfate (pH 9.3), 2 ml 0.5 M EDTA (pH 9.0), and 0.5 ml β-mercaptoethanol for 15 minutes. The cells are then collected followed by incubation at 37°C in 1 M sorbitol containing a 10 mM citrate–phosphate buffer (pH 5.8) and 0.1 M EDTA (pH 5.8) in the presence of the enzyme. Usually 0.1 ml of the enzyme solution is added per gram of cells. Spheroplast formation is monitored during this incubation, as with zymolyase.

Glusulase, however, contains DNase activity, hence it is desirable to wash (suspend, and then pellet at 4000 g for 5 minutes) the spheroplasts two to three times with sorbitol–EDTA before they are lysed to minimize degradation of the DNA. In general, Glusulase is less effective than zymolyase in converting stationary-phase yeast cells into spheroplasts. However, it does not seem to cause lysis of spheroplasts, even on extended treatment.

A nonenzymatic method of disrupting yeast cells is to incubate cells pretreated with β-mercaptoethanol in 1 volume of a solution containing 0.2 M EDTA (pH 10), 0.15 M NaCl and 4% SDS for 1 hour at room temperature (modified from Bicknell and Douglas, 1970).

5. The spheroplasts are pelleted by centrifugation at 4000 g for 5 minutes, and the pellet is then washed by dispersing it in 20 ml of sorbitol–EDTA and repeating the centrifugation.

6. The washed spheroplast pellet is next suspended in saline–EDTA at 2 ml/gm of cells used initially. It is important at this stage to break up all clumps of cells in order subsequently to obtain total lysis.

7. Proteinase K [Proteinase K (fungal) lyophilized, 15 mAnson units/mg, E.M. Laboratories, Inc., Elmsford, N.Y.] is dissolved in 1 ml of saline–EDTA immediately prior to use and is added to the cell suspension to a final concentration of 50–100 μg/ml. The spheroplasts are then lysed by making the suspension 1% (w/v) with respect to SDS by slowly adding an appropriate volume of a stock solution of 25% SDS. (To obtain high-molecular-weight DNA in this procedure, the preparation should be handled gently from the stage of lysis onward, e.g., wide-bore pipettes should be used and mechanical mixers should be avoided.) The lysate is incubated at 37°C for 3–4 hours.

8. The lysate is next heated to 60°C for 30 minutes, and then cooled to room temperature.

9. Deproteinization and extraction of lipids are accomplished by adding an equal volume of chloroform containing isoamyl alcohol (24:1, v/v) and gently shaking until a white, homogeneous emulsion is formed.

10. The emulsion is broken, and the phases separated by centrifugation at 12,000 g for 20 minutes. The aqueous upper phase contains the nucleic acids and is transferred to a 150-ml beaker. The lower phase and the precipitated protein which collects between the liquid phases are discarded. In occasional preparations, especially when the cell concentration is too high, some DNA may be occluded by the precipitated protein at the interface. This can be recovered by removing the precipitate, shaking gently with 5 ml 0.1 × SSC [0.15 M NaCl, 0.015 M trisodium citrate (pH 7.0)] until all lumps of protein are dispersed, and centrifuging at 12,000 g for 20 minutes. The supernatant is pooled with the aqueous phase of the previous centrifugation, and the pellet is discarded.

11. Next, 2 vol of 95% ethanol are layered over the aqueous phase, and the DNA threads are collected by slowly stirring with a glass rod. [The precipitated DNA is recognizable at this stage as a fibrous precipitate. Since much of the RNA and some residual protein coprecipitates with the DNA in this step, or if the DNA is of low molecular weight, the precipitate may not adhere readily to the stirring rod. In this case the DNA can be collected by a short low-speed centrifugation (e.g., 2000 g for 1 minute).] Stirring is continued until both layers are thoroughly mixed.

12. The DNA-containing precipitate is dissolved in a small volume of 0.1× SSC (about 15–20 ml for the DNA from 10 gm of cells). Then the solution is made up to 1× SSC by the addition of an appropriate volume of 10× SSC. DNA is more soluble in solutions of low ionic strength (i.e., 0.1 ×

SSC) but is more stable at higher ionic strength (i.e., $1\times$ SSC); thus the rationale for this step. Nonspecific DNA losses in subsequent steps of the purification are minimized if the DNA is kept at reasonably high concentrations (at least 100 μg/ml), so the DNA at this step should be dissolved in a minimal volume. DNA should be dissolved gently and, if it is of high molecular weight, may require several hours to dissolve. (The DNA solution at this stage of the procedure is sufficiently pure to be examined in isopycnic cesium chloride gradients in an analytical ultracentrifuge.)

13. Pancreatic RNase [Sigma, 1 mg/ml, in 10 mM sodium acetate (pH 7.0); this stock solution is prepared by heating to 100°C for 20 minutes at pH 5 and then neutralizing; aliquots are stored frozen at -20°C] is added to a final concentration of 40–50 μg/ml. T1 RNase [Sigma, 1000 units/mg, 340,000 units/mg, in 10 mM sodium acetate (pH 7.0); this stock solution is prepared by heating to 80°C for 10 minutes at pH 5 and then neutralizing; aliquots are stored frozen at -20°C] is also added to give a final concentration of 50–100 units/ml. The DNA solution is then incubated at 37°C for 1 hour.

14. The DNA solution is again deproteinized and ethanol-precipitated, and the DNA is redissolved (steps 9 to 12).

15. Repetition of the RNase digestion (step 13) and subsequently step 14 is usually desirable at this stage, because of the large amount of RNA initially isolated with the DNA.

16. Contaminating polysaccharides result in DNA solutions that are opalescent at this point. These may be either cleared by high-speed centrifugation [e.g., 45,000 rpm (200,000 g) for 30 minutes at 4°C in a Spinco 60 Ti rotor; Hennig, 1968], or digested by including α-amylase (Worthington Biochemical Corporation, 36.7 mg/ml, 717 units/mg) at a final concentration of 0.03 unit/ml during an RNase digestion step.

If a reasonable amount of DNA (0.3 mg or more) has been obtained at this stage, the following isopropanol steps allow a more selective precipitation of DNA from preparations that may still contain some RNA digestion products and polysaccharides. [The usual yield at this point is at least five times this amount. For an accurate DNA determination, however, a chemical method should be used (e.g., Burton, 1968).]

17. The DNA should be ethanol-precipitated as in step 11 and then dissolved in $0.1\times$ SSC (or simply kept in $0.1\times$ SSC in the repetition of step 12) at a concentration of at least 100–200 μg/ml. The solution is made 0.3 M with respect to sodium acetate using a stock solution of 3 M sodium acetate containing 1 mM EDTA (pH 7).

18. Isopropanol (0.54 vol) is added dropwise as the solution is stirred with a glass rod. If the DNA is of a reasonably high molecular weight, the precipitated DNA will wrap around the stirring rod. The DNA is finally dissolved in $0.1\times$ SSC and made up to $1\times$ SSC as described in step 12. Due to the high initial ratio of RNA to DNA, traces of RNA may still remain

at this stage but can be removed by cesium chloride gradient centrifugation or hydroxyapatite adsorption. The DNA solution is stored over a few drops of chloroform at 4° C to prevent the growth of microorganisms. The procedure can be interrupted at several points without adverse effects. The washed spheroplast pellet (from step 5) can be stored for several days at −20° C. The preparation can be stored at 4° C after any deproteinization, either before centrifugation (step 9) or after centrifugation (step 10). Precipitated DNA (step 11) can be stored in 70% ethanol, but it is often advantageous instead to allow a precipitate to dissolve slowly overnight in $0.1 \times$ SSC (step 12) in the presence of chloroform. Nucleic acid solutions in $1 \times$ SSC prepared by the above method can be stored for long periods of time at 4° C over a small amount of chloroform in screw-cap vials. The concentration of pure, native DNA, free of RNA, can be estimated from the absorbance at 260 nm using the relationship OD $1.0 = 50 \ \mu g/ml$.

Usually 0.20–0.25 mg DNA/gm of cells is obtained by this procedure. The molecular weight is generally 10 to 20×10^6. The DNA concentration is determined according to Burton (1968) using calf thymus DNA as a standard.

ACKNOWLEDGMENTS

This research was supported by NIH Grant CA 12410. Dennis R. Cryer was a medical scientist trainee supported by PHS Grant 5T5 GM 1674 from the National Institute of General Medical Sciences. Partial salary support for Julius Marmur was derived from NIH Grant GM 19100.

REFERENCES

Anderson, F. B., and Millbank, J. W. (1966). *Biochem. J.* **99**, 682–687.

Ballou, C. E. (1974). *Advan. Enzymol.* **40**, 239.

Bernardi, G. (1971). *In* "Methods in Enzymology" (L. Grossman and K. Moldave, eds.), Vol. 21, Part D, pp. 95–139. Academic Press, New York.

Bernardi, G., Piperno, G., and Fonty, G. (1972). *J. Mol. Biol.* **65**, 173–189.

Bicknell, J. N., and Douglas, H. C. (1970). *J. Bacteriol.* **101**, 505.

Britten, R. J., Pavich, M., and Smith, J. (1969). *Carnegie Inst. Wash. Yearb.* **68**, 400–402.

Burton, K. (1968). *In* "Methods in Enzymology" (L. Grossman and K. Moldave, eds.), Vol. 12, Part B, pp. 163–166. Academic Press, New York.

Clark-Walker, G. D., and Miklos, G. L. G. (1974). *Eur. J. Biochem.* **41**, 359–365.

Goldring, E. S., Grossman, L. I., Krupnick, D., Cryer, D. R., and Marmur, J. (1970). *J. Mol. Biol.* **52**, 323.

Grossman, L. I., Goldring, E. S., and Marmur, J. (1969). *J. Mol. Biol.* **46**, 367–376.

Guerineau, M., Grandchamp, C., Paoletti, C., and Slonimski, P. (1971). *Biochem. Biophys. Res. Commun.* **42**, 550–557.

Hennig, W. (1968). *J. Mol. Biol.* **38**, 227–239.

Kankeo, T., Kitamura, K., and Yamamoto, Y. (1973). *Agr. Biol. Chem.* **37**, 2295–2302.

Kitamura, K., and Yamamoto, Y. (1972). *Arch. Biochem. Biophys.* **153**, 403–406.

Marmur, J. (1961). *J. Mol. Biol.* **3**, 208–218.

Radloff, R., Bauer, W., and Vinograd, J. (1967). *Proc. Nat. Acad. Sci. U.S.* **57**, 1514–1521.

Retel, J., and Planta, R. J. (1972). *Biochim. Biophys. Acta* **281**, 299–309.

Chapter 4

Preparation of RNA and Ribosomes from Yeast

GERALD M. RUBIN[1]

*Medical Research Council Laboratory of Molecular Biology,
Cambridge, England*

I. Introduction

Most methods used for the preparation of RNA and ribosomes from extracts of mammalian cells or bacteria can be directly applied either to

[1] *Present address*: Department of Biochemistry, Stanford University, Stanford, California.

yeast spheroplasts or to disrupted yeast cells. These methods have been
described and evaluated in Volume VII of this series (Brawerman, 1973;
Muramatsu, 1973; Kates, 1973; Traugh and Traut, 1973). The following discussion is therefore limited to methods specifically designed for yeasts or to
methods we have found particularly useful in our own work.

II. Cell Growth and Disruption

Haploid *Saccharomyces cerevisiae* cells are grown in 1% Bacto yeast
extract, 2% Bacto peptone, and 2% glucose (YEPD) to midlog phase at
30°C, unless otherwise specified. The cells are harvested by centrifugation
and are washed once with distilled water. We have not found the storage of
cells at $-20°$ for several days or the method of cell disruption to affect
significantly the yield of material except in the preparation of polyribosomes. The methods commonly used for the disruption of yeast cells are
described by Fink (1970).

III. Radioactive Labeling

Yeast RNA can be radioactively labeled *in vivo* with purine and pyrimidine
bases, with phosphate-^{32}P, or with methyl-labeled methionine.

A. Labeling with Purines and Pyrimidines

Yeasts are able to utilize purines and pyrimidines added to the medium,
although the free bases are not normal intermediates in the biosynthesis of
of nucleotides (de Robichon-Szulmajster and Surdin-Kerian, 1971). Uracil
and adenine are the bases most commonly used to label RNA. It is difficult
to evaluate the usefulness for preparative purposes of the various labeling
protocols, since the specific activities of the labeled RNAs are rarely determined. RNA of specific activity on the order of 10^4 cmp/μg can be conveniently obtained by growing a culture of yeasts for several generations
in YEPD medium to which adenine-^{3}H (20 Ci/mmole) has been added to a
final concentration of 2 μCi/ml (Sogin *et al.*, 1972). Much higher specific
activities can probably be obtained if the labeling is done in a minimal
medium using a mutant defective in adenine biosynthesis.

B.   Labeling with Phosphate-[32]P

Low-phosphate medium can be prepared from YEPD medium by the selective precipitation of inorganic phosphate as $MgNH_4PO_4$ (Rubin, 1973). Ten milliliters of 1 M $MgSO_4$ and 10 ml of concentrated aqueous ammonia are added to 1 liter of 1% Bacto yeast extract and 2% Bacto peptone. The phosphates are allowed to precipitate at room temperature for 30 minutes, and the precipitate is removed by filtration through Whatman No. 1 filter paper. The filtrate is adjusted to pH 5.8 with HCl and autoclaved. Sterile glucose is added to a final concentration of 2%. This procedure gives a medium with a low inorganic phosphate concentration (of the order of 10^{-4} M as estimated by isotope dilution), but containing a relatively high proportion of organic phosphates. Since $S.$ $cerevisiae$ has an inducible acid phosphomonoesterase (Günther and Kattner, 1968), it can readily grow in this medium. Nevertheless, inorganic phosphate is used preferentially as a source of phosphate; over 90% of carrier-free phosphoric-^{32}P acid added to the medium at a concentration of 1 mCi/20 ml is found to be precipitable with the cells in less than half the generation time. Phosphate-^{32}P can be added at concentrations of 1 mCi/5 ml without significantly decreasing the generation time of the cells (Rubin, 1974). RNA of specific activity in excess of 10^6 cpm/μg is routinely obtained. Perhaps most important, when yeast is grown in this medium, inorganic polyphosphates do not accumulate within the cells. This is an advantage over media previously used for ^{32}P-labeling of yeasts containing only inorganic phosphates as a source of phosphate (Schweizer et al., 1969; Harold, 1966).

C.   Labeling with Methyl-Labeled Methionine

Methylation of high-molecular-weight RNA species is primarily limited to rRNAs and their precursors (Greenberg and Penman, 1966). Among the low-molecular-weight RNAs tRNA is heavily methylated. These methylated RNA species can be specifically labeled with methyl-labeled methionine if guanine and adenine are included in the medium at a concentration of 2 mM to prevent labeling of the purine ring via the one-carbon pool (Retèl et al., 1969; Udem et al., 1971; Smitt et al., 1972).

IV.   Preparation of Ribosomes

A.   Differential Centrifugation

Ribosomes are readily prepared from disrupted yeast cells by differential centrifugation. The cells are suspended in two to four times their wet weight of sodium–magnesium–Tris buffer [0.1 M NaCl, 0.03 M MgCl$_2$, 0.01 M tris–HCl (pH 7.4)] and passed through a French press twice at 16,000 lb/in^2. Bentonite (Watts and Mathias, 1967) can be added to a final concentration of 2 mg/ml immediately after cell disruption to inhibit RNase activity. The cell debris and bentonite are removed by centrifugation at 20,000 rpm (26,000 g) for 20 minutes at 2°C in a Beckman 50 Ti rotor. The ribosomes are sedimented from the 20,000-rpm supernatant by further centrifugation through a 3-ml layer of 5% (NH$_4$)$_2$SO$_4$, 15% sucrose in sodium–magnesium–Tris buffer at 40,000 rpm (105,000 g) for 6 hours (Warner, 1971). This concentration of (NH$_4$)$_2$SO$_4$ is reported to cause the most efficient release of loosely bound protein (Warner 1971).

B. Chromatography on DEAE-Cellulose

Stanley and Wahba (1967) have described a purification procedure for *Escherichia coli* ribosomes based on chromatography on DEAE-cellulose. This procedure removes components, such as RNase I and II, which are found to contaminate *E. coli* ribosomes prepared by differential centrifugation. Likewise, McPhie *et al.* (1966) found that the stability of yeast ribosomes was very greatly improved after chromatography on DEAE-cellulose. The ability of yeast ribosomes prepared by this method to function in *in vitro* protein synthesis has not been determined.

We have found the following modification of the Stanley and Wahba protocol to be convenient for large-scale preparations of *S. cerevisiae* ribosomes. Yeast cells [about 20 gm (wet weight)] are suspended in an equal weight of Tris–magnesium–0.05 M NH$_4$ buffer [0.05 M NH$_4$Cl, 0.01 M magnesium acetate, 0.01 M Tris–HCl (pH 7.4)] and passed through a French press twice at 16,000 lb/in^2. Bentonite is added to a final concentration of 2 mg/ml, and the suspension is centrifuged at 18,000 rpm (30,000 g) for 30 minutes at 2°C in a Sorvall SS34 rotor. The ribosomes are sedimented from the 18,000-rpm supernatant by further centrifugation at 40,000 rpm (105,000 g) for 90 minutes at 4°C in a Beckman 50 Ti rotor. The pellet is resuspended in Tris–magnesium–0.05 M NH$_4$ buffer, and the ribosomes are sedimented as before. The pellet is suspended in Tris–magnesium–0.25 M NH$_4$ [0.25 M NH$_4$Cl, 0.01 M magnesium acetate, 0.01 M Tris–HCl, (pH 7.4)] to give a concentration of 80 OD$_{260}$ units/ml. The ribosomal suspension is clarified by centrifugation at 5000 rpm (2000 g) for 10 minutes in a Sorvall SS34 rotor. The clarified ribosomal suspension (approximately 25 ml) is adsorbed at 4°C to a 2.5 × 50 cm column of DE23 (Whatman) which has been equilibrated with Tris–magnesium–0.25 M NH$_4$ buffer. The column is washed with 1 liter of Tris–magnesium–0.25 M NH$_4$ buffer at a flow rate of 300 ml per hour. The ribosomes are then eluted with Tris–

magnesium–0.60 M NH$_4$ buffer [0.60 M NH$_4$Cl, 0.01 M magnesium acetate, 0.01 M Tris–HCl (pH 7.4)] at a flow rate of 200 ml per hour. The portion of the column eluate containing the ribosomes is easily recognized by its bluish opalescence. More than 80% of the OD$_{260}$ units are recovered in less than 50 ml of eluate.

V. Preparation of Ribosomal Subunits

Yeast ribosomes, like those of other organisms, dissociate into two un-equal subunits when exposed to low concentrations of magnesium ions. Ribosomes are suspended in or dialyzed in 10^{-5} M MgCl$_2$, 0.1 M NaCl, and 0.01 M Tris–HCl (pH 7.4), which causes them to dissociate (Warner, 1971). The derived subunits can then be separated by centrifugation in a 10–25% (w/v) sucrose gradient in this buffer for 10 hours at 29,000 rpm (110,000 g) in an MSE SW 35 rotor at 3°C (Fig. 1). A disadvantage of this procedure is that 5 S RNA is lost from the large ribosomal subunit (Udem et al., 1971). Battaner and Vazquez (1971) have reported, however, that, when S. cere-visiae ribosomes are dissociated by dialysis into 0.2 mM magnesium acetate, 5 mM mercaptoethanol, 50 mM NH$_4$Cl, and 10 mM Tris–HCl (pH 7.4), the

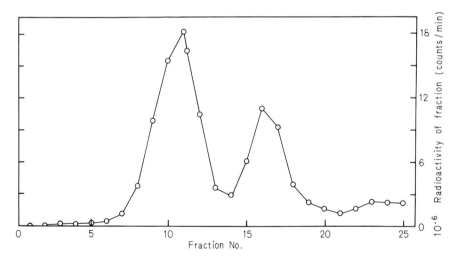

FIG. 1. Separation of ^{32}P-labeled ribosomal subunits. Derived subunits were centri-fuged in a 10–25% (w/v) linear sucrose gradient in 10^{-5} M MgCl$_2$, 0.1 M NaCl, and 0.01 M Tris–HCl (pH 7.4) for 10 hours at 29,000 rpm (110,000 g) in an MSE SW35 rotor at 3°C. Fractions were collected from the bottom of the tube.

resultant subunits are active in poly-U-directed protein synthesis and thus presumably the 5 S RNA has been retained.

Large ribosomal subunits containing a normal complement of 5 S RNA can also be isolated when the ribosomes are dissociated by high concentrations of KCl (Udem *et al.*, 1971; Martin and Wool, 1968). Dissociation by KCl appears to be specific for ribosomes that lack attached mRNA and nascent polypeptides (Martin and Hartwell, 1970). Thus, in cases in which maximal yield of subunits is crucial, it is necessary to increase the percentage of the cell's ribosomes that are in a susceptible state. This can be accomplished by treatment with puromycin to release nascent polypeptides (Martin *et al.*, 1969). Alternatively, Martin and Hartwell (1970) made use of the temperature-sensitive mutant *ts136*, in which mRNA formation becomes defective when the cells are transferred from the permissive to the restrictive temperature, to convert the majority of the cell's ribosomes to a KCl-dissociable form. To prepare subunits by this method, either a cell extract or a suspension of partially purified ribosomes is layered onto a 10–25% (w/v) sucrose gradient made in 0.8 M KCl, 0.1 M NaCl, 0.03 M MgCl$_2$, and 0.01 M Tris–HCl (pH 7.4). The gradient is centrifuged for 10 hours at 110,000 g at 4°C.

VI.   Preparation of Polyribosomes

Yeast polyribosomes are usually prepared from spheroplasts by sucrose gradient sedimentation. Procedures for the enzymatic conversion of yeast cells into spheroplasts are considered elsewhere in this volume and are not discussed here. Alternatively, Marcus and Halvorson (1967) have described a procedure for isolating polysomes from yeast cells disrupted by grinding with glass beads. The following procedure for the preparation of polyribosomes is that of Hartwell and McLaughlin (1968). Forty milliliters of spheroplast culture (prepared from approximately 4×10^8 cells) is harvested for each sucrose gradient by adding cycloheximide to a final concentration of 10^{-3} M and rapidly chilling the culture in an ice bath. The spheroplasts are collected by centrifugation at 4600 g for 5 minutes and lysed in 1 ml of buffer containing 0.01 M Tris–HCl, 0.1 M NaCl, 0.03 M MgCl$_2$, and 20 mg of poly-L-ornithine per milliliter (pH 7.4). An 0.1-ml amount of 5% sodium deoxycholate is added, and the mixture is held at 5°C for 5 minutes. Then, 0.15 ml of 5% Brij 58 is added, and the mixture is held for another 5 minutes at 5°C before it is layered onto 31 ml of a 10–60% (w/v) linear sucrose gradient in the above buffer minus poly-L-ornithine. Centrifugation is carried out in a Beckman SW 25.1 rotor at 60.000 g for 3–5 hours at 5°C.

VII.   Preparation of High-Molecular-Weight rRNAs

A.   Extraction of RNA from Ribosomes

Ribosomal pellets are resuspended in 50 mM NaCl, 30 mM MgCl$_2$, and 20 mM Tris–HCl (pH 7.4) containing 0.2% sodium dodecyl sulfate (SDS), and extracted three times at room temperature with equal volumes of water-saturated phenol. If ribosomes are prepared by the DEAE chromatography procedure, an equal volume of 0.4% SDS is simply added to the pooled fractions and the extraction is performed as described above. The RNA is allowed to precipitate from the final aqueous layer overnight at $-20°$C after the addition of 2 vol of ethanol. The precipitated RNA is collected by centrifugation, and the residual ethanol is removed from the pellet by drying *in vacuo* for 10 minutes. The RNA can then be dissolved in an appropriate buffer. Drying the pellet for more than 10 minutes results in removal of residual water as well as ethanol, and renders the pellet much more difficult to dissolve.

B.   Separation of High-Molecular-Weight rRNAs

The precipitated RNA is dissolved in (0.15 M NaCl and 0.015 M sodium citrate (pH 7.0) (SSC) at a concentration of 1 mg/ml (OD$_{260}$ = 20), and up to 1 ml is layered onto a 10–30% (w/v) linear sucrose gradient made in SSC. The gradient is centrifuged for 18 hours at 26,000 rpm (90,000 g) in a Beckman SW 27 rotor at 4°C (Fig. 2). Intact 18 and 28 S rRNA can be isolated from ribosomes prepared by either of the procedures given above. A sensitive measure of the intactness of the RNA is the ratio of the amount of 28 to 18 S RNA. This ratio should be greater than 2.

VIII.   Preparation of Low-Molecular-Weight RNAs

Holley *et al.* (1961) demonstrated over a decade ago that yeast cells release their "soluble RNA" when permeablized with phenol. Since that time many modifications of the basic phenol permeablization technique have been introduced. These methods have two major advantages over phenol extraction of disrupted yeast cells in the isolation of low-molecular-weight RNAs. First, the permeablization is highly selective for low-molecular-weight RNA, and its specificity can be varied by changing the conditions of the extraction. Second, cell disruption is not required; thus large amounts of material or highly radioactive material can be conveniently processed.

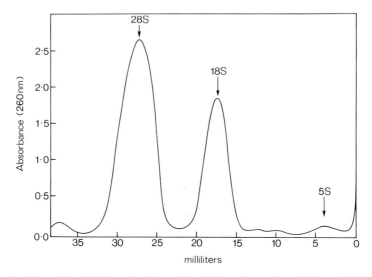

FIG. 2. Separation of high-molecular-weight rRNAs. Approximately 1 mg of RNA was sedimented for 18 hours at 26,000 rpm (90,000 g) in a Beckman SW 27 rotor at 4°C. The terms 18 and 28 S are used to denote the general classes of RNA. The actual sedimentation constants of these RNAs as isolated from yeasts are 18 and 25 S as determined by Udem and Warner (1972), or 17 and 26 S as determined by Retèl and Planta (1970b).

A. tRNA

Yeast cells are suspended in at least three times their volume of 0.1 M sodium acetate and 1% SDS (pH 5.0). One-half volume of water-saturated phenol is added, and the mixture is shaken vigorously at room temperature for 1–2 hours. The phases are separated by centrifugation for 10 minutes at 10,000 g. The yeast cells, which remain intact, sediment to the bottom of the tube. The aqueous phase is removed and reextracted with an equal volume of water-saturated phenol. RNA is allowed to precipitate from the reextracted aqueous phase overnight at $-20°C$ after the addition of 2 vol of ethanol. The precipitated RNA consists of tRNA with a trace of 5 S RNA. A discussion of tRNA fractionation is beyond the scope of this article. A list of specific tRNA species that have been purified from yeasts is given in Table I.

B. 5 and 5.8 S RNA

If the yeast cells are permeablized at room temperature as described above in a buffer containing EDTA [20 mM EDTA, 0.1 M sodium acetate, and 1% SDS (pH 5)], the full complement of 5 S RNA, as well as a trace of 5.8 S RNA, is released from the cells in addition to tRNA. If permeablization

TABLE I

PURIFICATION AND SEQUENCE DETERMINATION OF tRNAs[a]

tRNA Species	Reference
$\text{tRNA}_1^{\text{Ala}}$	Holley et al. (1965); Merrill (1968)
$\text{tRNA}_2^{\text{Arg}}$	Weissenbach et al. (1972)
$\text{tRNA}_3^{\text{Arg}}$	Kuntzel et al. (1972)
$\text{tRNA}_1^{\text{Asp}}$	Gangloff et al. (1972)
$\text{tRNA}_1^{\text{Gly}}$	Yoshida (1973)
$\text{tRNA}_3^{\text{Leu}}$	Kowalski et al. (1971); Chang et al. (1973)
$\text{tRNA}_1^{\text{Lys}}$	Smith et al. (1973)
$\text{tRNA}_2^{\text{Lys}}$	Madison et al. (1972)
$\text{tRNA}_f^{\text{Met}}$	Simsek and RajBhandary (1972)
tRNA^{Phe}	RajBhandary et al. (1968); Nakanishi et al. (1970)
$\text{tRNA}_{1,2}^{\text{Ser}}$	Zachau et al. (1966a,b)
tRNA^{Trp}	Keith et al. (1971)
tRNA^{Tyr}	Madison and Kung (1967)
$\text{tRNA}_1^{\text{Val}}$	Baev et al. (1967); Bonnet et al. (1971)

[a] The yeast tRNA species listed have been purified and their complete nucleotide sequences determined.

of the cells is carried out at a temperature of 33°C or greater, the molar amount of 5.8 S RNA released is equal to that of the 5 S RNA. Under these conditions the extraction is still specific for low-molecular-weight RNAs; at 33°C less than 1% of the cells' high-molecular-weight RNA is released from cells of *S. cerevisiae* strain S288C. The yield of 5 and 5.8 S RNA, however, is as great as that obtained when the cells are completely disrupted and phenol-extracted (G. Rubin, unpublished results). The lowest temperature at which the 5.8 S RNA is quantitatively released from the cells may depend on the strain or growth conditions, and should be determined empirically. At higher temperatures increasing amounts of high-molecular-weight RNA are released from the cells. If permeablization of the cells is carried out for 30 minutes at 65°C in a buffer containing 10 mM EDTA, 10 mM Tris–HCl and 0.5% SDS (pH 7.4), the majority of the cell's high-molecular-weight RNA is released (Rubin, 1973). Thus, by sequentially varying the permeablization conditions, RNA fractions very greatly enriched for tRNA or 5 or 5.8 S RNA can be obtained.

The 5 and 5.8 S RNAs are best separated and purified by polyacrylamide gel electrophoresis. Up to 500 μg each of 5 and 5.8 S RNA can be obtained from a single 4 mm × 15 cm × 20 cm slab gel when the starting material is RNA obtained by permeablization of yeast cells at 33°C. For larger amounts of material chromatography on Sephadex (Monier, 1971) or DEAE-Sephadex (Moriyama et al., 1970) can be carried out.

IX.   Polyacrylamide Gel Electrophoresis of Low-Molecular-Weight RNAs

A useful polyacrylamide gel system for the fractionation of low-molecular-weight RNAs is a modification of the discontinuous buffer system of Laemmli (1970). This system has several advantages for both analytical and preparative separations of low-molecular-weight RNAs. First, the bands are generally sharper, and consequently the resolution is better than with continuous buffer gel systems. Second, the low percentage of acrylamide stacking gel prevents any high-molecular-weight RNA present in the sample from clogging the pores at the top of the running gel before the low-molecular-weight RNA has entered it. In the absence of a stacking gel, such clogging can cause the low-molecular-weight RNA to streak badly; this is the limiting factor in the loading of continuous-buffer gel systems. This problem is especially severe in the case of phenol extracts of crude cell lysates or of ribosomes, in which the majority of the RNA in the sample is high molecular weight. Finally, this gel system can tolerate larger sample volumes and higher concentrations of salt in the sample than can continuous-buffer systems. Detailed descriptions of the continuous-buffer systems commonly used for the polyacrylamide gel electrophoresis of both high- and low-molecular-weight RNAs are given in DeWachter and Fiers (1971) and Dingman and Peacock (1971). Preparation of discontinuous-buffer system polyacrylamide gels is as follows.

A.   Preparation of the Gels

1.   STOCK SOLUTIONS

All solutions are filtered through 0.45-μm pore size filters (Millipore) to remove insoluble material. The solutions are stored at 4°C and, with the exception of the 5 M urea and 10% ammonium persulfate, are stable for at least 6 months.

1.5 M Tris–HCl (pH 8.8).
0.5 M Tris–HCl (pH 6.8).
Acrylamide stock solution 1: 30% acrylamide and 0.8% bisacrylamide.
Acrylamide stock solution 2: 20% acrylamide and 1.5% bisacrylamide.
5 M Urea (the urea is deionized with ion-exchange resin and is made fresh monthly).
10% Ammonium persulfate (made fresh daily).
Electrode buffer (dilute 1:10 before use): 30.3 gm Tris and 144.1 gm glycine; add distilled water to 1 liter and adjust to pH 8.4 with HCl if necessary.
N,N,N',N'-tetramethylethylenediamine (TEMED).

2. GEL APPARATUS

A simple and inexpensive slab gel apparatus is shown in Fig. 3a. Details of the construction and assembly of the apparatus are given in DeWachter and Fiers (1971). A 4-mm slab is excellent for preparative separations. For analytical separations a thinner slab (1.5 mm) can be used. If the gel is to be dried for autoradiography, the slab thickness should be less than 2 mm.

3. POURING THE GEL

To make 100 ml of running gel, combine in a filter flask 34 ml of distilled water, 26.5 ml of 1.5 M Tris–HCl (pH 8.8), 39 ml of acrylamide stock solution 1, and 0.05 ml of TEMED.

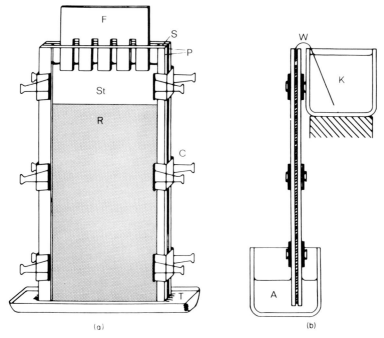

FIG. 3. Apparatus for slab gel electrophoresis. (a) Assembly of the cell. Two 40 cm × 20 cm × 4 mm glass plates (P) are kept at a suitable distance by two 40 cm × 15 cm × 4 mm Perspex strips (S). The assembly is made leakproof with grease and kept together with six steel clips (C). It is placed vertically in a trough (T), and 100 ml of running gel solution is allowed to polymerize in the trough to close the lower end of the gel mold. The running (R) and stacking gels (St) are then poured as described in the text. A sample well template (F) is used when more than one sample is to be applied to the gel. (b) Electrophoresis (transverse section). The lower end of the slab gel is in direct contact with the electrode buffer of the lower reservoir which contains the anode (A), a platinum wire about 20 cm long. The top of the gel is covered by about 2 cm of a buffer, and forms electrical contact with the upper reservoir containing the cathode (K) through a paper wick (W) soaked with buffer. (After DeWachter and Fiers, 1971).

Dissolved gases are removed by evacuating the flask for 10 minutes with a water aspirator; 0.5 ml of 10% ammonium persulfate is added, and the gel solution is poured into the gel mold. Water or water-saturated 2-butanol is layered over the gel solution to ensure a flat interface and to prevent oxygen in the air from interfering with the polymerization reaction. 2-Butanol is immiscible with the gel solution, and thus is much easier than water to apply without disturbing the gel (Burgess, personal communication). The mixture should set in about 10 minutes.

To make approximately 20 ml of stacking gel, combine 14 ml of 5 M urea, 2.5 ml of 0.5 M Tris–HCl (pH 6.8), 4 ml of acrylamide stock solution 2, and 0.04 ml of TEMED.

Dissolved gases are removed as described above. The top of the running gel should be rinsed thoroughly with water. 0.04 ml of 10% ammonium persulfate is added to the stacking gel solution, and the top of the running gel is rinsed once with the complete stacking gel solution. The remainder of the gel mold is filled with stacking gel solution, and the sample well template is inserted. If only one sample is to be applied to the gel, the sample well template can be omitted. In this case water or water-saturated 2-butanol should be layered over the stacking gel.

The stacking gel should polymerize in about 10 minutes. The sample well template may then be removed. The top of the gel is rinsed with electrode buffer, and the gel is then placed in the apparatus as shown in Fig. 3b. The gel is now ready for sample application; do not preelectrophorese. For best results the gel should be used within an hour, before the buffer zones diffuse.

B.   Running the Gel

If the RNA has been precipitated with ethanol, the precipitate is collected by centrifugation and the residual ethanol is removed from the pellet as described in Section VII,A. The RNA is then dissolved in sample buffer [10% (w/v) sucrose, 5 M urea, 0.5% SDS, and 0.05% bromphenol blue], heated 1 minute at 65°C, and carefully layered onto the gel. Alternatively, an aliquot of the aqueous phase of a phenol extraction can be combined with an equal volume of sample buffer, heated 1 minute at 65°C, and layered directly onto the gel. In general, the volume of the sample should be such that the height of the sample when applied to the gel is less than 5 mm.

The gel can be run either at room temperature or at 4°C. At room temperature a 1.5 mm $\times$ 15 cm $\times$ 16 cm slab gel will run in 9 hours at a constant voltage of 120 V. The separation of yeast RNAs obtained under these conditions is shown in Fig. 4.

Time (min)

2   4   6   8   10   13   16   20   30   45   60   90

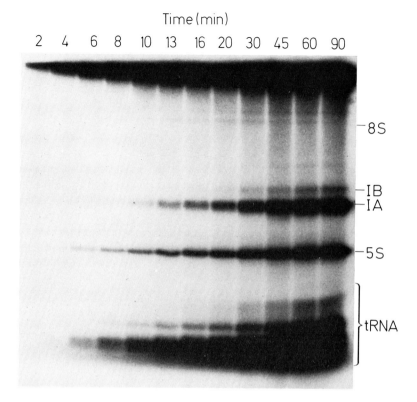

−8S

−IB
−IA

−5S

}tRNA

FIG. 4. Polyacrylamide gel electrophoresis of low-molecular-weight RNAs. One milli-curie of carrier-free phosphoric-^{32}P acid was added to a 5-ml midlog-phase culture of *S. cerevisiae* S288C, and 200-μl samples were removed at the indicated times. Each sample was combined with 400 μl of phenol [saturated with 0.01 M Tris–HCl$_2$ (pH 7.6)], 200 μl of distilled water, 100 μl of 0.01 M EDTA, and 5% SDS at 65°C. The mixture was shaken at 60°C for 30 minutes followed by 1 hour at room temperature. The phases were separated by centri-fugation, and a 10-μl portion of the aqueous phase was removed, combined with 10 μl of sample buffer, heated 1 minute at 65°C, and applied to a 1.5 mm × 15 cm × 16 cm slab gel.

There are three stable forms of the 5.8 S RNA in *S. cerevisiae*. The major form (band IA) constitutes approximately 90% of the total 5.8 S RNA, and the two minor forms (both found in band IB) each constitute approximately 5%. The nucleotide sequence of the major form begins pApApApCp. The sequences of the minor forms begin pUpApUpUpApApApAp-ApCp and pApUpApUpUpApApApApApCp. The remaining nucleotide sequences of all three forms appear to be identical (Rubin, 1974). The band designated 8 S shows kinetics expected for an unstable RNA species. The role of this RNA is unknown. From Rubin (1974) with permission of the Federation of European Biochemical Societies.

C.   Detection of Fractionated RNA

1.   AUTORADIOGRAPHY

Preparative gels of ^{32}P-labeled RNAs are autoradiographed directly after electrophoresis. One of the two glass plates of the gel mold is removed, and the gel is covered with a thin layer of plastic film (Saran wrap or an equivalent product). A sheet of x-ray film (Kodak Royal Blue Medical or Auto-Process AP54) is placed on top of the gel for a period of time sufficient to visualize the RNA bands. On a 4-mm slab gel a band with 10^6 cpm is visible after a 5-minute exposure. There are two convenient methods for aligning the x-ray film with the gel. Thin tape with radioactive ink markings can be placed on the corners of the gel before autoradiography. The ink markings and their autoradiographs are then used to align the film and gel. Alternatively, holes can be punched through the x-ray film into the gel with a 16-gauge needle during autoradiography. The holes in the developed film are then aligned with the holes in the gel.

Analytical gels of ^{32}P-labeled RNAs can be dried before autoradiography to prevent the RNA bands from diffusing during long periods of autoradiography. The efficiency of autoradiography is also greater in dried gels. Gels of ^{14}C-labeled RNAs must be dried before autoradiography. Gels can be conveniently dried in about 1 hour under vacuum using the apparatus shown in Fig. 5.

2.   STAINING

a. Methylene Blue.   RNA bands can be located in the gel by straining with methylene blue as described by Peacock and Dingman (1967). The gel is immersed in 1 M acetic acid for 30 minutes to lower the pH before staining

FIG. 5. Apparatus for drying slab gels for autoradiography. The wet slab gel (D) is transferred to a wet piece of Whatman 3 MM paper (C). Care is taken to remove air bubbles between the paper and gel. The gel and paper are placed paper side down on a piece of porous polyethylene (B) (Bel-Art Products, Pequannock, New Jersey; linear polyethylene, ⅛ inch thick F1255). The porous polyehtylene is supported and held flat by a wood block (A). A vacuum can be applied to the porous polyethylene via a series of air channels in the wood block. The gel is covered by a piece of plastic film (Saran wrap or an equivalent product), and the entire assembly is enclosed in a polythene bag. The bag is then evacuated, and the apparatus is placed gel side down on a steel plate maintained at 100°C by a boiling water bath. The gel should dry in about 1 hour.

with 0.2% methylene blue in a mixture of equal volumes of 0.4 M sodium acetate and 0.4 M acetic acid. The gel is destained with successive changes of distilled water. If the RNA is to be recovered from the gel, the staining should be done with 0.2% methylene blue in distilled water. When RNA is extracted from the gel as described in Section IX,E, the methylene blue is removed from the RNA and is extracted into the phenol phase.

b. Ethidium Bromide. The fluorescent dye ethidium bromide, which has been used extensively for staining DNA in polyacrylamide gels, also stains RNA under appropriate conditions. An advantage of ethidium bromide is that its fluorescence is greatly enhanced when it binds to DNA or RNA (Cantor and Tao, 1971). The gel is immersed in a 20 μg/ml solution of ethidium bromide in 1 mM EDTA (pH 7) for 1 hour at room temperature, followed by destaining for 2–3 hours with successive changes of 1 mM EDTA. The fluorescent bands can be observed by illuminating the gel with ultraviolet light in a darkened room. Maximum sensitivity, however, is only obtained by integrating the emitted light photographically. The gel is illuminated with ultraviolet light in a darkroom and is photographed through a red filter using Kodak Tri-X film. The filter screens out the incident ultraviolet light, and thus only the light emitted by the ethidium is recorded. This procedure is at least an order of magnitude more sensitive than staining with methylene blue. 0.01 μg of RNA in a band 1 cm $\times$ 1 mm $\times$ 1.5 mm can be detected after a 5-minute exposure (G. Rubin, unpublished observation). Much longer exposures are also feasible.

D. Further Purification of RNA

Although a single polyacrylamide gel electrophoresis routinely yields both 5 and 5.8 S RNA in greater than 90% purity, techniques such as RNA–DNA hybridization require RNA of higher purity. 5 and 5.8 S RNA suitable for RNA–DNA hybridization or other procedures sensitive to minor contaminants can be obtained by a second gel electrophoresis (Rubin and Sulston, 1973). This can be accomplished without eluting the material from the first gel. The portion of the gel corresponding to the RNA in question is cut out of the gel with a scalpel and partially desalted by immersion in distilled water at 0°C for 30 minutes. The gel slice is then applied to the second gel as shown in Fig. 6. Electrophoresis is carried out in the same manner as for the first gel.

E. Recovery of RNA from Preparative Polyacrylamide Gels

The portion of the gel containing the RNA to be eluted is homogenized in a motor-driven Potter-Elvehjem homogenizer with three times its volume

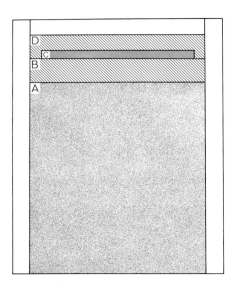

Fig. 6. Reelectrophoresis of RNA without prior elution of the RNA from the gel. The running gel (A) and a stacking gel (B) are poured as described in Section IX,A.

The partially desalted gel slice (C) is pushed into the gel mold onto the surface of the stacking gel (B). Another stacking gel (D), identical to the first stacking gel except for the addition of bromphenol blue to a final concentration of 0.01%, is poured over the gel slice. Water or water-saturated 2-butanol is layered over this gel as described in Section IX,A. After the stacking gel polymerizes, the gel is electrophoresed in the apparatus as shown in Fig. 3b. The bromphenol blue should concentrate and enter the running gel (A) as a single sharp band. Likewise, the RNA is eluted from the gel slice (C) and concentrated so that it also enters the running gel (A) as a sharp band.

of 0.5 M NaCl, 0.1 M Tris–HCl, and 10 mM EDTA (pH 9.1) and two times its volume of water-saturated phenol. The homogenizer and all the glassware used in the elution procedure are pretreated with repelcote (Hopkin & Williams, Essex, England) to siliconize their surfaces, as described by Barrell (1971). Nonradioactive carrier RNA can be added at a concentration of 1 A_{260} unit/ml of buffer. The homogenized mixture is allowed to stand on ice for 30 minutes. The phases are separated by centrifugation for 15 minutes in a clinical centrifuge at 4°C. The aqueous phase is removed without disturbing the particles of gel at the phenol–water interface. A volume of buffer equal to the volume removed is added to the lower phase, and the tube is mixed on a vortex mixer and allowed to stand for 10 minutes on ice. The phases are again separated, and the pooled aqueous phases are centrifuged for 20 minutes at 10,000 g to remove any remaining particles of gel. The eluted RNA is then precipitated by the addition of 2 vol of ethanol at −20°C.

In some circumstances, such as when the RNA is to be used for DNA–RNA hybridizations, it is undesirable to add carrier RNA. If the concentration of RNA in the aqueous phase is less than 20 μg/ml, it is necessary to concentrate the pooled aqueous phases in order to achieve quantitative precipitation of the RNA with ethanol. This can be accomplished with complete recovery of the RNA by extracting water from the aqueous phase with phenol. Solid phenol is melted and added to the pooled aqueous phases. The phenol extracts approximately 40% of its volume of water and thus concentrates the RNA. Alternatively, the aqueous phase can be extracted with ether to remove dissolved phenol and then concentrated by evaporation.

X. High-Molecular Weight rRNA Precursors

Yeasts, like other eukaryotes, transcribe the two high-molecular-weight RNAs and the 5.8 S RNA as a single precursor molecule which contains one of each of the RNA components and some excess nonribosomal RNA. The precursor is subsequently methylated and cleaved in several steps. The intermediates in this process have been isolated by fractionation of pulse-labeled total cell RNA (Taber and Vincent, 1969; Retèl and Planta, 1970a; Van den Bos et al., 1971; Klootwijk et al., 1972; Smitt et al., 1972; Van den Bos and Planta, 1973; Udem and Warner, 1972). Total cell RNA can be prepared by phenol–SDS extraction of disrupted whole cells (Van den Bos and Planta, 1971) or of spheroplasts (Udem and Warner, 1972). Fractionation of the various precursor RNAs can be accomplished either by sucrose gradient sedimentation (Fig. 7a) or by polyacrylamide gel electrophoresis (Fig. 7b).

XI. mRNA

Many yeast mRNAs contain a polyadenylate sequence covalently linked to their 3′ ends (McLaughlin et al., 1973; Reed and Wintersberger, 1973). These mRNAs can be purified on an analytical or preparative scale by binding to immobilized poly dT or poly rU. The relevant procedures are described in detail by Kates (1973) in Volume VII of this series. An alternative analytical procedure, based on selective binding of ribosomal subunits to Millipore filters, was used by Hartwell et al. (1970) to distinguish mRNA from rRNA.

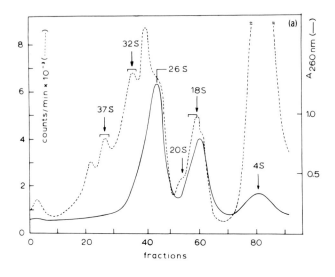

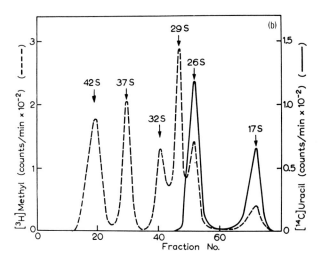

FIG. 7. Separation of rRNA precursors on sucrose gradients and polyacrylamide gels.
(a) Sucrose gradient sedimentation of rapidly labeled yeast RNA. RNA, pulse-labeled for
4 minutes with methionine-methyl-³H was prepared from whole cells. The RNA was fractionated by centrifugation through a linear 5–20% sucrose gradient made in 0.1 M NaCl
and 0.01 M EDTA (pH 5.2) in a Beckman SW 27.1 rotor at 23,000 rpm for 16 hours.
Fractions were collected, and the absorbance at 260 nm (solid line) and radioactivity
(broken line) were determined. From Van den Bos *et al.* (1972) with the permission of the
Federation of European Biochemical Societies. (b) Polyacrylamide gel electrophoresis of the
methionine-methyl-³H labeled nuclear RNA fraction of yeast spheroplasts. The nuclear RNA
fraction was mixed with an appropriate amount of ¹⁴C-labeled rRNA and electrophoresed on
a 2% polyacrylamide gel containing 1% agarose. After electrophoresis the gel was sliced, and
the radioactivity in each slice determined. From Retèl and Planta (1970a) with permission of
Elsevier Scientific Publishing Company.

REFERENCES

Baev, A. A., Venkstern, T. V. Mirzabekov, A. D., Krutilina, A. I., Li, L., and Aksel'rod, V. D. (1967). *Mol. Biol.* **1**, 754.

Barrell, B. G. (1971) *Procedures Nucl. Acid Res.* **2**, 751.

Battaner, E., and Vazquez, D. (1971). In "Methods in Enzymology" (K. Moldave and L. Grossman, eds.), Vol. 20, Part C, p. 446. Academic Press, New York.

Bonnet, J., Ebel, J. P., and Dirheimer, G. (1971). *FEBS (Fed. Eur. Biochem. Soc.) Lett.* **15**, 286.

Brawerman, G. (1973). In "Methods in Cell Biology" (D. M. Prescott, ed.), Vol. VII, p. 2. Academic Press, New York.

Cantor, C. R., and Tao, T. (1971). *Procedures Nucl. Acid Res.* **2**, 31.

Chang, S. H., Kuo, S., Hawkins, E., and Miller, N. R. (1973). *Biochem. Biophys. Res. Commun.* **51**, 951.

deRobichon-Szulmajster, H., and Surdin-Kerian, Y. (1971). In "The Yeasts" (A. H. Rose and J. S. Harrison, eds.), Vol. 2, p. 335. Academic Pres, New York.

De Wachter, R., and Fiers, W. (1971). In "Methods in Enzymology" (K. Moldave and L. Grossman, eds.), Vol. 21, Part D, p. 167. Academic Press, New York.

Dingman, C. W., and Peacock, A. C. (1971). *Procedures Nucl. Acid Res.* **2**, 623.

Fink, G. R. (1970). In "Methods in Enzymology" (H. Tabor and C. W. Tabor, eds.), Vol. 17A, p. 59. Academic Press, New York.

Gangloff, J., Keith, G., Ebel, J. P., and Dirheimer, G. (1972). *Biochim. Biophys. Acta* **259**, 198.

Greenberg, H., and Penman, S. (1966). *J. Mol. Biol.* **21**, 527.

Günther, T., and Kattner, W. (1968). *Z. Naturforsch. B* **23**, 77.

Harold, F. M. (1966). *Bacteriol. Rev.* **30**, 772.

Hartwell, L. H., and McLaughlin, C. S. (1968). *J. Bacteriol.* **96**, 1664.

Hartwell, L. H., McLaughlin, C. S., and Warner, J. R. (1970). *Mol. Gen. Genet.* **109**, 42.

Holley, R. W., Apgar, J., Doctor, B. P., Farrow, J., Marini, M. A., and Merrill, S. H. (1961). *J. Biol. Chem.* **236**, 200.

Holley, R. W., Apgar, J., Everett, G. A., Madison, J. I., Marquisee, M., Merrill, S. H., Penswick, J. R., and Zamir, A. (1965). *Science* **147**, 1462.

Kates, J. (1973). In "Methods in Cell Biology" (D. M. Prescott, ed.), Vol. VII, p. 53. Academic Press, New York.

Keith, G., Roy, A., Ebel, J. P., and Dirheimer, G. (1971). *FEBS (Fed. Eur. Biochem. Soc.) Lett.* **17**, 306.

Klootwijk. J., Van den Bos, R. C., and Planta, R. J. (1972). *FEBS (Fed. Eur. Biochem. Soc.) Lett.* **27**, 102.

Kowalski, S., Yamane, T., and Fresco, J. R. (1971). *Science* **172**, 285.

Kuntzel, B., Weissenbach, J., and Dirheimer, G. (1972). *FEBS (Fed. Eur. Biochem. Soc.) Lett.* **25**, 189.

Laemmli, U. K. (1970). *Nature (London)* **227**, 680.

McLaughlin, C. S., Warner, J. R., Edmonds, M., Nakazato, H., and Vaughan, M. H. (1973). *J. Biol. Chem.* **248**, 1466.

McPhie, P., Hounsell, J., and Gralzer, W. (1966). *Biochemistry* **5**, 988.

Madison, J. T., and Kung, H. (1967). *J. Biol. Chem.* **242**, 1324.

Madison, J. T., Boguslawski, S. J., and Teetor, G. H. (1972). *Science* **176**, 687.

Marcus, L., and Halvorson, H. O. (1967). In "Methods in Enzymology" (L. Grossman and K. Moldave, eds.), Vol. 12, Part A, p. 498. Academic Press, New York.

Martin, T. E., and Hartwell, L. H. (1970). *J. Biol. Chem.* **245**, 1504.

Martin, T. E., and Wool, I. (1968). *Proc. Nat. Acad. Sci. U.S.* **60**, 569.

Martin, T. E., Rolleston, F. S., Low, R. B., and Wool, I. G. (1969). *J. Mol. Biol.* **43**, 135.

Merrill, C. R. (1968). *Biopolymers* **6**, 1727.

Monier, R. (1971). *Procedures Nucl. Acid Res.* **2**, 618.

Moriyama, Y., Ip, P., and Busch, H. (1970). *Biochim. Biophys. Acta* **209**, 161.

Muramatsu, M. (1973). *In* "Methods in Cell Biology" (D. M. Prescott, ed.), Vol. VII, p. 24. Academic Press, New York.

Nakanishi, K., Furutachi, N., Funamizo, M., Grunberger, D., and Weinstein, I. B. (1970). *J. Amer. Chem. Soc.* **92**, 7617.

Peacock, A. C., and Dingman, C. W. (1967). *Biochemistry* **6**, 1818.

RajBhandary, U. L., Chang, S. H., Stuart, A., Faulkner, R. D., Hoskinson, R. M., and Khorana, H. G. (1968). *J. Biol. Chem.* **243**, 556.

Reed, J., and Wintersberger, E. (1973). *FEBS (Fed. Eur. Biochem. Soc.) Lett.* **32**, 213.

Retèl, J., and Planta, R. J. (1970a). *Biochim. Biophys. Acta* **199**, 286.

Retèl, J., and Planta, R. J. (1970b). *Biochim. Biophys. Acta* **224**, 458–469.

Retèl, J., Van den Bos, R. C., and Planta, R. J. (1969). *Biochim. Biophys. Acta* **195**, 370.

Rubin, G. M. (1973). *J. Biol. Chem.* **248**, 3860–3875.

Rubin, G. M. (1974). *Eur. J. Biochem.* **41**, 197.

Rubin, G. M., and Sulston, J. E. (1973). *J. Mol. Biol.* **79**, 521.

Schweizer, E., MacKechnie, C., and Halvorson, H. O. (1969). *J. Mol. Biol.* **40**, 261.

Simsek, M., and RajBhandary, U. L. (1972). *Biochem. Biophys. Res. Commun.* **49**, 508.

Smith, C. J., Teh, H.-S., Ley, A. N., and O'Obrenan, P. (1973). *J. Biol. Chem.* **248**, 4475.

Smitt, W. W., Vlak, J. M., Schiphof, R., and Rozijn, M. H. (1972). *Exp. Cell Res.* **71**, 33.

Sogin, S. J., Haber, J. E., and Halvorson, H. O. (1972). *J. Bacteriol.* **112**, 806.

Stanley, W. M., Jr., and Wahba, A. J. (1967). *In* "Methods in Enzymology" (L. Grossman and K. Moldave, eds.), Vol. 12, Part A, p. 524. Academic Press, New York.

Taber, R. L., and Vincent, W. S. (1969). *Biochim. Biophys. Acta* **186**, 317.

Traugh, J. A., and Traut, R. R. (1973). *In* "Methods in Cell Biology" (D. M. Prescott, ed.), Vol. VII, p. 68. Academic Press, New York.

Udem, S. A., and Warner, J. R. (1972). *J. Mol. Biol.* **65**, 227.

Udem, S. A., Kaufman, K., and Warner, J. R. (1971). *J. Bacteriol.* **105**, 101.

Van den Bos, R. C., and Planta, R. J. (1971). *Biochim. Biophys. Acta* **247**, 175.

Van den Bos, R. C., and Planta, R. J. (1973). *Biochim. Biophys. Acta* **294**, 464.

Van den Bos, R. C., Retèl, J., and Planta, R. J. (1971). *Biochim. Biophys. Acta* **232**, 494.

Van den Bos, R. C., Klootwijk, J., and Planta, R. J. (1972). *FEBS (Fed. Eur. Biochem. Soc.) Lett.* **24**, 93.

Warner, J. R. (1971). *J. Biol. Chem.* **246**, 447.

Watts, R. L., and Mathias, A. P. (1967). *Biochim. Biophys. Acta* **145**, 828.

Weissenbach, J., Martin, R., and Dirheimer, G. (1972). *FEBS (Fed. Eur. Biochem. Soc.) Lett.* **28**, 353.

Yoshida, M. (1973). *Biochem. Biophys. Res. Commun.* **50**, 779.

Zachau, H. G., Dütting, D., and Feldmann, H. (1966a). *Angew. Chem., Int. Ed. Engl.* **5**, 422.

Zachau, H. G., Dütting, D., and Feldmann, H. (1966b). *Hoppe-Seyler's Z. Physiol. Chem.* **347**, 212.

Chapter 5

DNA-Dependent RNA Polymerases from Yeasts

H. PONTA AND U. PONTA

Max-Planck-Institut für Molekulare Genetik,
Berlin-Dahlem, Germany

AND E. WINTERSBERGER

Physiologisch-Chemisches Institut der Universität Würzburg,
Würzburg, West Germany

I. Introduction

All eukaryotic cells investigated so far contain multiple forms of DNA-dependent RNA polymerases (Roeder and Rutter, 1969, 1970a,b; Roeder *et al.*, 1970; Strain *et al.*, 1971; Doenecke *et al.*, 1972; Mondal *et al.*, 1972; Seifart *et al.*, 1972; Roeder, 1974). These enzymes are localized in different compartments. One enzyme is found in the nucleolus, and its function seems to be the synthesis of rRNA. Another RNA polymerase can be isolated from the nucleoplasm, and it is believed to synthesize the bulk of nucleoplasmic RNA, including mRNA. A third enzyme activity has been obtained from the nucleus in some cases, but its role is still unknown. These different RNA polymerases can be clearly distinguished by their chromatographic behavior, their response to the action of α-amanitin, their subunit structure, and their sedimentation value (Kedinger *et al.*, 1970, 1971; Blatti *et al.*, 1970; Chambon *et al.*, 1970; Mandel and Chambon, 1971; Chesterton and Butterworth, 1971; Weaver *et al.*, 1971; Seifart and Sekeris, 1969).

Other DNA-dependent RNA polymerases are localized in the mitochondria and in the chloroplasts of eukaryotic cells. Mitochondrial and chloroplast enzymes differ from nuclear enzymes in that they are smaller and are solubilized only by methods different from those used for nuclear enzymes (Tsai *et al.*, 1971; Kützel and Schäfer, 1971; Bottomley *et al.*, 1971; Wintersberger, 1972; Scragg, 1971).

In this article we describe methods for the purification of DNA-dependent RNA polymerases from yeasts.

II. Purification Procedures

A. Materials

Two different strains of *Saccharomyces cerevisiae* are used. One is a haploid leucine-requiring mutant. The other is a commercially obtained diploid wild-type strain. Essentially identical results are obtained with both strains.

1. COLUMN MATERIALS

DEAE-cellulose (Whatman DE-52), preswollen
DEAE-Sephadex A-25 (Pharmacia)
Phosphocellulose (Serva)
DNA-agarose (prepared according to Schaller *et al.*, 1972)

2. BUFFER SOLUTIONS

Buffer A: 50 mM Tris–HCl (pH 7.9), 5 mM MgCl$_2$, 0.1 mM EDTA, 1mM dithioerythritol, and glycerol. Except where specified, the glycerol concentration is 10%. The buffer also contains a saturating amount of the protease inhibitor phenylmethylsulfonyl fluoride.

Buffer B: 20 mM Tris–HCl (pH 7.5), 1 mM 2-mercaptoethanol, 10% glycerol, 100 mM NaCl, and 1 mM EDTA.

Buffer C: 0.25 M mannitol, 20 mM Tris–HCl (pH 7.5), and 1 mM EDTA.

Buffer D: 50 mM Tris–HCl (pH 7.9), 0.1 mM EDTA, 5 mM 2-mercapto-ethanol, and 10% glycerol.

Buffer E: 10 mM Tris–HCl (pH 7.9), 10 mM MgCl$_2$, 0.1 mM EDTA, 0.1 mM dithioerythritol, and 5% glycerol.

3. DETERMINATION OF RNA POLYMERASE ACTIVITY

The standard assay mixture contains in a total volume of 0.125 ml: 0.05 M Tris–HCl (pH 7.9), 1.6 mM MnCl$_2$, 1 mM EDTA, 1 mM 2-mercapto-ethanol, 100 μM each of ATP, GTP, and CTP, 25 μM UTP plus 0.25 μC of UTP-^{3}H (specific activity 3–13 Ci/mmole), 12.5 μg heat-denatured calf thymus DNA and enzyme. After incubation at 35° C for 15 minutes, samples are cooled in ice and 100-μl aliquots are applied to Whatman GF/C filters (25 mm ϕ) placed in cold 5% trichloroacetic acid (TCA) plus 1% PP for 10 minutes and then washed thoroughly with cold TCA on a Büchner funnel. Finally, the filters are washed with a methanol–ether (1:1) mixture and dried at 90°–100° C. The filters are placed in a toluene scintillator and counted.

One unit of RNA polymerase is that amount of enzyme that incorporates 1 nmole UMP at 35°C in 15 minutes into acid-insoluble product. Specific activity is expressed as units per milligram of protein. Protein is assayed according to the procedure of Lowry et al. (1951).

B. Separation of RNA Polymerases A and B

All operations are carried out at 0°–4°C. The procedure is given for 500 gm (wet weight) of yeast cells.

1. SOLUBILIZATION OF RNA POLYMERASES A AND B

The cells are washed with water, suspended in 500 ml of buffer C, and disrupted by passing the suspension twice for 1 minute through a Manton Gaulin homogenizer at a pressure of 550 atm, with cooling of the suspension between the two cycles for 10 minutes. DNA in the homogenate is digested with pancreatic DNase (10 μg/ml) for 30 minutes. Whole cells, cell debris, mitochondria, and ribosomes are removed by stepwise centrifugation (5 minutes at 10,000 g, 20 minutes at 45,000 g, and 2 hours at 90,000 g).

2. ADSORPTION OF THE RNA POLYMERASES TO DEAE-CELLULOSE

The clear supernatant (700 ml) is stirred with DEAE-cellulose [200 gm (wet weight) equilibrated with buffer A] for 20 minutes. The filtrate is discarded, and the cellulose is washed with buffer A. The resin is finally resuspended in 700 ml of buffer A containing 0.7 M (NH$_4$)$_2$SO$_4$, stirred for 20 minutes, and filtered again. 42 gm/100 ml of (NH$_4$)$_2$SO$_4$ is added to the filtrate and the precipitate is centrifuged at 50,000 g for 20 minutes. The pellet containing the RNA polymerases is dissolved in 30 ml of buffer A and dialyzed against 4 liters of the same buffer for 3 hours to give fraction 1 (total RNA polymerase activity, 420 to 480 units; specific activity, 0.2 units/mg of protein).

3. SEPARATION OF RNA POLYMERASES A AND B ON DEAE-CELLULOSE

Fraction 1 (60 ml, 35 to 40 mg protein/ml) is diluted with 200 ml of buffer A and chromatographed on DEAE-cellulose (35 $\times$ 2.5 cm). A large amount of proteins without RNA polymerase activity can be removed by washing the column with 300 ml of buffer A. Adsorbed protein is eluted with a linear gradient from 0 to 0.5 M $(NH_4)_2SO_4$ in buffer A (1400 ml) and collected in 18 ml-fractions. Two peaks of RNA polymerase activity are eluted, polymerase A at 0.15–0.18 M $(NH_4)_2SO_4$ and polymerase B at 0.20–0.26 M $(NH_4)_2SO_4$. Active fractions of the two peaks are collected separately and concentrated by precipitation with $(NH_4)_2SO_4$ (42 gm/100 ml). The pellets are dissolved in buffer A and dialyzed extensively against the same buffer to give fractions 2 A and 2 B, respectively (specific activity for fraction 2 A, 0.05 units/mg of protein; specific activity for fraction 2 B; 9.4 units/mg of protein).

C.   Further Purification of RNA Polymerase A

1.   DEAE-CELLULOSE

Fraction 2A (12 ml, 6–8 mg protein/ml) is diluted to a final volume of 50 ml and rechromatographed on a DEAE-cellulose column (15 $\times$ 3 cm). Five-hundred milliliters of a linear gradient from 0 to 0.4 M $(NH_4)_2SO_4$ in buffer A is used for the elution of the adsorbed proteins (flow rate 50 ml/per hour), and fractions of 5 ml are collected. Fractions with polymerase-A activity (50 ml) are pooled and dialyzed against 2 liters of buffer A containing 25% glycerol. The specific activity of this fraction is 0.09 units/mg of protein.

2.   DEAE-SEPHADEX

The dialyzed solution from the preceding purification step (1–1.5 mg/ml of protein) is further purified on a DEAE-Sephadex column (25 $\times$ 0.8 cm). The column is washed with 100 ml of buffer A containing 25% glycerol and 50 mM $(NH_4)_2SO_4$. Enzyme activity is eluted with 300 ml of a linear gradient of the same buffer containing 0.05–0.4 M $(NH_4)_2SO_4$. Fractions of 3 ml are collected at a flow rate of 15 ml per hour. Polymerase A is eluted at 0.17 M $(NH_4)_2SO_4$. Active fractions (15 ml) are pooled and dialyzed against buffer D for 2 hours (specific activity 0.8 units/mg protein).

3. Phosphocellulose

The RNA polymerase-A–containing fractions (0.5–1 mg protein/ml) are finally chromatographed on phosphocellulose (5 × 0.5 cm). Polymerase A is eluted with 200 ml of a linear gradient ranging from 0 to 0.5 M (NH$_4$)$_2$SO$_4$ in buffer D (flow rate of 40 ml per hour). Fractions of 2 ml are collected. Polymerase A is eluted at 0.3 M (NH$_4$)$_2$SO$_4$. Active fractions are combined, dialyzed against buffer A, and concentrated by adsorbing the enzyme on a small (0.5 × 0.5 cm) DEAE-cellulose column followed by elution with a small volume of 0.7 M (NH$_4$)$_2$SO$_4$ in buffer A. Protein-containing fractions are dialyzed against buffer A containing 50% glycerol and stored at −15°C (total protein, 0.8–1 mg; specific activity, 1 unit/mg of protein).

D. Further Purification of RNA Polymerase B

1. DEAE-Cellulose

Fraction 2B (10 ml, 4–6 mg protein/ml) is diluted with buffer A to 60 ml and rechromatographed on DEAE-cellulose (9 × 1 cm). The column is washed with 100 ml of buffer A, and the enzyme is eluted with 500 ml of a linear gradient from 0 to 0.4 M (NH$_4$)$_2$SO$_4$ in buffer A (5-ml fractions). Fractions containing RNA polymerase B are pooled and concentrated by precipitation with (NH$_4$)$_2$SO$_4$ (42 gm/100 ml). The pellet is dissolved in buffer A containing 25% glycerol and dialyzed against the same buffer for 2 hours (8–12 mg protein in 10 ml; specific activity, 28 units/mg of protein).

2. DEAE-Sephadex

The solution of RNA polymerase B is adsorbed to DEAE-Sephadex (10 × 0.5 cm). The column is washed with 50 ml of buffer A containing 25% glycerol. The protein is eluted with a linear gradient (200 ml) from 0 to 0.5 M (NH$_4$)$_2$SO$_4$ in the same buffer. Three-milliliter fractions are collected (flow rate 20 ml per hour). RNA polymerase B is eluted at about 0.3 M (NH$_4$)$_2$SO$_4$. Active fractions are combined, dialyzed, and concentrated by adsorption on and elution from a very small DEAE-cellulose column as described in Section II,C,3 (specific activity, 240 units/mg of protein).

Further purification of RNA polymerase B can be achieved by either one of the procedures described below.

a. DNA-Agarose. 0.1 mg of polymerase B is dialyzed against buffer B and chromatographed on a DNA–agarose column (7 × 0.8 cm). The column is washed first with 20 ml of buffer B, and the protein is eluted with 100 ml of a linear gradient from 0.1 to 0.7 M NaCl in buffer B. Fractions of 2 ml each are collected. Enzyme activity is eluted at 0.45 M NaCl. Fractions containing the bulk of activity are pooled, dialyzed, and concentrated as described above.

b. Phosphocellulose Column. One milligram of RNA polymerase B (see Section II,D,2) dialyzed against buffer D is applied to a phosphocellulose column (3 × 0.5 cm). The column is washed with 20 ml of buffer D, and 100 ml of a linear gradient from 0 to 0.5 M $(NH_4)_2SO_4$ in buffer D is used for elution (2-ml fractions; flow rate 40 ml per hour). The enzyme is eluted at 0.18 M $(NH_4)_2SO_4$. Fractions containing activity are pooled and concentrated on a small DEAE-cellulose column as described above.

c. Chromatography on Biogel A. Two milligrams of polymerase B (see Section II,D,2) dialyzed against buffer A containing 0.5 M KCl is chromatographed on a Biogel A 1.5-m column (45 × 2 cm). The enzyme is eluted with buffer A containing 0.5 M KCl at a flow rate of 20 ml per hour. Fractions containing polymerase activity are pooled, and the protein is concentrated as above. The specific activity for RNA polymerase B after one of the purification steps (Section II,D,3) is 320 to 350 units/mg of protein.

E.   Isolation and Purification of RNA Polymerase C

The large-scale purification method described above is not suitable for the simultaneous purification of RNA polymerase C. Polymerase C is isolated from extracts of yeast by chromatography on DEAE-Sephadex A-25.

1.   Solubilization

Fifty grams of cells (washed and suspended in 100 ml of buffer A with 15% glycerol) are disrupted by shaking with glass beads in a Braun homogenizer for 40 seconds. After decantation from the glass beads, the homogenate is sonified (MSE Model 3000, maximum output, 20 seconds). $(NH_4)_2SO_4$ is added to a final concentration of 0.3 M, and the sonication procedure is repeated. The suspension is mixed with 2 vol of buffer A containing 25% glycerol and centrifuged at 90,000 g for 2 hours. The supernatant is mixed with $(NH_4)_2SO_4$ (42 gm/100 ml). The precipitate is centrifuged at 50,000 g for 20 minutes and resuspended in buffer A containing 25% glycerol.

2.   DEAE-Sephadex

The solution (300–350 mg of protein) is dialyzed against buffer A containing 25% glycerol for several hours and chromatographed on DEAE-Sephadex (30 × 5 cm). After washing the column with 300 ml of buffer A, a linear gradient of 1500 ml of 0.1–0.5 M $(NH_4)_2SO_4$ is applied. Three peaks of activity are obtained. The first elutes at 0.17 M (RNA polymerase A), the second at 0.28 M (RNA polymerase B), and the third at 0.35 M (RNA polymerase C) $(NH_4)_2SO_4$ concentration. The fractions containing RNA polymerase activity are pooled and brought to 60% saturation with $(NH_4)_2SO_4$. The partially pure enzyme C (8 mg) is dissolved in buffer A (25% glycerol) and dialyzed. This fraction is rechromatographed on a second DEAE-Sephadex column (15 × 0.8 cm). The protein is eluted with a linear gradient from 0.1 to 0.5 M $(NH_4)_2SO_4$ in the same buffer (80 ml; fractions of 2 ml are collected). Fractions containing RNA polymerase C activity are combined, and the protein is concentrated as described in Section II,C,3.

F. Purification of Mitochondrial RNA Polymerase

The three enzymes described, RNA polymerases A, B, and C, are localized in the nucleus and/or in the cytoplasm, but not in the mitochondria (Ponta *et al.*, 1971). The solubilization of mitochondrial RNA polymerase requires an approach different from the method used above.

1. PREPARATION OF MITOCHONDRIA

Washed yeast cells (200 gm) are suspended in cold buffer C to a final volume of 300 ml. The cells are homogenized with glass beads (diameter 0.45–0.50 mm) in a Braun homogenizer (Braun, Melsungen, Germany) by shaking at 0°C at 4000 cycles per minute for 25 seconds. The homogenate is decanted from the glass beads and centrifuged for 15 minutes at 1500 g. Centrifugation of the supernatant for 30 minutes at 30,000 g results in the separation of a crude mitochondrial pellet. The mitochondria are washed once by resuspension in buffer C and centrifuged as before. The washed mitochondrial pellet is suspended in 80 ml of buffer C, and $MgCl_2$ and pancreatic DNase are added to 10 mM and 100 μg/ml, respectively. The suspension is incubated at 0°C for 30 minutes, and the mitochondria are then reisolated by centrifugation (30 minutes at 40,000 g) and washed once with buffer C.

2. SOLUBILIZATION OF MITOCHONDRIAL RNA POLYMERASE

Two grams (wet weight) of mitochondrial pellet is suspended in 10 ml of buffer E and homogenized in an Ultra-Turrax three times for 15 seconds (0°C, maximal speed) (type TP 18/2, Jankel & Kunkel KG, Staufen im Breisgau, West Germany). After dilution with an equal volume of the same buffer, mitochondrial DNA is digested with DNase (20 μg/ml, 30 minutes, 0°). An extraction with 0.5 M KCl (30 minutes, 0°) follows. After centrifugation at 30,000 rpm for 90 minutes, solid $(NH_4)_2SO_4$ is added to the clear extract to 50% saturation. The precipitate is collected by centrifugation, dissolved in 4 ml of buffer A (25% glycerol) containing 50 mM (NH$_4$)$_2$SO$_4$, and dialyzed against the same buffer for 2 hours.

3. CHROMATOGRAPHY ON DEAE-CELLULOSE

The dialyzed solution is applied to a DEAE-cellulose column [1.8 × 20 cm, for chromatography of material obtained from 4–6 gm (wet weight) of mitochondria] which has previously been well equilibrated with buffer A. The column is first washed with 30 ml of buffer A containing 50 mM (NH$_4$)$_2$SO$_4$, and the RNA polymerase is then eluted by 160 ml of a linear gradient from 0.05 to 0.5 M (NH$_4$)$_2$SO$_4$ in buffer A. Fractions of 3.5 ml each are collected and tested for RNA polymerase activity. Fractions with enzyme activity are combined and concentrated 10-fold by either (NH$_4$)$_2$SO$_4$ precipitation or by ultrafiltration using an Amicon ultrafiltration cell. Concentrated enzyme solutions are dialyzed against buffer A.

4. GLYCEROL-GRADIENT CENTRIFUGATION

The dialyzed enzyme solution is layered onto 25 ml of a 15–30% glycerol gradient (in buffer A) and centrifuged at 23,000 rpm at 0°C for 24 hours (SW 25 rotor of a Spinco centrifuge). Peak fractions are combined and dialyzed against buffer A containing 50% glycerol and 50 mM (NH$_4$)$_2$SO$_4$. This enzyme preparation can be stored at −20°C for several weeks; loss of activity is about 20% in 1 month.

An alternative to the method described above has recently been used to prepare homogeneous mitochondrial RNA polymerase from yeasts (Rogall and Wintersberger, 1974). It involves preparation of mitochondria from yeast spheroplasts by a modification of the procedure of Kováč et al. (1968), and purification of the mitochondria by DNase treatment and repeated differential centrifugation. Solubilization of RNA polymerase is achieved using the Ultra-Turrax homogenizer as described above, or with the aid of the detergent Nonidet P-40 (final concentration 0.5%) in the presence of 0.5 M KCl. Purification of the polymerase is again carried out by chromatography of DEAE-cellulose, followed by glycerol gradient centrifugation in the presence or absence of 0.5 M KCl (Küntzel and Schäfer, 1971).

III.   Characterization of the RNA Polymerases of Yeasts

The final enzyme preparations obtained for RNA polymerases A and B are fairly homogeneous. Both enzymes contain two large subunits with molecular weights of about 190,000 and 135,000 for polymerase A, and 175,000 and 140,000 for enzyme B, in a 1:1 ratio (Ponta et al., 1972; Dezelee and Sentenac, 1973). The sedimentation constants of RNA polymerases A and B are found to be 18 and 13 S, respectively, in the presence of 0.5 M $(NH_2)_4SO_4$, and 24 and 21 S, respectively, in the absence of the salt (Ponta et al., 1972). Polymerase C has not yet been sufficiently purified to determine molecular weight and sedimentation constants. Yeast mitochondrial RNA polymerase consists of subunits with a molecular weight of 67,000 (Rogall and Wintersberger, 1974) which very readily form high-molecular-weight aggregates.

Only RNA polymerase B is sensitive to α-amanitin, while all nuclear enzymes are insensitive to rifampicin (Ponta et al., 1971; Dezelee et al., 1972); reports on the mitochondrial enzyme have been conflicting (Tsai et al., 1971; Wintersberger, 1972; Rogall and Wintersberger, 1974; Scragg, 1971).

All polymerases prefer denatured to native calf thymus DNA. Mitochondrial DNA is the best template tested for the mitochondrial enzyme, and nuclear enzymes are stimulated by this DNA only to a low degree. Native yeast DNA prepared in the usual manner is a very poor template for yeast RNA polymerases. A method for the preparation of a yeast DNA that is used with higher efficiency as template by the yeast RNA polymerases is described in Chapter 6.

IV. Preparation of Yeast DNA Active as Template for Yeast RNA Polymerase

Yeast DNA was prepared either from yeast nuclei (Bhargava and Halvorson, 1971) using Marmur's procedure (Marmur, 1961) or according to the following modification: Nuclei from 20 gm (wet weight) of cells are lysed in 1 M NaCl–0.1 M EDTA (pH 8) at 55°C by the addition of sodium N-lauroyl sarcosinate to a final concentration of 2%. The lysate is deproteinized once by shaking at room temperature with an equal volume of chloroform isoamyl alcohol (24:1) for 60 minutes. The phases are separated, and protein is sedimented by centrifugation at 3000 g for 20 minutes. DNA is precipitated from the clear aqueous phase by the addition of 2 vol of ethanol, wound onto a glass rod, and dissolved in 0.1 × SSC. The DNA solution is transferred to a dialysis bag together with amylase (100 μg/ml), pancreatic RNase (100 μg/ml), and T1 RNase (100 units/ml). The DNA is dialyzed at 4°C against 1 × SCC overnight. Cesium chloride is then added to give a final density of 1.7, and the DNA is centrifuged at 35,000 rpm and 20°C for 48 hours (50 Ti rotor or a Spinco centrifuge). The gradients are fractionated, the absorbance of each fraction is determined, and the peak fraction containing the DNA is pooled and dialyzed extensively against 20 mM Tris–HCl (pH 7.9). To avoid contamination of the DNA preparation with a very powerful inhibitor (possibly a polyphosphate, Ponta *et al.* 1974, it is useful to dialyze the fractions of the cesium chloride gradient and to test for inhibition of transcription. The inhibiting activity bands at higher density than the DNA does. DNA actively transcribed by yeast RNA polymerases A and B can also be obtained by the method of Bhargava *et al.* (1972).

V. Discussion

The procedures for purification of yeast RNA polymerases outlined here can be conveniently used to purify RNA polymerases A and B from up to 1 kg of yeast. While enzyme B can be readily obtained in soluble form, as reported earlier (Frederick *et al.*, 1969; Dezelee *et al.*, 1970; Sebastian *et al.*, 1973), solubilization of polymerase A requires incubation with DNase. Polymerase C can only be solubilized by sonication of the cell homogenate under controlled conditions (Ponta *et al.*, 1971; Brogt and Planta, 1972; Adman *et al.*, 1972), and solubilization of the mitochondrial polymerase is possible only by use of the Ultra-Turrax and high-salt extraction (Tsai *et al.*, 1971; Wintersberger, 1972). These latter facts have, up to now, excluded large-scale purification of polymerase C and mitochondrial polymerase.

RNA polymerases from yeasts have many properties in common with the corresponding enzymes of multicellular organisms. It seems that RNA polymerase A corresponds to the nucleolar enzyme and therefore most likely functions in the synthesis of rRNA in *S. cerevisiae*, and that RNA polymerase B corresponds to the nucleoplasmic polymerase of animal cells and might therefore be engaged in synthesis of mRNA. This correlation of special functions of the different RNA polymerases has not yet been proven directly. The use of homologous native yeast DNA in *in vitro* synthesis of RNA by yeast RNA polymerases, and competitive hybridization experiments with these RNAs, should be powerful tools to test these assumptions. Furthermore, the use of a homologous template for transcription by yeast RNA polymerases allows investigation of the molecular function of the different polymerases in the cell.

One should keep in mind that, because of the complexity of the eukaryotic transcription machinery, the conditions of *in vitro* RNA synthesis may still be incorrect, e.g., the RNA polymerases may have lost essential factors during isolation, and the question which yeast DNA preparation is an appropriate template has not been answered.

ACKNOWLEDGMENT

We thank Dr. Peter Herrlich for critically reading the manuscript.

REFERENCES

Adman, R., Schultz, L. D., and Hall, B. D. (1972). *Proc. Nat. Acad. Sci. U.S.* **69**, 1702.

Bhargava, M. M., and Halvorson, H. O. (1971). *J. Cell Biol.* **49**, 423.

Bhargava, M. M., Cramer, J. H., and Halvorson, H. O. (1972). *Anal. Biochem.* **49**, 276.

Blatti, S. P., Ingles, C. J., Lindell, T. J., Morris, P. W., Weaver, R. F., Weinberg, F., and Rutter, W. J. (1970). *Cold Spring Harbor Symp. Quant. Biol.* **35**, 649.

Bottomley, W., Smith, H. J., and Bogorad, L. (1971). *Proc. Nat. Acad. Sci. U.S.* **68**, 2412.

Brogt, T. M., and Planta, R. J. (1972). *FEBS (Fed. Eur. Biochem. Soc.) Lett.* **20**, 47.

Chambon, P., Gissinger, F., Mandel, J. L., Jr., Kedinger, C., Gniazdowski, M., and Meihlac, M. (1970). *Cold Spring Harbor Symp. Quant. Biol.* **35**, 693.

Chesterton, C. J., and Butterworth, P. H. W. (1971). *FEBS (Fed. Eur. Biochem. Soc.) Lett.* **15**, 181.

Dezelee, S., and Sentenac, A. (1973). *Eur. J. Biochem.* **34**, 41.

Dezelee, S., Sentenac, A., and Fromageot, P. (1970). *FEBS (Fed. Eur. Biochem. Soc.) Lett.* **7**, 220.

Dezelee, S., Sentenac, A., and Fromageot, P. (1972). *FEBS (Fed. Eur. Biochem. Soc.) Lett.* **21**, 1.

Doenecke, D., Pfeiffer, C., and Sekeris, C. E. (1972). *FEBS (Fed. Eur. Biochem. Soc.) Lett.* **21**, 237.

Frederick, E. W., Maitra, U., and Hurwitz, J. (1969). *J. Biol. Chem.* **244**, 413.

Kedinger, C., Gniazdowski, M., Mandel, J. L., Gissinger, F., and Chambon, P. (1970). *Biochem. Biophys. Res. Commun.* **38**, 165.

Kedinger, C., Nuret, P., Chambon, P. (1971). *FEBS (Fed. Eur. Biochem. Soc.) Lett.* **15**, 169.

Kováĉ, L., Bednárová, H., and Greksák, M. (1968). *Biochim. Biophys. Acta* **153**, 32.

Küntzel, H., and Schäfer, K. P. (1971). *Nature (London), New Biol.* **231**, 265.

Lowry, O. H., Rosebrough, N. J., Farr, A. L., and Randall, R. J. (1951). *J. Biol. Chem.* **193**, 265.

Mandel, J. L., and Chambon, P. (1971). *FEBS (Fed. Eur. Biochem. Soc.) Lett.* **15**, 175.

Marmur, J. (1961). *J. Mol. Biol.* **3**, 208.

Mondal, H., Ganguly, A., Das, A., Mandal, R. K., and Biswas, B. B. (1972). *Eur. J. Biochem.* **28**, 143.

Ponta, H., Ponta, U., and Wintersberger, E. (1971). *FEBS (Fed. Eur. Biochem. Soc.) Lett.* **18**, 204.

Ponta, H., Ponta, U., and Wintersberger, E. (1972). *Eur. J. Biochem.* **29**, 110.

Ponta, H., Ponta, U., and Wintersberger, E. (1974). *Eur. J. Biochem.* (in press).

Roeder, R. G. (1974). *J. Biol. Chem.* **249**, 241.

Roeder, R. G., and Rutter, W. J. (1969). *Nature (London)* **224**, 234.

Roeder, R. G., and Rutter, W. J. (1970a). *Biochemistry* **9**, 2543.

Roeder, R. G., and Rutter, W. J. (1970b). *Proc. Nat. Acad. Sci. U.S.* **65**, 675.

Roeder, R. G., Reeder, R. H., and Brown, D. D. (1970). *Cold Spring Harbor Symp. Quant. Biol.* **35**, 727.

Rogall, H., and Wintersberger, E. (1974). In preparation.

Schaller, H., Nusslein, C., Bonhoeffer, F. J., Kurz, C., and Nietzschmann, J. (1972). *Eur. J. Biochem.* **26**, 474.

Scragg, A. H. (1971). *Biochem. Biophys. Res. Commun.* **45**, 701.

Sebastian, J., Bhargava, M. M., and Halvorson, H. O. (1973). *J. Bacteriol.* **114**, 1.

Seifart, K. H., and Sekeris, C. E. (1969). *Eur. J. Biochem.* **7**, 408.
Seifart, K. H., Benecke, B. J., and Juhasz, P. P. (1972). *Arch. Biochem. Biophys.* **151**, 519.
Strain, G. C., Mullinix, K. P., and Bogorad, L. (1971). *Proc. Nat. Acad. Sci. U.S.* **68**, 2647.
Tsai, M., Michaelis, G., and Criddle, R. S. (1971). *Proc. Nat. Acad. Sci. U.S.* **68**, 473.
Weaver, R. F., Blatti, S. P., and Rutter, W. J. (1971). *Proc. Nat. Acad. Sci. U.S.* **68**, 2994.
Wintersberger, E. (1972). *Biochem. Biophys. Res. Commun.* **48**, 1287.

Chapter 6

The Isolation of Yeast Nuclei and Methods to Study Their Properties

JOHN H. DUFFUS

*Department of Brewing and Biological Sciences,
Heriot-Watt University, Edinburgh,
Scotland, United Kingdom*

I. Introduction

Light and electron microscopy of intact yeast cells has in recent years produced a fairly clear picture of the structure of the yeast nucleus (Matile *et al.*, 1969). However, our understanding of the properties and functioning of the nucleus in biochemical terms has lagged behind. This seems to reflect a lack of awareness of the techniques available. As late as 1970, J. D. Watson, in *Molecular Biology of the Gene*, stated that good techniques had not yet been developed for the isolation of intact yeast nuclei, much of the difficulty being due to the problem of breaking the yeast cell wall. Considering that satisfactory techniques for preparing yeast protoplasts had been known since 1957 (Eddy and Williamson, 1957), and that Rozijn and Tonino had published their method for isolating nuclei from protoplasts in 1964, Watson's statement highlights the need for the current review.

Two main approaches have been made to the isolation of yeast nuclei. The longer established starts with yeast protoplasts. This method was pioneered by Eddy (1959), but his technique of sonication has proved difficult to repeat. Accordingly, the fundamental technique here must be regarded as that of Rozijn and Tonino (1964), although the subsequent work of May (1971) and Wintersberger *et al.* (1973) suggests that this may have to be modified appreciably with different strains and species of yeasts.

More recently it has proved possible to isolate nuclei from normal yeast cells using an Eaton press (Duffus, 1969; Bhargava and Halvorson, 1971), a modified French press (Bhargava and Halvorson, 1971; Sajdel-Sulkowska *et al.*, 1974), or a Biotec X-Press (Wintersberger *et al.*, 1973). This has the advantage that the cells, and consequently the nuclei, are initially in a normal state, but it has the disadvantages that yields are variable and that the nuclei may suffer considerable damage.

II.   Isolation of Nuclei from Yeast Protoplasts

The first reported method for the isolation of nuclei from yeast protoplasts was that of Rozijn and Tonino (1964). Although it seems likely that the method of May (1971) is preferable, nuclei prepared by the latter method have not yet been quite so thoroughly characterized biochemically, and so both methods are described here. The snail gut enzyme used for obtaining protoplasts may be prepared according to Rost and Venner (1964), or obtained commercially from Industrie Biologique Française S.A., 35–49 Quai du Moulin de Cage, Gennevilliers (Seine), France (Helicase), or from Endo Laboratories, Garden City, L.I., New York (Glusulase). From the former, the enzyme is available in solution or in freeze-dried form. One milliliter of solution is roughly equivalent to 0.1 gm of the freeze-dried enzyme. However, in our experience, the solution is the more effective preparation, although it may contain preservatives which should be removed by ultrafiltration with a noncellulosic filter.

A. Isolation of Nuclei from *Saccharomyces carlsbergensis* NCYC 74 by the Method of Rozijn and Tonino (1964) modified according to Sillevis-Smitt *et al.* (1972a)

1. Protoplasts are prepared from exponentially growing cells (6×10^6 cells/ml) growing in malt extract medium (Wickerham, 1951) The cells are harvested by centrifuging at $2000\,g_{av}$ for 5 minutes and washed by suspension in distilled water and recentrifuging. The yeast pellet is suspended at 10^8 cells/ml in 0.005 M citrate–phosphate buffer (pH 5.8) containing 10% mannitol and 10 mg snail gut enzyme/ml. After about 1 hour in a water bath at 30°C, nearly all the cells should be converted to protoplasts. The suspension is centrifuged ($2000\,g_{av}$ for 5 minutes). The resultant pellet is washed three times with ice-cold acetate buffer (pH 6.0) containing 10% mannitol.

2. The protoplasts are lysed at 4°C by suspending them in 5 vol of a solution (0.02% w/v) of Triton X-100 in PVP medium [8% polyvinylpyrrolidone (MW 40,000), 1.0 mM $MgCl_2$–0.02 M KH_2PO_4 (pH 6.5)]. The lysate is homogenized in a Potter-Elvehjem homogenizer by one or two movements of the tube along the rotating pestle (500 rpm). The tube is kept surrounded with crushed ice.

3. The homogenates are diluted with an equal volume of PVP medium containing 0.06 M sucrose. They are then centrifuged at $3000\,g_{av}$ for 7 minutes at 4°C. The supernatant is discarded, and the crude nuclear sediment is suspended in a small volume of 0.6 M in PVP medium.

4, One milliliter of the crude nuclear suspension is layered onto the following gradient of sucrose solutions in PVP medium: 2.0 ml 2.0 M, 1.0 ml 1.8 M, and 1.0 ml 1.5 M. The gradient is centrifuged at $120,000\,g_{av}$ for 60 minutes. All the supernatant is discarded by careful pipetting or other means. The residual pellet represents purified nuclei. Occasionally, an intact yeast cell may be seen, but the total amount of contaminating material is estimated by Rozijn and Tonino at 1% or less of the nuclear mass. Figure 1 shows an electron micrograph of the nuclear pellet, and the analysis of DNA, RNA, and protein is shown in Table I.

B. Isolation of Nuclei by the Method of May (1971, 1972) from *Kluyveromyces fragilis* H_{32}, formerly *Saccharomyces fragilis* (Van der Walt, 1970) and also from *Saccharomyces carlsbergensis* H_{60}.

1. Protoplasts are prepared from cells growing exponentially at 28°C in the medium described by Ohnishi *et al.* (1966). These cells are harvested by centrifuging at $2000\,g_{av}$ for 2 minutes and washed twice by suspension in distilled water. The method of protoplasting suggested by May is based on

JOHN H. DUFFUS

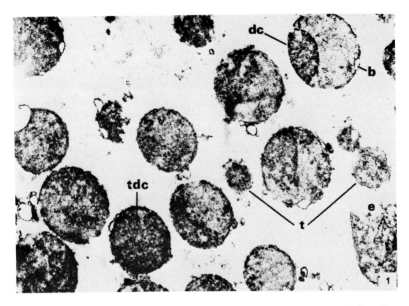

FIG. 1. Section through a pellet of yeast nuclei prepared by the method of Rozijn and Tonino (1964). t, Tangentially sectioned nucleus; b, "bleb" between the nuclear membranes; e, extrusion of nuclear content; dc, dense crescent; tdc, nucleus with a tangentially sectioned dense crescent. ×12,000 (Molenaar *et al.*, 1970, with permission of Academic Press, Inc.).

Rost and Venner (1964), Suomalainen *et al.* (1966), and Schwencke *et al.* (1969). The washed yeast cells [3 gm wet weight] are suspended in 10 ml Tris–EDTA medium (100 mM Tris, 100 mM disodium, EDTA adjusted to pH 8.0 with KOH) and shaken for 10 minutes at 30°C. The cells are then harvested by centrifuging and washed twice with distilled water as before. The final pellet is suspended in 40 ml of protoplast preparation medium (PPM), i.e., phosphate–citrate buffer (McIlvaine, 1921) (pH 6.0), 0.6 M $(NH_4)_2SO_4$, and 0.02 M 2-mercaptoethanol. This suspension is incubated at 30°C with shaking. After 5 minutes, 480 mg of snail gut enzyme dissolved in 10 ml of PPM is added. After a further 30 minutes, protoplasting should be complete. The protoplasts are washed three times with a 1.4 M sorbitol solution by centrifuging (2000 g_{av} for 4 minutes), and the final protoplast pellet is cooled in ice to 4°C.

2. The protoplasts are lysed at 4°C by suspending 1 gm of protoplasts (wet weight) in 10 ml of polymer solution [15–18%, w/v, Ficoll in 0.02 M phosphate buffer (pH 6.5) (Sørensen, 1912), or 0.02 M potassium maleate buffer (pH 6.5)] and homogenized in a Potter-Elvehjem homogenizer with a well-fitting pestle by one or two movements of the tube along the pestle rotating at 600–800 rpm. The tube is surrounded with crushed ice.

3. The homogenate is transferred to a centrifuge tube and centrifuged

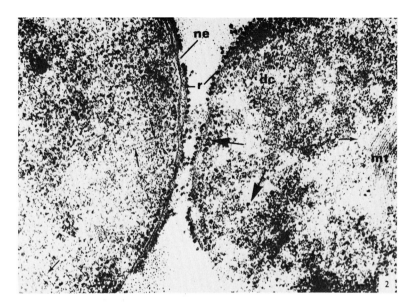

FIG. 2. Parts of two yeast nuclei prepared by the method of Rozijn and Tonino. The ribosomes (r) on the nuclear envelope (ne) are very obvious; dc, dense crescent; mt, microtubules. The small arrows point to 100-Å strands, and the large arrows to 125-Å particles. ×80,000. (Molenaar *et al.*, 1970, with permission of Academic Press, Inc.).

at 3000 g_{av} for 15 minutes at 4°C. The supernatant suspension is carefully removed and layered onto a solution of 30% Ficoll in 0.02 M KH_2PO_4 (pH 6.5) in a suitable centrifuge tube. The resultant step gradient is centrifuged at 100,000 g_{av} for 45 minutes at 4°C. Nuclei are pelleted. Protoplasts remain in suspension at the interface between the two Ficoll solutions, and mitochondria and vacuoles float to the top of the tube. The nuclei obtained are similar in structure to those obtained by Rozijn and Tonino (Figs. 1 and 2), except that they seem to have more invaginations of the nuclear envelope (May, 1973).

The nuclear pellet may be resuspended in the appropriate polymer solution. Electron micrographs of the nuclear preparation, following fixation with 1.5% glutaraldehyde in the polymer solution and postfixing in 1% aqueous osmic acid in 0.1 M KH_2PO_4 buffer (pH 7.0) for 1 hour, show that the nuclear envelope is generally preserved and that the ultrastructure of the mitotic apparatus is similar to that of nuclei *in situ* (May, 1972). Cytoplasmic microtubules attached to the extranuclear part of the kinetochore equivalent, however, are absent. Invaginations of the nuclear envelope are observed in nuclei of *K. fragilis* but are never found in those of *S. carlsbergensis*.

TABLE I

PROPORTIONAL CHEMICAL COMPOSITION OF YEAST NUCLEAR AND CHROMATIN
PREPARATIONS COMPARED WITH WHOLE CELLS

Yeast[a]	Cell			Nucleus			Chromatin		
	DNA	RNA	Pro-tein	DNA	RNA	Pro-tein	DNA	RNA	Pro-tein
Saccharomyces carlsbergensis NCYC 74 (1)[b]	1	109	391	1	3.7	35.3	—	—	—
Saccharomyces cerevisiae (commercial) (2)	1	26	173	—	—	—	1	0.3	4.2
Saccharomyces cerevisiae Y55 (3)	1	82	189	1	2.6	15.6	—	—	—
Saccharomyces cerevisiae (wild type, diploid) (4) [b]	1	89	252	1	4.2	17.3	1	3.8	9.7
Schizosaccharomyces pombe NCYC 132 (5)	1	121	227	1	9.4	115	1	3.5	16

[a]Numbers in parentheses refer to the following: (1) Rozijn and Tonino (1964); (2) Tonino and Rozijn (1966a); (3) Bhargava and Halvorson (1971); (4) Wintersberger et al. (1973) (5) Duffus (1969) and J. H. Duffus (unpublished).

[b]Protoplasts.

C.   Comparison of the Two Methods

Although the method of Rozijn and Tonino (1964) has been available for 10 years, it does not appear to have been exploited by workers outside the originators' research group. One can only speculate that this may reflect some variable, not so far elucidated, which makes the method applicable only to the strain of *S. carlsbergensis* used and possibly only under the described conditions of growth and protoplasting. May (1971) investigated the method as applied to his strain of *K. fragilis* and found that the 8% (w/v) concentration of PVP used by Rozijn and Tonino caused stabilization of the protoplasts, making them somewhat resistant to homogenization. A concentration of 15–20% seemed to be optimal for stabilizing nuclei with minimal stabilization of intact protoplasts. Rozijn and Tonino, however, reported that concentrations of more than 10% PVP made the protoplasts resistant to disruption.

May considered Ficoll (from Pharmacia) or dextran (MW about 80,000) (also from Pharmacia) preferable to PVP, because of the high viscosity of 15–20% PVP solutions. Calcium and magnesium ions were found by May to have a slight stabilizing effect at the concentrations of 0.5 and 2 mM used.

Triton X-100 and Tween 80 have very little effect, either of a beneficial or a harmful nature. This contrasts markedly with the observations of Rozijn and Tonino. They found that the presence of Mg^{2+} in the homogenizing medium was essential, and that protoplasts suspended in PVP medium without Triton X-100 swelled and eventually burst explosively, damaging extruded nuclei. May did not observe this phenomenon. Both Rozijn and Tonino and May observed similar effects of pH on isolated nuclei, but the "capped nuclei" observed at pH 5.0 in preparations from *S carlsbergensis* were not observed in preparations from *K. fragilis* at the same pH. In general terms, May (1972) has defined three fundamental methods for stabilizing nuclei: (1) by buffers with a high H^+ concentration (pH < 6.5), (2) by metal ions (Mg^{2+}, Ca^{2+}, Cd^{2+}, e.g., 10 mM in distilled water), and (3) by unbuffered polymer solutions [e.g., > 25% (w/v) Ficoll in distilled water]. It will be seen that the medium quoted in the description of his technique incorporates two of these factors. Finally, it may be noted that May uses as one of his main criteria for "morphologically intact" nuclei their aspherical form, similar to that of nuclei in intact cells of *K. fragilis*. Where applicable, this appears to be a most important criterion, as any swelling of the nuclei caused by the medium tends to make the nuclei spherical.

The method of nuclear preparation described by May (1971, 1972) is much simpler than that of Rozijn and Tonino (1964). Both methods can be used for complete cell fractionation, but possible deleterious effects of the media used must be borne in mind. Mitochondrial swelling occurs in the medium recommended by May and, although the swelling is reversible (R. May, unpublished), other associated effects may be irreversible. Some of the possible disadvantages of the Rozijn and Tonino medium have already been cited. Two others remain to be mentioned, namely, the presence of Triton X-100, which is known to solubilize many cell components, and the presence of sucrose, which, in the absence of polymers, causes expansion and lysis of *K. fragilis* nuclei (May, 1971) and tends to cause cosedimentation of nuclei and mitochondria (May, 1972). On balance, therefore, the method of May (1971, 1972) seems preferable, and recently Wintersberger *et al.* (1973) successfully applied May's method to a strain of *S. cerevisiae*. An analysis of their preparations is shown in Table I. U. Wintersberger (personal communication) recommends 20% Ficoll in phosphate buffer for the isolation of nuclei from protoplasts of log-phase cells, giving a yield of nuclei of 50% as judged by DNA recovery. Nuclei from protoplasts of stationary-phase cells tend to float in this medium, and so a slightly lower Ficoll concentration must be used. However, the Ficoll concentration is crucial and must not be altered excessively.

III. Isolation of Nuclei from Normal Yeast Cells

This technique was originally devised for use on the fission yeast *Schizosaccharomyces pombe* (Duffus, 1969), but was subsequently modified to make it applicable to *Saccharomyces cerevisiae* (Bhargava and Halvorson, 1971; Sajdel-Sulkowska *et al.*, 1974). The modified technique is probably more generally applicable and gives a purer preparation (Table I). It is therefore the one described here.

A. Isolation of Nuclei from *S. cerevisiae* Y55 by the Method of Bhargava and Halvorson (1971)

1. The yeast cells are grown with continuous shaking at 30°C to an optical density (660 nm) of between 2 and 10 in YEP medium (1% yeast extract, 2% Bacto peptone, and 2% glucose).

2. The cells are harvested by centrifuging at 3300 g_{av} for 5 minutes at 2°–4°C and washed by suspending in 100 vol of distilled water and re-centrifuging.

3. The resultant pellet is suspended in 5 vol of the homogenizing medium (HM) (1 M sorbitol, 20% glycerol, and 5% PVP, average MW 40,000).

4. The suspension is poured into a French press with its outlet tube replaced by a stainless-steel tube, about 2 cm long and with an internal diameter of 0.95 mm. The piston is inserted, and the press placed in an equilibrated mixture of Dry Ice and ethanol. It is allowed to cool for about 7 minutes. The mixture is forced through the outlet at a pressure of 20,000 lb/in².

5. The homogenate is thawed and centrifuged at 5500 g_{av} for 5 minutes. The supernatant (A) is carefully removed with a Pasteur-type pipette and centrifuged at 10,600 g_{av} for 10 minutes. The pellet obtained is a crude preparation of nuclei.

6. The crude preparation is suspended in 10 ml of HM and layered onto a step gradient consisting of 7 ml of 3.2 M sorbitol and 8 ml of 2.6 M sorbitol, both solutions containing 20% glycerol and 5% PVP. The gradient is centrifuged at 13,000 g_{av} for 20 minutes in a swing-out rotor. The nuclei sediment to the top of the 3.2 M sorbitol layer. This nuclear layer is carefully removed with a Pasteur-type pipette, diluted 1:4 with 5% PVP–20% glycerol solution, and centrifuged at 6350 g_{av} for 10 minutes. The pellet consists mainly of nuclei. An analysis of the pellet for DNA, RNA, and protein is shown in Table I. Morphologically, the nuclei are similar to those shown in Fig. 1 and 2, but the percentage that micrographs show to be intact can be very low.

Recently, Sajdel-Sulkowska *et al.* (1974) have described what they claim to be an improvement of the above method in terms of increased RNA polymerase activity in the isolated nuclei. PVP, which inhibits RNA polymerase, is replaced by Ficoll in the homogenizing medium. The fractionation scheme is also altered.

a. The yeast is grown, harvested, and washed as in steps 1 and 2.

b. The washed cell pellet is suspended in 2–5 vol of modified homogenizing medium (MHM) ($1M$ sorbitol, 20% glycerol, and 7% Ficoll). This suspension is passed through a modified French press as described in step 4.

c. The homogenate is thawed and made $1:5$ (w/v) with MHM at $4°C$. The temperature is maintained at $4°C$ in all subsequent steps.

d. The diluted homogenate is centrifuged at $6000\,g_{max}$ for 10 minutes to remove unbroken cells and cell walls. The supernatant from this spin is centrifuged at $18,000\,g_{max}$ for 10 minutes to pellet broken cells and cell walls.

e. The cell wall–free supernatant from step d is centrifuged at $22,000\,g_{max}$ for 20 minutes to pellet the nuclei. Contaminating mitochondria are removed by resuspending the pellet in MHM and centrifuging again at $22,000\,g_{max}$ for 20 minutes. This washing is repeated once. The nuclear pellet is indistinguishable under phase-contrast microscopy from that obtained by the original method of Bhargava and Halvorson. The yield of nuclei on a wet weight or DNA basis is also similar (40–50%).

As an alternative to washing in MHM to remove mitochondria from the nuclear pellet, the suspension of nuclei and mitochondria in MHM is layered on top of $2\,M$ sorbitol containing 30% glycerol and 7% Ficoll. This gradient is centrifuged at $13,000\,g_{av}$ for 20 minutes in a swing-out rotor. The mitochondria remain in the $1\,M$ sorbitol layer, while the nuclei are pelleted at the bottom of the tube.

B. Comments on This Method and Comparison with Preparations from Protoplasts

Although Bhargava and Halvorson claim that a modified French press as described above gives the best results, satisfactory preparations can be obtained using an Eaton press (Eaton, 1962) or an X-Press (Biotec, Sweden) (Wintersberger *et al.*, 1973). Care must be taken to avoid excessive freezing of the cell suspension, and ideally it should be pressed immediately after freezing of the medium is complete. Excessive freezing reduces the yield of nuclei drastically.

In isolating nuclei from *S. cerevisiae*, the presence of PVP-40 or Ficoll throughout the process is reported to be essential to maintain the morphological integrity of the nuclei and, in the later stages of purification, to prevent clumping of mitochondria (Bhargava and Halvorson, 1971).

Above 8% PVP causes the nuclei to become pycnotic. This agrees with the observations of Rozijn and Tonino (1964), but not with those of May (1971). Perhaps it is an effect produced only in nuclei exposed to Triton X-100 or to sorbitol, the latter of which, like sucrose, was shown by May (1971), to cause lysis of the nuclei of *K. fragilis.*

The addition of 3 mM Ca$^+$ to the homogenizing medium makes very little difference, probably because of the high ratio of cells to medium used, ensuring that the cells themselves provide sufficient divalent ions.

The method described by Duffus (1969) for the isolation of nuclei from *Schizo. pombe* differs from the above in omitting PVP or Ficoll from the homogenizing medium. PVP was found to have little effect at the 5% level. Perhaps part of the difference from budding yeasts that this reflects is due to the absence of a vacuole of any size in *S. pombe.* Some experiments suggested that dimethyl sulfoxide might be used to replace glycerol in the homogenizing medium. The gradient used by Duffus also differs slightly in consisting of 5-ml layers of 4.0, 3.0, and 2.0 M sorbitol centrifuged at 62,000 g_{av} for 20 minutes in a swing-out head. The nuclei settle on top of the 4.0 M layer. The layers are separated with a Pasteur pipette, and the nuclear layer, diluted with 5 vol of cold distilled water, is spun at 2000 g_{av} for 20 minutes. The pellet is washed by resuspension in cold 0.5 M sorbitol, followed by further centrifuging. This is repeated until the supernatant is clear. The final pellet represents a partially purified preparation of nuclei with contamination from damaged cells cell walls and membranes. Attempts to improve the purity by washing with Triton X-100 or Tween 80 were unsuccessful, although 1% acetic acid seemed to remove at least some of the contaminants. The nuclei were best visualized following Giemsa staining (Ganesan and Swaminathan, 1958). Electron microscopy frequently did not show intact nuclei, but material similar to nuclear contents without the characteristic doubled membrane and in a diffuse state. This contrasts with the clearly defined nuclei seen by light microscopy and suggests some inadequacy in the techniques used, although similar techniques have been applied successfully to isolated nuclei from budding yeasts (Molenaar *et al.*, 1970). The overall yield of nuclei varies between 10 and 70%, depending primarily on careful control of the initial freezing stage and the number of purification stages employed. With the advent of a reliable method for protoplasting *S. pombe* (Mitchison *et al.*, 1973), a comparison of nuclei prepared as above with nuclei from protoplasts should soon be possible.

Wintersberger *et al.* (1973) compared nuclei prepared by the method of May (1971) from protoplasts with those prepared by the method of Bhargava and Halvorson (1971). They concluded that the nuclei obtained were very similar with regard to DNA, RNA, and protein content. However, the yield with the Bhargava and Halvorson method varied from 10% with one strain of *S. cerevisiae* to 40% with another, as compared to 50% claimed by the original investigators for their strain. The method of May was more reliable. The homogenizing medium of Bhargava and Halvorson was found to be unsuitable for preparing nuclei from protoplasts, as it seemed to make them resistant to homogenization.

A final point worth considering in the comparison of yeast nuclear preparations is that differing methods of analysis of DNA, RNA, and protein may not be entirely comparable. For a study of this in relation to whole yeast cells, see Sokurova (1967).

IV. Selection of a Method for the Isolation of Yeast Nuclei

There can be little doubt that, for precise characterization, nuclei obtained from protoplasts by one of the methods described should be used, as they suffer little or no damage during the isolation procedure and can be obtained in high yield. However, the physiological state of these nuclei is difficult to assess. The conditions involved in preparing protoplasts require exposure of the cells to conditions that must precipitate entry into stationary phase. Structural changes in nuclei after protoplasting have been reported (May *et al.*, 1972; May, 1974). Although the potential of protoplasts for cell wall regeneration and subsequent normal division shows that the changes occurring are reversible, there is a great need for precise biochemical characterization of the cells. For instance, what changes occur in the enzyme content of the cells during protoplasting? What is the state of protoplasted cells with respect to the cell cycle? Since they do not divide until the cell wall is regenerated, one may guess that they may be synchronized to some extent. If so, this may have a profound effect on the nucleus.

It seems likely that, of the two methods for isolating nuclei from protoplasts, that of May (1971) is to be preferred, since it is much simpler than that of Rozijn and Tonino and since the effects of various possible additives on the morphology of the nuclei have been carefully studied and the best selected. However, there seems good reason to suppose that different species and strains of yeasts require at least slightly different methods of nuclear isolation to provide optimum conditions (see Haff and Keller, 1973; Udem and Warner, 1973; Wintersberger *et al.*, 1973). The biochemical characterization of the nuclear preparation of Rozijn and Tonino is more complete than that of May, but the work of Wintersberger *et al.* (1973) on nuclei prepared by May's method shows no obvious biochemical aberrations.

If nuclei must be isolated from cells in precisely defined physiological states, the method of Bharghava and Halvorson (1971) or of Sajdel-

Sulkowska *et al.* (1974) must be used, but it seems likely that nuclei obtained in this way will have suffered varying degrees of damage although their gross properties are similar to those of nuclei prepared from protoplasts (Winters-berger *et al.*, 1973). Besides mechanical damage during freezing and homogenization, there is the possibility of biochemical damage due to the medium. R. May (unpublished) observed the formation of crystalline structures in cells treated with the Bhargava and Halvorson medium. Further characterization of physiological effects of homogenizing media is required before evaluation of the properties of these nuclei can be under-taken satisfactorily.

V. Preparation of a Dense Crescent (Nucleolar) Fraction from Yeast Nuclei (Sillevis-Smitt *et al.*, 1973)

1. Freshly isolated nuclei (2×10^9) are suspended at 4°C in 1.4 ml of a 35% (w/v, 1.0 M) sucrose solution in PVP medium (see above).
2. To the suspension of nuclei, 0.1 ml of a solution of DNase ($100 \mu g/0.1$ ml, Worthington, RNase-free) is added, and the suspension is incubated at 27°C for 12 minutes with occasional shaking.
3. After incubation, the suspension is rapidly chilled and centrifuged at 32,000 g_{av} for 7 minutes at 2°C in a swing-out rotor. The pellet obtained is enriched in dense crescent material (Fig. 3).

It can be seen from Fig. 3 that this preparation, while undoubtedly en-riched in dense crescent material, still contains most of the nuclear mem-brane and attached ribosomes, together with other less easily defined contaminants. There is thus scope for further improvement of this prepara-tion. Possibly, careful treatment with a lipid solvent such as acetone at the right dilution and at about 0°C or slightly below would solubilize the mem-branes and allow the released ribosomes to be separated by centrifugation.

VI. Electron Microscopy of Isolated Yeast Nuclei

All the methods used to prepare isolated nuclei for electron microscopy are very similar (Molenaar *et al.*, 1970; Bhargava and Halvorson, 1971; May *et al.*, 1972; May, 1973). The one presented below is that of Molenaar *et al.* (1970). All the operations described are carried out at room temperature.
1. The nuclear pellet is fixed in 2–5% glutaraldehyde in 0.1 M sodium cacodylate buffer (pH 7.4) for 2 hours.

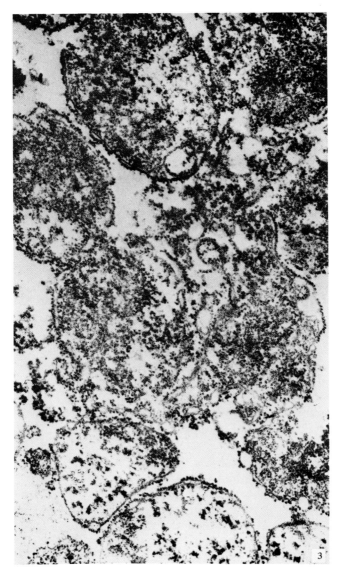

FIG. 3.   Section through a pellet of DNase-treated, residual nuclei. The pellet consists mainly of dense crescents attached to, or partially encircled by, nuclear membranes with attached ribosomes. ×26,000. (Sillevis-Smitt *et al.*, 1973, with permission of Academic Press, Inc.).

2. The pellet is washed five times with 0.2 *M* sucrose in the buffer used for the glutaraldehyde solution.

3. The pellet is postfixed for 1 hour with 1% osmic acid in Palade buffer (Palade, 1952).

4. The pellet is washed five times with physiological saline solution (0.95%) (pH 7.4).

5. The pellet is dehydrated with alcohol and embedded in Epon by the standard procedure (Glauert, 1965).

6. The final light-gold or silver sections of the embedded nuclei are post-stained with uranyl acetate and lead citrate (Glauert, 1965).

Figures 1 and 2 are electron micrographs of isolated protoplast nuclei that have been treated as above. Yeast nuclei, fixed and stained as described, have also been studied by high-resolution autoradiography to investigate the distribution of newly formed RNA (Sillevis-Smitt et al., 1972a).

VII. Isolation and Characterization of Chromatin from Yeast Nuclei

Chromatin may be defined structurally as the chemical complex of which the chromosomes are composed. Its biochemical definition is imprecise, being highly dependent on the techniques used in its isolation. Two methods of preparing yeast chromatin have been described. The first (Tonino and Rozijn, 1966a,b; Van der Vliet et al., 1969) starts with homogenates of whole cells of S. cerevisiae. This gives a preparation similar in DNA, RNA, and protein composition to those from higher organisms. The second (Wintersberger et al., 1973) starts with isolated nuclei. Clearly, this should give a preparation approximating more closely to structural chromatin than the first. However, the overall DNA, RNA, and protein analysis bears very little resemblance to corresponding analyses of well-defined chromatin preparations from higher organisms (Bonner et al., 1968), and differs from that of nuclei only in the loss of about half of the nuclear protein (Table I). This may be the true picture, but it is difficult to reconcile the results of the two groups, especially as both preparations are from strains of S. cerevisiae. In this context, it may be relevant to cite results obtained during attempts to isolate chromatin from nuclei of Schizo. pombe (J. H. Duffus, unpublished). The method used was that described by Bonner et al. (1968) for liver chromatin. The analysis of the product obtained was similar in terms of proportional composition to that of Wintersberger et al. The same problems in terms of appreciable loss of DNA during attempts at further purification were encountered. It was concluded that the product was probably not chromatin alone, since it still contained much of the nuclear RNA. Electron micrographs of the chromatin preparations might be helpful in resolving these doubts.

The material Wintersberger et al. describe as chromatin is prepared as follows:

1. Nuclei are suspended in 3–4 vol of cold $0.025M$ NaCl containing 10 mM EDTA (pH 7.5).

2. Three milliliters of this suspension is layered onto 35 ml of 2.6 M sorbitol in the same salt–EDTA solution and centrifuged for 60 minutes at 80,000 g_{av} and 0°C in a swing-out rotor. The pellet obtained is taken to be chromatin.

This chromatin preparation has been characterized mainly with respect to the acid-extractable proteins it contains and with regard to its transcription capacity. The acid-extractable proteins (extracting in 0.25 N or 0.4 N H_2SO_4 or 0.2 N HCl) have been analyzed by acrylamide gel electrophoresis. Transcription by isolated yeast RNA polymerases A and B is poor. Increasing salt concentrations cause removal of proteins from the preparation, allowing endogenous RNA polymerases to function. α-Amanitin-resistant polymerases are activated in this way at lower salt concentrations than α-amanitin-sensitive polymerases. The techniques used are standard and are fully described by Wintersberger et al. Unfortunately, the preparation has not been characterized with regard to its sedimentation properties and melting profiles under different conditions. This would have permitted a better comparison with the chromatin preparation of Van der Vliet et al. (1969), which was so characterized.

VIII.   Isolation and Characterization of DNA from Yeast Nuclei

The first report of the isolation and partial characterization of DNA from yeast nuclei is that of Bostock (1969) who worked with a nuclear preparation from Schizo. pombe. Subsequently, Bhargava et al. (1972) described a method for the isolation of high-molecular-weight DNA from nuclei of S. cerevisiae as follows:

1. The dialysis tubing to be used is boiled for 30 minutes, and the RNase and Pronase solutions (see below) are heated to 85°C for 10 minutes to inactivate any DNase that may be present.

2. Nuclei from 15–20 gm of cells (wet weight) are suspended gently in 2 ml of a solution containing 1 M sorbitol and 20% glycerol and transferred to a dialysis bag. Recrystallized sodium dodecyl sulfate (SDS) is added to a final concentration of 2%. The dialysis bag is placed in a beaker containing 2 liters of 1 M NaCl–0.01 M EDTA (pH 7) preheated to 55°C. One end of the dialysis bag is twisted and taped to the side of the beaker to permit subsequent addition of enzymes.

3. After 20 minutes at 55°C the dialysis bag is transferred to a beaker containing 2 liters of 0.15 M NaCl–0.015 M sodium citrate (pH 7) (SSC), and kept at 37°C. Pronase (Calbiochem, 10 mg/ml in SSC) is added to the dialysis bag to a final concentration of 1 mg/ml. Pronase digestion is allowed to take place for 8–16 hours against three changes of SSC in the beaker. It is monitored by following increased absorbance at 280 nm in the dialysate. Sometimes the further addition of the same amount of Pronase may be necessary if clumps of denatured protein are still visible in the sample.

4. Once Pronase digestion is complete, pancreatic RNase and T_1 RNase (Calbiochem, 2.5 mg/ml and 2500 units/ml, respectively, in SSC) are added to the dialysis bag to give final concentrations of 100 μg/ml and 100 units/ml. Digestion of RNA is allowed to occur at 37°C against two to three changes of SSC until no further increase in absorbance at 260 nm in the dialysate is observed.

5. The Pronase treatment is repeated.

6. Residual protein is removed by the method of Massie and Zimm (1965). The DNA suspension is transferred carefully to a glass centrifuge tube which is then placed in a water bath at 55°C. An equal volume of freshly distilled phenol (equilibrated with SSC at 55°C) is slowly dripped into the suspension. The mixture is left at 55°C for 5 minutes, and then centrifuged at 20,000 g_{av} for 15 minutes at 2°C; the supernatant is then withdrawn carefully with a wide-mouth pipette. The phenol is removed by dialysis at 4°C against several changes of SSC. Note: Stirring the dialysate during any of the steps or shaking the tube during the phenol treatment can result in considerable degradation of the DNA.

The DNA obtained is characterized by cesium chloride equilibrium density gradient centrifugation (Bostock, 1969; Bhargava *et al.*, 1972) and by analytical zone sedimentation and molecular-weight determination (Bhargava *et al.*, 1972). The γ Satellite DNA may be isolated following controlled shearing in a Virtis homogenizer by centrifugation on Hg^{2+}–cesium sulfate density gradients (Cramer *et al.*, 1972).

IX. Isolation and Characterization of RNA from Yeast Nuclei

Work on nuclear RNA from yeasts seems to be entirely associated with Rozijn and his co-workers (Rozijn *et al.*, 1964; Molenaar *et al.*, 1970; Sillevis-Smitt *et al.*, 1970, 1972a,b, 1973; Rozijn, 1974). The methods of extraction are standard and fully described (Rozijn *et al.*, 1964; Sillevis-Smitt *et al.*, 1970). Sillevis-Smitt *et al.* (1970, 1972b) characterized yeast nuclear RNA as to its sedimentation properties in sucrose gradients, and also as to its separation on acrylamide gels. They further showed (Sillevis-Smitt *et al.*, 1972a, 1973) that almost all the 37 and 28 S ribosomal precursors are confined to the dense crescent area of the nucleus. This work has been reviewed by Rozijn (1974).

In essence, the method used by Rozijn and his co-workers to extract RNA from nuclei is as described below (Sillevis-Smitt *et al.*, 1972b).

1. The nuclear pellet (about 2×10^9 nuclei) is suspended in 0.01 M sodium acetate buffer (pH 5.1) containing 0.02% polyvinyl sulfate. This suspension is incubated with 100 μg DNase (Worthington, RNAase-free) for 12 minutes at 27°C.

2. The suspension is then made 1% (w/v) in triisopropylnaphthalene sulfonate (TIPNS) (Kodak Ltd., Kirkby, Liverpool, T3513), 6% (w/v) in 4-aminosalicylate (PAS), and 6% (v/v) in 2-butanol, and shaken with an equal volume of phenol–*m*-cresol–water–8-hydroxyquinoline (500:70:55:0.5, by volume) (Kirby, 1965) for 10 minutes at 20°C. The phases are separated by centrifuging.

3. The phenol phase and interphase are then extracted, together, with an equal volume of a solution containing 1% TIPNS, 6% PAS, and 6% 2-butanol. Again the phases are separated by centrifuging.

4. The combined water phases are made 3% with respect to NaCl, and the phenol extraction is repeated as described above.

5. RNA is precipitated from the combined aqueous extracts by the addition of 2.5 vol 96% ethanol containing 2% potassium acetate. After storage overnight at −20°C, the precipitated RNA is washed twice with 75% ethanol containing 0.1 M NaCl (Loening *et al.*, 1969). tRNA may be removed with LiCl by the method of Schweizer *et al.* (1969).

The RNA obtained may be analyzed by gel electrophoresis according to Loening (1969), and by sucrose gradient centrifugation (Sillevis-Smitt *et al.*, 1970).

X. Isolation and Characterization of Nuclear Proteins

The first attempted characterization of yeast nuclear proteins was the work done by Tonino and Rozijn (1960a,b) on the acid-extractable proteins from a chromatin preparation from whole cells of pressed bakers' yeast, *S. cerevisiae*. This of course is open to criticism on the grounds of possible contamination with cytoplasmic proteins. Subsequently, Bhargava and Halvorson (1971) extracted acid-soluble proteins from yeast nuclei by the method of Lee and Scherbaum (1966). Duffus (1971) examined a similar fraction from *Schizo. pombe* nuclei and whole cells. The nuclei were prepared from normal cells using an Eaton press (Duffus, 1969), and the acid-extractable proteins were obtained by the pH titration technique of Murray *et al.* (1968). Subsequently, the same technique was applied to nuclei and cells of *K. fragilis* (Duffus and Penman, 1973). Again the nuclei were isolated using an Eaton press. Work in progress (J. H. Duffus and C. S. Penman, unpublished) suggests that the pH titration technique as previously used may result in much neutral protein being included in the acid-extractable fractions, since these fractions can be greatly reduced in quantity by adding an initial distilled-water extraction step to the technique. The absence of such a step in previous work may explain the low content of basic amino acids in the acid-extractable fractions. Another point to be borne in mind is that the outer membrane of isolated yeast nuclei carries a large number of tightly bound ribosomes (Fig. 2), and that these most certainly contribute to fractions of supposed intranuclear protein. Whether these ribosomes contribute to the acid-extractable proteins obtained by Wintersberger *et al.* (1973) from their chromatin preparation must be a matter of speculation, as no electron micrographs of this preparation are available. All these acid-extractable proteins have been characterized by polyacrylamide gel electrophoresis, but the detailed techniques differ and so comparison is difficult. Tonino and Rozijn (1966a,b) have published amino acid analyses of their fractions, as have Duffus (1971) and Duffus and Penman (1973). Presumably, these fractions contain proteins that may be considered yeast histones, but the evidence for the existence of yeast histones must still be considered equivocal when one allows for the possibilities of the presence of ribosomal and other protein indicated above.

So far, no attempts appear ot have been made to characterize other nuclear proteins, except by their enzymatic activity (see Section XI).

XI.   Yeast Nuclear Enzymes

Very little has so far been published about the enzymes of the yeast nucleus. The presence of RNA polymerases (EC 2.7.7.6) in nuclei from *S. carlsbergensis* (Rozijn and Tonino, 1964), *Schizo. pombe* (Duffus, 1969), *K. fragilis* (May, 1972), and *S. cerevisiae* (Bhargava and Halvorson, 1971; Sebastian *et al.*, 1973) has been demonstrated. The role of these enzymes in the synthesis of rRNA has been elucidated by the work of Sillevis-Smitt *et al.* (1970, 1972a,b, 1973). Their properties when part of yeast chromatin have been partially characterized by Wintersberger *et al.* (1973), particularly with respect to effects of varying salt concentrations and of α-amanitin.

Sebastian *et al.* (1973) separated out two nuclear DNA–dependent RNA polymerases (designated I and II) from *S. cerevisiae*. RNA polymerase I is active only in the presence of Mn^{2+} or Mg^{2+} and has a higher affinity for poly (dAdT) than for denatured DNA. RNA polymerase II shows a higher affinity for denatured DNA and is α-amanitin-sensitive. However, Hildebrandt *et al.* (1973) have shown that these two polymerases are immunologically related. They show independent regulation during the cell cycle (Sebastian *et al.*, 1974).

Duffus (1969) reported the presence of NAD pyrophosphorylase (EC 2.7.7.1) in *Schizo. pombe* nuclei. This enzyme has not been further characterized, but recent work has indicated that it is also localized in the nucleus of *K. fragilis* (R. May, unpublished).

Haff and Keller (1973) separated two distinct poly-A polymerases from *S. cerevisiae* nuclei, which differ in primer requirements. One can initiate poly-A chains on messengerlike RNA but extends the chains slowly. The other cannot use messengerlike RNA as a primer, but extends preexisting poly-A chains. It therefore appears that these two enzymes are complementary.

XII.   Concluding Remarks

There can be no doubt that yeast nuclear preparations still leave something to be desired. A major improvement would be a modification to remove the ribosomes from the outer layer of the nuclear envelope. So long as these are present, clear distinction between intranuclear and extranuclear ribosomal components is impossible. Another development that should be pursued is a nonaqueous technique for the isolation of yeast nuclei. It is well established that water-soluble components of nuclei leach out during aqueous isolation procedures, resulting in considerable loss of material. The application of nonaqueous techniques to protoplasts should be fairly straightforward. Their application to normal cells poses more problems, since the methods used for isolation of nuclei from intact cells are based on homogenization in the presence of ice (Eaton press, etc).

Despite their shortcomings, much can still be done with the yeast nuclear preparations now available. In many ways they are insufficiently characterized. Nothing is known of their lipid constituents. One may surmise that the lipids present are largely components of the nuclear envelope, but this will be uncertain until a nuclear envelope preparation is available. Similarly, nothing is known of the purely carbohydrate constituents or of that mysterious compound poly ADP-ribose (see Shall, 1972). Even the acid-soluble proteins have been characterized only slightly. The presence of histones has not yet been established. Are there any proteins corresponding to the acidic nonhistone proteins of mammalian cells? What percentage of the nuclear protein is microtubule protein? What precisely is the enzyme profile of a yeast nucleus in terms of activities and amounts? These and other questions remain to be answered. In fact, at the biochemical level only the nucleic acids have been characterized to any extent (see Sections VIII and IX).

The availability of methods for the isolation of yeast nuclei has opened the way to a clearer understanding of nuclear function in a simple eukaryotic system for which there exists a large fund of biochemical and genetic knowledge. As yet these methods have not been fully exploited. It is hoped that this article will provide some impetus for such exploitation and further development of these techniques.

ACKNOWLEDGMENTS

I am deeply indebted to Prof. J. M. Mitchison for directing my attention to the yeast nucleus and for invaluable guidance. My work cited here was supported by grants from the Science Research Council. I am grateful to Vanora Hereward and Colin Penman for their help and advice. Particular thanks must go to Drs. R. May and U. Wintersberger for their help with unpublished observations, and to Dr. Th. H. Rozijn for permission to use Fig. 1–3.

REFERENCES

Bhargava, M. M., and Halvorson, H. O. (1971). *J. Cell Biol.* **49**, 423.
Bhargava, M. M., Cramer, J. H., and Halvorson, H. O. (1972). *Anal Biochem.* **49**, 276.
Bonner, J., Chalkley, G. R., Dahmus, M., Fambrough, D., Fujimura, F., Huang, R. C., Hubermann, J., Jensen, R., Marushige, K., Ohlenbusch, H., Olivera, B., and Widholm, J. (1968). *In* "Methods in Enzymology" (L. Grossman and K. Moldave, eds.), Vol. 12, Part B, p. 3. Academic Press, New York.
Bostock, C. J. (1969). *Biochim. Biophys. Acta* **195**, 579.
Cramer, J. H., Bhargava, M. M., and Halvorson, H. O. (1972). *J. Mol. Biol.* **71**, 11.
Duffus, J. H. (1969). *Biochim. Biophys. Acta* **195**, 230.
Duffus, J. H. (1971). *Biochim. Biophys. Acta* **228**, 627.
Duffus, J. H., and Penman, C. S. (1973). *J. Gen. Microbiol.* **79**, 189.
Eddy, A. A. (1959). *Exp. Cell Res.* **17**, 447.
Eddy, A. A., and Williamson, D. H. (1957). *Nature (London)* **179**, 1252.
Ganesan, A. T., and Swaminathan, M. S. (1958). *Stain Technol.* **33**, 115.
Glauert, A. M. (1965). *In* "Techniques for Electron Microscopy" (D. H. Kay, ed.), pp. 194 and 254. Blackwell, Oxford.
Haff, L. A., and Keller, E. B. (1973). *Biochem. Biophys. Res. Commun.* **51**, 704.
Hildebrandt, A., Sebastian, J., and Halvorson, H. O. (1973). *Nature (London)* **246**, 73.
Kirby, K. S. (1965). *Biochem. J.* **96**, 266.
Lee, Y. C., and Scherbaum, O. H. (1966). *Biochemistry* **5**, 2067.
Loening, U. E. (1969). *Biochem. J.* **113**, 131.
Loening, U. E., Jones, K. W., and Birnstiel, M. L. (1969). *J. Mol. Biol.* **45**, 353.

McIlvaine, T. C. (1921). *J. Biol. Chem.* **49**, 183.

Massie, H. R., and Zimm, B. H. (1965). *Proc. Nat. Acad. Sci. U.S.* **54**, 1635.

Matile, P., Moor, H., and Robinow, C. F. (1969). *In* "The Yeasts" (A. H. Rose, ed.), Vol. 1, p. 219. Academic Press, New York.

May, R. (1971). *Z. Allg. Mikrobiol.* **11**, 131.

May, R. (1972). *In* Symposium on "Replication and Segregation of Genetic Material in Micro-organisms," held at the Castle of Reinhardsbrunn, GDR.

May, R. (1973). *Z. Allg. Mikrobiol.* **13**, 529.

May, R. (1974). *Z. Allg. Mikrobiol.* **14**, 409.

May, R., Unger, E., and Strunk, C. (1972). *Protoplasma* **75**, 195.

Mitchison, J. M., Creanor, J., and Sartirana, M. L. (1973). *In* "Yeast, Mould and Plant Protoplasts" (J. R. Villanueva *et al.*, eds.), pp. 229–245. Academic Press, New York.

Molenaar, I., Sillevis-Smitt, W. W., Rozijn, T. H., and Tonino, G. J. M. (1970). *Exp. Cell Res.* **60**, 148.

Murray, K., Vidali, G., and Neelin, J. M. (1968). *Biochem. J.* **107**, 207.

Ohnishi, T., Kawaguchi, K., and Hagihara, B. (1966). *J. Biol. Chem.* **241**, 1797.

Palade, G. E. (1952). *J. Exp. Med.* **95**, 285.

Rost, K., and Venner, H. (1964). *Hoppe-Seyler's Z. Physiol. Chem.* **339**, 230.

Rozijn, T. H. (1974). *Biochem. Soc. Symp.* **37**, 83.

Rozijn, T. H., and Tonino, G. J. M. (1964). *Biochim. Biophys. Acta* **91**, 105

Rozijn, T. H., Tonino, G. J. M., Frens-v. d. Bilt, E. M. B., Bloemers, H. P. J., and Koningsberger, V. V. (1964). *Biochim. Biophys. Acta* **80**, 675.

Sajdel-Sulkowska, E. M., Bhargava, M. M., Arnaud, M. V., and Halvorson, H. O. (1974). *Biochem. Biophys. Res. Commun.* **56**, 496.

Schweizer, E., MacKechnie, C., and Halvorson, H. O. (1969). *J. Mol. Biol.* **40**, 261.

Schwencke, J., Gonzalez, G., and Farias, G. (1969). *J. Inst. Brew.* **75**, 15.

Sebastian, J., Bhargava, M. M., and Halvorson, H. O. (1973). *J. Bacteriol.* **114**, 1.

Sebastian, J., Takano, I., and Halvorson, H. O. (1974). *Proc. Nat. Acad. Sci. U.S.* **71**, 769.

Shall, S. (1972). *FEBS (Fed. Eur. Biochem. Soc.) Lett.* **24**, 1.

Sillevis-Smitt, W. W., Nanni, G., Rozijn, T. H., and Tonino, G. J. M. (1970). *Exp. Cell Res.* **59**, 440.

Sillevis-Smitt, W. W., Vermeulen, C. A., Vlak, J. M., Rozijn, T. H., and Molenaar, I. (1972a). *Exp. Cell. Res.* **70**, 140.

Sillevis-Smitt, W. W., Vlak, J. M., Schiphof, R., and Rozijn, T. H. (1972b). *Exp. Cell Res.* **71**, 33.

Sillevis-Smitt, W. W., Vlak, J. M., Molenaar, I., and Rozijn, T. H. (1973). *Exp. Cell Res.* **80**, 313.

Sokurova, E. N. (1967). *Biokhimiya* **32**, 936.

Sørensen, S. P. L. (1912). *Ergeb. Physiol.* **12**, 393.

Suomalainen, H., Nurminen, T., and Oura, E. (1966). *Arch. Biochem. Biophys.* **118**, 219.

Tonino, G. J. M., and Rozijn, T. H. (1966a). *Biochim. Biophys. Acta* **124**, 427.

Tonino, G. J. M., and Rozijn, T. H. (1966b). *In* "The Cell Nucleus—Metabolism and Radiosensitivity" (J. A. Cohen, ed.), p. 125. Taylor & Francis, London.

Udem, S. A., and Warner, J. R. (1973). *J. Biol. Chem.* **248**, 1412.

Van der Vliet, P. C., Tonino, G. J. M., and Rozijn, T. H. (1969). *Biochim. Biophys. Acta* **195**, 473.

Van der Walt, J. P. (1970). *In* "The Yeasts, a Taxonomic Study" (J. Lodder, ed.), p. 316. North-Holland Publ., Amsterdam.

Watson, J. D. (1970). "Molecular Biology of the Gene," 2nd ed. p. 520. Benjamin, New York.

Wintersberger, U., Smith, P., and Letnansky, K. (1973). *Eur J. Biochem.* **33**, 123.

Chapter 7

Isolation of Vacuoles from Yeasts

ANDRES WIEMKEN

Department of General Botany,
Swiss Federal Institute of Technology,
Zürich, Switzerland

I. Introduction

Vacuoles are components of an elaborate membrane system which also includes the endoplasmic reticulum and the Golgi complex. Since the components of this system have manifold interconnections and are involved in a continuous process of interconversion, it is not always possible to distinguish clearly among them. Consequently, de Duve (1969) adopted the old cytological term "vacuole" and applied it to the whole of this membrane system.

Obvious difficulties arise in defining the term "vacuome" for cells of higher plants. The meristematic cells contain numerous small vesicles which are probably derived from the endoplasmic reticulum. These vesicles enlarge and fuse as the cells differentiate, while other vesicles, apparently derived from the Golgi complex, are engulfed (Matile and Moor, 1968).

This process of progressive vacuolation finally results in the formation of the large vacuole proper, known as the central vacuole. The cytoplasm is pressed to the wall by the central vacuole and there forms only a thin lining (Matile, 1969; Buvat, 1971).

In fungi vacuolation proceeds similarly. Small vacuoles in the young cells of growing hyphal tips give rise to a few large vacuoles in the aged cells located a distance from the tips (Park and Robinson, 1967; Iten and Matile, 1970). Also, in yeast cells the numerous small vacuoles present in the stage of bud emergence enlarge and coalesce into one or two large vacuoles as the cells progress in the budding cycle (Wiemken, 1969; Wiemken et al., 1970).

The reverse process, i.e., shrinkage and division of the large vacuoles into numerous small vesicles, takes place in yeasts at the moment of bud initiation. This also occurs in hyphae at sites of lateral branch formation (Park and Robinson, 1967), and in higher plants during the course of division of vacuolate cells (Sinnott and Bloch, 1940; Jensen, 1968).

Thus it becomes clear that the term "vacuole" is only descriptive or operational, and does not stand for a clearly delimited physiological unit as, for example, the term "mitochondrion" does.

In this article methods are described by which vacuoles can be isolated from yeast cells. All these methods involve an initial transformation of the yeast cells into spheroplasts. This procedure results in spheroplasts metabolically similar to cells in the stationary growth phase. Furthermore, the spheroplasts typically contain only one or two large vacuoles, whereas a multitude of small vacuoles is often present in growing yeast cells (Wiemken et al., 1970). This greatly facilitates isolation, since the large vacuoles can easily be identified, using a light microscope, throughout the entire isolation procedure. However, it also limits the validity of the results obtained from studies on these vacuoles. They are only representative of certain metabolic states of the cell.

New methods must therefore be developed in order to investigate the whole spectrum of vacuoles occurring in cells in different metabolic states. This is of particular interest, since the degree of vacuolation is known to undergo profound changes during the processes of cell differentiation such as the cell cycle (Wiemken et al., 1970; Nurse and Wiemken, 1974) and sporulation (McClary, 1964; Svihla et al., 1964; Croes, 1967).

II.   Preparation of Spheroplasts

Vacuoles are inevitably ruptured when the tough cells are disintegrated by the mechanical methods presently available. Therefore the transformation of the cells to fragile spheroplasts by degradation of the cell walls is a prerequisite for isolating intact vacuoles. As the techniques used for preparing spheroplasts are reviewed elsewhere in this volume, only the principle and the procedure applied in our laboratory are described here.

The cell walls are degraded in an isotonic solution with enzymes from snail gut or from microorganisms. Supplementary treatments with reducing and chelating agents, or with deoxyglucose (an analog of glucose, which interferes with cell wall synthesis), are often essential to render the cell walls susceptible to enzymatic attack. The facility of formation of spheroplasts greatly depends on the yeast strain, the composition of the culture medium, and the growth phase from which the cells are taken. With the following procedure we obtained high yields of spheroplasts from both *Saccharomyces cerevisiae* and *Candida utilis* which we successfully used for the isolation of vacuoles.

Saccharomyces cerevisiae (strain LBGH 1022) and *C. utilis* (NCYC 737) are grown in standard media containing glucose or lactate as carbon sources and an ammonium salt as nitrogen source. The cultures are harvested during growth and centrifuged and washed twice with distilled water at $0°-4°C$. The wet weight of the cells is determined on a pellet (5 minutes at 2000 g). The cells are then transferred to a solution containing 0.6 M sorbitol and 10 mM sodium citrate buffer (pH 6.8) (SOB). In this solution supplemented with 0.14 M cysteamine–HCl, they are incubated for 20 minutes at 28°C (approximately 5 ml/gm wet weight of cells). They are washed free of cysteamine with SOB and then incubated with snail enzyme: 25–35 mg of lyophilized Helicase (Industrie Biologique Française S.A., Gennevilliers, France) per milliliter of SOB and gram wet weight of cells. After 1–2 hours at 28°C, all the cells are transformed to spheroplasts. Because many of the cells are ruptured if they are washed free of the Helicase by repeated sedimentation and resuspension, it is preferable to wash them by a single sedimentation through a density gradient. The spheroplast–Helicase suspension is diluted with SOB (3×) and layered on top of a denser but isotonic solution containing 0.6 M sucrose and 10 mM sodium citrate buffer (pH 6.8) (SUB); the spheroplasts are then purified efficiently by sedimentation through the layer of SUB at 2000 g for 15 minutes.

III. Extraction of Intact Vacuoles from Spheroplasts

A. Osmotic Lysis

Spheroplasts that are transferred to hypotonic solutions swell and finally burst. Observation of the process of spheroplast lysis by means of phase-contrast microscopy shows that the vacuoles are released and remain intact (Garcia-Acha *et al.*, 1967; Wiemken and Nurse, 1973b). Moreover, inspection with the ultraviolet microscope reveals that *S*-adenosylmethionine accumulated in the vacuoles *in vivo* can be detected in the free vacuoles after lysis (Nakamura, 1973). This demonstrates that the vacuoles to be isolated in the spheroplast lysate are identical to those *in vivo* and are not artefacts of the lytic process.

The critical tonicity below which the spheroplasts are lysed is highly variable and depends on the species of yeast and on the previous treatment. Generally, it lies between 0.2 and 0.4 *M*. However, it appears that it is not necessary to establish this concentration exactly in each case, as the vacuoles can stand solutions of very low tonicity. For instance, they remain perfectly stable if the spheroplasts are lysed in a solution containing only 8% (w/v). Ficoll (MW 400,000; Pharmacia, Uppsala, Sweden) and 10 m*M* sodium citrate buffer (pH 6.8). Under these conditions the vacuoles swell and loose their content of low-molecular-weight substances. However, they still retain high-molecular-weight substances such as soluble enzymes. Also, the Ficoll is kept out of the vacuoles. It is interesting that large vacuoles, which can be compared to bags, do not collapse but preserve their spherical shape even after having lost the low-molecular-weight substances which maintain turgor. This resistance to collapse probably results from the presence of a slimy macromolecular substance within, possibly composed of mannan (P. Matile, unpublished results).

Most investigators try to reduce the stretching of the tonoplast by lowering the tonicity only just to the critical point at which the lysis of spheroplasts is induced. The resistance of the spheroplasts to lysis by osmotic shock can be lowered by chelating agents, by low pH values (Indge, 1968a), or by traces of Triton X-100 (Matile and Wiemken, 1967). Nakamura (1973) isolated vacuoles that retained accumulated *S*-adenosylmethionine; to rupture the spheroplasts he lowered the tonicity from 0.8 to 0.4 *M* sorbitol for only 3–5 minutes at 25°C. Generally, however, osmotic shock results in leakage of micromolecules from the vacuoles. It is therefore advisable to use this method only for investigations on enzymes and other high-molecular-weight substances retained within the vacuoles.

B.   Lysis under Isotonic Conditions

1.   METABOLIC LYSIS

This is a convenient method of rupturing spheroplasts under isotonic conditions (Indge, 1968b; Cabib *et al*, 1973). The procedure involves incubation of the spheroplasts 2×10^7/ml) at 30°C in a medium containing 8.5% (w/v) mannitol, 0.5% (w/v) glucose, 10 mM imidazole chloride (pH 6.4), and 6 mM citrate as Tris salt at the same pH, or 5 mM EDTA. Lysis of the spheroplasts is virtually complete after 20 minutes. The advantage of this method is that up to 75% of the spheroplasts release intact vacuoles (Indge, 1968c). However, a disadvantage of metabolic lysis lies in the requirement for chelating agents and prolonged incubation at temperatures allowing high metabolic activity, conditions that are unfavorable for the integrity of membranes and the stability of metabolic intermediates.

2. MECHANICAL DISINTEGRATION

Mechanical disintegration of spheroplasts is a useful alternative (Wiemken and Dürr, 1974). The following procedure yields intact vacuoles. A centrifuge tube of 12 ml capacity is filled with a 0.5-ml pellet of packed spheroplasts, 1 ml of SOB or SUB solution, and 3.5 ml of glass beads (diameter 0.45–0.5 mm). These components are mixed vigorously for 1–2 minutes at 0°–4°C by means of a 0.5-cm-thick glass rod fixed on a vibrating shaker (Chemap AG, Männedorf, Switzerland) with the tip immersed in the mixture. About 50–70% of the spheroplasts are disrupted by this treatment, and many vacuoles remain intact. The obvious disadvantage of this method is the relatively small yield of vacuoles. The advantage is that it proceeds rapidly and at a low temperature.

3. ULTRAVIOLET IRRADIATION

Ultraviolet light (6 × 10⁴ ergs/mm² energy of 253 nm) has been used to rupture spheroplasts for isolating vacuoles (Svihla et al., 1961).

4. TREATMENT WITH BASIC PROTEIN SOLUTIONS

This is another possible method of specifically disrupting the plasmalemma while keeping the tonoplast intact (Schlenk et al., 1970; Wiemken and Nurse, 1973a).

C. Isolation of Vacuoles

After lysis of the spheroplasts the vacuoles often remain entangled within other membranous elements. Further, the lipid granules tend to stick to the tonoplast. In order to liberate the vacuoles from such attachments, gentle shearing in a Potter-Elvehjem homogenizer or vigorous stirring (e.g., with a Sorvall Omni-Mixer, Norwalk, Conn.) may be resorted to without risk of disruption of too many vacuoles.

The procedures for isolating vacuoles all rely on two features. The vacuoles of spheroplasts are very large, and their specific density is generally not much higher than that of water. It is therefore not very difficult to design density gradients in which vacuoles can be flotated or sedimentated rapidly. All the methods described for isolating vacuoles are based on differential flotation, or sedimentation, or both consecutively (Matile and Wiemken, 1967; Indge, 1968c; Wiemken, 1969; Beteta and Gascon, 1971; Holley and Kidby, 1973; Cabib et al., 1973; Meyer, 1973; Wiemken and Nurse, 1973b; Nakamura, 1973; Nakamura and Schlenk, 1974; Wiemken and Dürr, 1974). Density gradients such as those depicted in Figs. 1 and 2 yielded very pure vacuoles from S. cerevisiae.

Vacuoles extracted from spheroplasts by osmotic shock are permeable to micromolecules. Hence the density of the medium can be raised above that of the vacuoles only if solutions of macromolecular substances, such as Ficoll, that do not penetrate the tonoplast are used. Lysis by osmotic shock followed by flotation in a Ficoll gradient (Fig. 1) has been found to be the simplest and quickest procedure for obtaining large numbers of highly purified vacuoles.

Vacuoles extracted from spheroplasts under isotonic conditions are impermeable to small molecules such as sucrose. Therefore sucrose can be used in addition to the Ficoll in order to increase the density of the medium so that the vacuoles float. Strictly isotonic density gradients can be constructed by combining sorbitol, sucrose, and Ficoll (Fig. 2).

It must be emphasized, however, that isolation schemes involving density gradients can be adopted in principle only. Modifications are probably necessary for isolating vacuoles from other yeast strains, or from yeasts grown under different culture conditions. Vacuoles from S. cerevisiae grown on arginine as sole nitrogen source, for example, were found to float in SUB solution only when 15% (w/v) Ficoll was added (Wiemken and Dürr, 1974). Thus a wide variation in vacuolar density must be expected, especially when the vacuoles are isolated isotonically.

IV. Characterization of Isolated Vacuoles

A. Enzymes

Table I presents a compilation of the activities of enzymes measured in the vacuolar fraction of S. cerevisiae (Wiemken, 1969; Matile et al., 1971; Cortat et al., 1972; van der Wilden et al., 1973; Meyer, 1973; Lenney et al., 1974). The yeast was cultivated under derepressing conditions for the respective enzymes. Vacuoles were extracted from the spheroplasts by osmotic shock and isolated by means of a Ficoll gradient essentially like that shown in Fig. 1.

FIGS. 1 and 2. Density gradients for the isolation of vacuoles. The left half of the density gradient diagrams presents their construction, and the right half presents the distribution of particles after centrifugation. The conditions of centrifugation are indicated above the gradients. Open circles, vacuoles; small solid circles, lipid granules; large solid circles, spheroplasts; hatching, soluble fraction. SOB, 0.6 M sorbitol with 10 mM sodium citrate buffer (pH 6.8); SUB, 0.6 M sucrose with 10 mM sodium citrate buffer (pH 6.8).

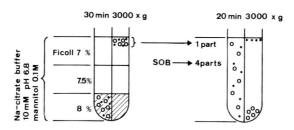

FIG. 1. Hypotonic density gradient.

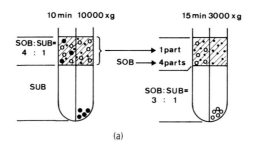

(a)

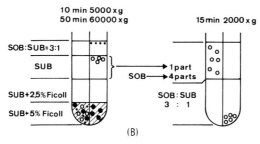

(B)

FIG. 2. Isotonic gradients. (The isotonic isolation of vacuoles was worked out in collaboration with M. Dürr, J. Meyer, G. Schaffner, and B. Jans.)

TABLE I

ENZYME ACTIVITIES IN ISOLATED VACUOLES AS COMPARED WITH
THE ENZYME ACTIVITIES IN THE TOTAL SPHEROPLAST LYSATE[a]

Enzyme	Relative activity[a]
Protease A (hemoglobin, pH 3)	20.7
Protease B (hide powder azure, pH 7)	40.2
Carboxypeptidase (benzoyl-tyrosine-p-nitroanilide)	24.0
Aminopeptidase (leucyl-p-nitroanilide)	7.7
RNase	19.2
Invertase	20.1
α-Mannosidase (p-nitrophenyl-α-mannoside)	72.6
β-1, 3-Glucanase (laminarin)	27.8
Phosphatase, alkaline	40.0
Phosphatase, acid	15.0
ATPase, Mg^{2+}-dependent	9.0
NADH$_2$-dichlorphenol-indophenol diaphorase[b]	0.638
NADH$_2$-cytochrome-c oxidoreductase[b]	0.016
Cytochrome-c oxidase[c]	<0.01
Succinate dehydrogenase[c]	<0.01
Succinate-cytochrome-c oxidoreductase[c]	<0.01
Glucose-6-phosphate dehydrogenase[d]	<0.01
Ethanol dehydrogenase[d]	<0.01
α-Glucosidase (p-nitrophenyl-α-glucoside[d]	<0.01

[a] Ratio of activity (per protein) in vacuoles and activity (per protein) in total spheroplast lysate.
[b] Indicative of endoplasmic reticulum.
[c] Indicative of mitochondria.
[d] Indicative of soluble fraction.

Activities of indicator enzymes for the mitochondrial and the soluble fractions are barely detectable in the isolated vacuoles. This demonstrates that the separation of the vacuoles is very good. The traces of two enzyme activities in the vacuolar fraction, which are indicators for the endoplasmic reticulum, may reflect the relationship of this membrane system with the tonoplast. This relationship is known from cells of higher plants (Matile, 1969).

High activities of hydrolytic enzymes are characteristic of vacuoles (Table I). Yeast vacuoles are therefore lysosomes (Matile and Wiemken, 1967; Wiemken, 1969). As most of these enzymes are soluble when the vacuoles are disrupted, they are not specific for the tonoplast but only for intact vacuoles. α-Mannosidase is an exception; it is firmly bound to the tonoplast (van der Wilden et al., 1973). Hydrolytic enzymes have also been found in vacuolar fractions by Beteta and Gascon (1971), Holley and Kidby (1973), and Cabib et al. (1973).

B.   Content of Low-Molecular-Weight Substances

Vacuoles isolated under isotonic conditions contain large amounts of soluble amino acids (Wiemken and Dürr, 1974). Up to 240 mmoles of amino acids (α-NH_2 nitrogen) per gram of protein has been found in the vacuolar fraction as compared with 3.8 mmoles/gm of protein in the total spheroplast lysate. Basic amino acids and glutamine are accumulated preferentially in the vacuoles. It has been estimated that more than 60% of the total soluble amino acid pool of spheroplasts is located in the vacuoles. The localization of soluble amino acids here is further supported by the results of a method that does not involve the isolation of vacuoles. Basic proteins disrupt the plasmalemma specifically, while the tonoplast remains intact (Yphantis *et al.*, 1967; Svihla *et al.*, 1969). In this way a cytoplasmic and a vacuolar extract can be obtained. In using this method it appeared that as much as 80–90% of the total pool of soluble amino acids in yeast is located in the vacuoles (Wiemken and Nurse, 1973a).

In addition to amino acids, which normally constitute the bulk of soluble intermediates of yeasts (Gancedo and Gancedo, 1973), various purine derivatives are known under certain conditions to accumulate in large quantities in the vacuoles. As these substances may precipitate within the vacuoles (e.g., uric acid) or strongly absorb ultraviolet light (*S*-adenosyl-methionine), their vacuolar localization was demonstrated even *in vivo* by means of light microscopy (Roush, 1961; Svihla *et al.*, 1963). *S*-Adenosylmethionine has also been found in isolated vacuoles of *C. utilis* (Nakamura, 1973).

These results show that vacuoles play a central role in the compartmentation of metabolites. They serve as a repository for intermediates which temporarily are not metabolized. Clearly, the lengthy procedures of preparing spheroplasts and vacuoles prohibit the investigation of this role in more detail.

C.   Content of High-Molecular-Weight Substances

Vacuoles are found to contain polyphosphate (Indge, 1968d) and mannan (P. Matile, unpublished results). These substances are probably responsible for the amorphous aggregates in the vacuoles visible in many electron micrographs of thin sections. The metachromatic granules in vacuoles that are stainable with neutral red (reviewed by McClary, 1964) are thought to represent polyphosphate (Indge, 1968d).

D.   The Tonoplast

Little is known about the tonoplast, i.e., the single unit membrane that confines the vacuoles. When the technique of freeze-fracturing is used, the membrane generally splits and an inner fracture face is seen in the electron micrograph (Hereward and Northcote, 1972). This face reveals a distinct pattern of particles (Moor and Mühlethaler, 1963; Matile *et al.*, 1969), which is characteristic of vacuoles. The structural aspect of yeast membranes is dealt with elsewhere in this volume.

Note added in proof: Hasilik *et al.* (1974) and Matern *et al.* (1974) have found proteinase A and B and carboxypeptidase Y in their preparations of vacuoles. The three specific inhibitors of these enzymes were absent from the vacuolar fractions; they were only found in the soluble fraction. This is in accordance with our observations (Matile and Wiemken, 1967; Lenney *et al.*, 1974) but contrasts with a result of Cabib *et al.* (1973) indicating an extravacuolar location of proteinase A.

Vacuoles isolated from mechanically disrupted spheroplasts of *S. cerevisiae* according to the scheme of Fig. 2 were successfully used for studies on amino acid transport. A specific system for arginine transport was found (Boller *et al.*, 1975).

ACKNOWLEDGMENTS

I wish to thank S. Türler, G. Caroll, and Ph. Matile for a critical reading of the manuscript.

REFERENCES

Beteta, P., and Gascon, S. (1971). *FEBS (Fed. Eur. Biochem. Soc.) Lett.* **13**, 297–300.
Boller, T., Dürr, M., and Wiemken, A. (1975). *Eur. J. Biochem.* **54**, 81–91.
Buvat, R. (1971). *In* "Results and Problems in Cell Differentiation" (J. Reinert and H. Ursprung, eds.), Vol. II, pp. 127–157. Springer-Verlag, Berlin and New York.
Cabib, E., Ulane, R, and Bowers, B. (1973). *J. Biol. Chem.* **248**, 1451–1458.
Cortat, M., Matile, P., and Wiemken, A. (1972). *Arch. Mikrobiol.* **82**, 189–205.
Croes, A. F. (1967). *Planta* **76**, 209–226.
de Duve, C. (1969). *In* "Lysosomes in Biology and Pathology" (J. T. Dingle and H. B. Fell, eds.), Vol. I, pp. 3–40. North-Holland Publ., Amsterdam.
Gancedo, J. M., and Gancedo, C. (1973). *Biochimie* **55**, 205–211.
Garcia-Acha, F., Lopez-Belmonte, F., and Villanueva, J. R. (1967). *Can. J. Microbiol.* **13**, 433–437.
Hasilik, A., Müller, H., and Holzer, H. (1974). *Eur. J. Biochem.* **48**, 111.
Hereward, F. V., and Northcote, D. H. (1972). *J. Cell Sci.* **10**, 555–561.
Holley, R. A., and Kidby, D. K. (1973). *Can. J. Microbiol.* **19**, 113–117.
Indge, K. J. (1968a). *J. Gen. Microbiol.* **51**, 425–432.
Indge, K. J. (1968b). *J. Gen. Microbiol.* **51**, 433–440.
Indge, K. J. (1968c). *J. Gen. Microbiol.* **51**, 441–446.
Indge, K. J. (1968d). *J. Gen. Microbiol.* **51**, 447–445.
Iten, W., and Matile, P. (1970). *J. Gen. Microbiol.* **61**, 301–309.
Jensen, W. A. (1968). *Planta* **79**, 346–366.
Lenney, J. F., Matile, P., Wiemken, A., Schellenberg, M., and Meyer, J. (1974). *Biochem. Biophys. Res. Commun.* **60**, 1378–1883.
McClary, D. O. (1964). *Bot. Rev.* **30**, 167–225.
Matern, H., Betz, H., and Holzer H. (1974). *Biochem. Biophys. Res. Commun.* **60**, 1051.

Matile, P. (1969). *In* "Lysosomes in Biology and Pathology" (J. T. Dingle and H. B. Fell, eds.), Vol. I, pp. 406–430. North-Holland Publ., Amsterdam.

Matile, P., Moor, H., and Robinow, C. F. (1969). *In* "The Yeasts" (A. H. Rose and J. S. Harrison, eds.), Vol. I, pp. 220–302. Academic Press, New York.

Matile, P., and Moor, H. (1968). *Planta* **80**, 159–175.

Matile, P., and Wiemken, A. (1967). *Arch. Mikrobiol.* **56**, 148–155.

Matile, P., Wiemken, A., and Guyer W. (1971). *Planta* **96**, 43–53.

Meyer, J. (1973). Ph.D. thesis No. 5121, ETHZ, Zürich, Switzerland.

Moor, H., and Mühlethaler, K. (1963). *J. Cell Biol.* **17**, 609–628.

Nakamura, K. D. (1973). *Prep. Biochem.* **3**, 553–561.

Nakamura, K. D., and Schlenk, F. (1974). *J. Bacteriol.* **118**, 314–316.

Nurse, P., and Wiemken, A. (1974). *J. Bacteriol.* **117**, 1108–1116.

Park, D., and Robinson, P. M. (1967). *Symp. Soc. Exp. Biol.* **21**, 323–336.

Roush, A. H. (1961). *Nature (London)* **190**, 449.

Schlenk, F., Dainko, J. L., and Svihla, G. (1970). *Arch. Biochem. Biophys.* **140**, 228–236.

Sinnott, E. W., and Bloch, R. (1940). *Proc. Nat. Acad. Sci. U.S.* **26**, 223–227.

Svihla, G., Schlenk, F., and Dainko, J. L. (1961). *J. Bacteriol.* **82**, 808–814.

Svihla, G., Dainko, J. L., and Schlenk, F. (1963). *J. Bacteriol.* **85**, 399–409.

Svihla, G., Dainko, J. L., and Schlenk, F. (1964). *J. Bacteriol.* **88**, 449–456.

Svihla, G., Dainko, J. L., and Schlenk, F. (1969). *J. Bacteriol.* **100**, 498–504.

van der Wilden, W., Matile, P., Schellenberg, M., Meyer, J., and Wiemken, A. (1973). *Z. Naturforsch. C* **28**, 416–421.

Wiemken, A. (1969). Ph.D. thesis No. 4340, ETHZ, Zürich, Switzerland.

Wiemken, A., and Dürr, M. (1974). *Arch. Mikrobiol.* **101**, 45–57.

Wiemken, A., and Nurse, P. (1973a). *Planta* **109**, 293–306.

Wiemken, A., and Nurse, P. (1973b). *Proc. Int. Spec. Symp. Yeasts*, 3rd, 1973, pp. 331–347.

Wiemken, A., Matile, P., and Moor, H. (1970). *Arch. Mikrobiol.* **70**, 89–103.

Yphantis, D. A., Dainko, J. L., and Schlenk, F. (1967). *J. Bacteriol.* **94**, 1509–1515.

Chapter 8

Analytical Methods for Yeasts

P. R. STEWART

Department of Biochemistry, Faculty of Science,
Australian National University,
Canberra, Australia

I. Introduction

Yeast cells present particular difficulties in the application of standard extractive and analytical methods. The thick and chemically refractile cell wall of the yeast cell often has a decided propensity to behave as a molecular filter, allowing only particular sizes of molecules in or out even when cell membrane integrity has been destroyed completely. Moreover, it might reasonably be suspected that the properties of the cell wall, like those of the membranes of the organism, may alter with the physiological status of the cell. This tends to make slow-growing cells or glucose-repressed cells, for example, behave differently during extraction and disruption when compared with cells from fast-growing or derepressed cultures. Clearly, the compromise that must be made between efficacy of extraction and recovery of extracted substance is more critical with yeasts than with many other organisms used by cell biologists. Analytical methods applied to yeasts must therefore provide a carefully struck balance between yield and reproducibility, together with reasonable facility. The methods of quantitative analysis of the major components of yeast cells described in this chapter have been selected because they provide a workable combination of these requirements. The particular methods mentioned or described are those that appear to be most suitable for comparative studies on the composition of yeast cells and the substructures within these cells.

While the extraction procedures and assay methods described can be developed to a very high level of reproducibility, it is always possible that differences between cell types may well be the result of differences due to (1) altered strength of binding of components to cellular substructures, (2) interference by other substances that have been cofractionated, or (3) chemical alteration or degradation during extraction. Optimum extraction conditions may thus need to be established for different cell types, or for different subcellular fractions, in order that observed differences may be considered significant. Moreover, when organelle or other subcellular fractions are analyzed, it may often be the case that procedures of extraction less vigorous than those adopted with intact cells are needed. Parameters, especially of time, temperature, and concentration of extractants, ought to be examined to obtain optimum yield and satisfactory reproducibility with the analytical method ultimately adopted.

The scope of this survey of analytical methods applicable to yeasts is limited, for obvious reasons. A working description of even a single representative method for each class of compounds that has been analyzed in yeast cells would encompass the whole of this volume. In the space of a chapter, therefore, it has been necessary to restrict detailed descriptions to those substances that are quantitatively the most important. In the case of the minor components of the cell, although they may be crucial to the normal biochemical and physiological behavior of the cell, only an outline at most of general procedures has been given, together with references which should provide the reader with access to a detailed practical description of methods that have previously been successfully adopted.

II.   Macromolecular Components of Yeast Cells

A.   Protein

A comprehensive and critical review of methods of analysis of the macromolecular composition of microbial cells in general is provided by

Herbert *et al.* (1971). The chemical bases of certain of the extractive and assay procedures described below are discussed in detail in that article.

1. QUANTITATIVE ASSAY OF THE PROTEIN CONTENT OF WHOLE CELLS AND SUBCELLULAR FRACTIONS

Protein accounts for approximately 40–60% of the dry weight of most yeasts. In order to render all this protein soluble, hot-alkali extraction of intact cells, or the complete disruption of cells followed by extraction with a detergent solution, is necessary. The former procedure is simpler and provides results similar to those of the second method; it is thus preferred for routine analyses. Heating cell suspensions in hot alkali results in disruption of cell wall and cell membrane structures, such that macromolecules within the cell are able to diffuse out into the extracting solution. Hot alkali serves also to solubilize proteins within cellular substructure (cell wall, ribosomes, membranes), so that extraction of the total protein of the cell may be achieved.

A method that incorporates the Folin assay method of Lowry *et al.* (1951) and which is suitable for most yeasts is as follows (the procedure is essentially that of Herbert *et al.* (1971).

Reagents:

i. 1.0 M KOH

ii. 2% Na_2CO_3

iii. 0.125% $CuSO_4 \cdot 5H_2O$ in 0.25% Sodium potassium tartrate

iv. Mixture of Solutions *ii* and *iii* in a proportion of 50:4, prepared on day of use

v. Folin-Ciocalteu phenol reagent (available commercially from several supply houses); this reagent should be 1.0 M with respect to acid when titrated with KOH to a phenolphthalein end point (add water as appropriate)

vi. Protein standard (e.g., bovine serum albumin), 200 μg/ml in water (a stock solution of 2–20 mg/ml in 5 mM NaOH is stable for long periods at room temperature and can be diluted when needed)

Procedure: To 0.5 ml of cell suspension [100–400 μg (dry weight)/ml in water or buffer] add 0.5 ml of 1.0 M KOH and heat tubes in a boiling water bath for 15 minutes. Cool, add 2.0 ml of reagent *iv* and incubate at room temperature for at least 10 minutes. Add and rapidly mix in 0.5 ml of Folin-Ciocalteu reagent. The development of a stable blue color requires at least 30 minutes at room temperature. Measure the absorbance at 750 nm versus a reagent blank, and determine the protein content by a comparison with standards (containing 20, 50, and 100 μg of protein) that have been heated and assayed in the same way. With each cell type examined, it is desirable to determine the minimum heating time necessary to obtain the maximum release of protein from the cell.

In cases in which turbidity due to cell residues remains after color development, a centrifugation step may need to be incorporated immediately prior to measuring absorbance.

A major problem with alkali extraction and assay as described above is that substances other than protein may also contribute to the color reaction. These include aromatic amines and amino acids, peptides, and many reducing agents (Lowry *et al.*, 1951). One means of avoiding this problem is to use the biuret assay procedure (Herbert *et al.*, 1971), or to carry out a preliminary extraction of the cells with cold trichloracetic acid, or other small-molecule extractant, to remove interfering substances before hot-alkali extraction of the protein. $HClO_4$ should not be used, since precipitation of insoluble $KClO_4$ may occur during the assay.

The biuret method offers an added advantage of greater speed; it is however, very much less sensitive than the Folin-Lowry method.

Reagents:

i. 1 M NaOH

ii. $CuSO_4$, 2.5% (w/v)

iii. Protein standard (e.g., bovine serum albumin), 20 mg/ml in 5 mM NaOH

iv. KCN

Procedure: Resuspend cells at concentrations up to 10 mg/ml in 1 M NaOH and heat in a boiling water bath for 15 minutes. Cool, add an equal volume of 2.5% (w/v) $CuSO_4$, and mix thoroughly. After 5 minutes centrifuge and read the absorbance at 550 nm against a reagent blank; compare with protein standards (up to 8 mg per assay) that have been heated in alkali in the same way as the cell suspension. If centrifugation is not possible, or is unsatisfactory in clearing the residue from turbid solutions, solid KCN (approximately 10 mg per assay tube) is added and incubated with the samples for 10 minutes to discharge the blue color of the biuret reagent. The turbidity of the solution is then determined at the same wavelength, and this nonspecific absorbance is subtracted from the absorbance measured initially to give a value corrected for light scattering.

2. DETECTION OF SPECIFIC PROTEINS IN YEAST CELLS

Analysis of the total spectrum of proteins in yeast cells is rarely a meaningful exercise. More frequently, however, a situation arises in which analysis for a specific enzyme(s) or other protein(s) within the cell is sought. There are two useful general approaches to this problem. First, a rapid and effective means is available, using the technique of electrophoresis in agar, agarose, or starch gels coupled with detection methods for specific enzymes involving staining of the gel, to examine and compare the enzyme content and types in unfractionated homogenates of different cell types or subcellular fractions. Second, provided a specific protein can be obtained in pure form, it may be used as an antigen with which to prepare an antiserum. Subsequently, isolation and quantitative estimation of this protein can be carried out by immunoprecipitation from relatively crude fractions. Two examples, in outline only, are given below of these useful specific analytical approaches. They appear to be capable of more general application.

a. *Detection of Enzymes (or Isoenzymes) by Specific Reaction Tests.* Atzpo-dien *et al.* (1968) describe the extraction and separation of isoenzymes of malate dehydrogenase in cell homogenates and mitochondrial extracts prepared from *Saccharomyces cerevisiae.* The procedures are simple and rapid, and can be adapted to examine other enzymes capable of yielding an insoluble and/or colored end product as a consequence of their catalytic action. In essence, the procedure entails cell rupture (mechanical or enzymatic), removal of heterogeneous particulate matter, and/or isolation of specific subcellular fractions, followed by extraction or solubilization of enzymes from these fractions. The fractions containing the putative enzyme are loaded onto polyacrylamide or agar gels and subjected to electro-phoresis; the gels are then immersed in a reaction mixture formulated to provide visual evidence of the presence of the enzyme concerned. In the case of dehydrogenases or enzymes catalyzing oxidation–reduction re-actions, this is most readily achieved by coupling the reaction to formation of a colored, insoluble formazan produced by reduction of a tetrazolium

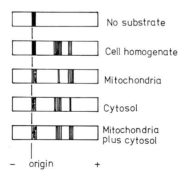

FIG. 1. Isoenzymes of malate dehydrogenase in *S. cerevisiae.* Electrophoresis was carried out as described by Wieme (1965), but using the electrode buffer and gel concentration of Atzpodien *et al.* (1968). The apparatus was somewhat larger than Wieme's, so that up to five samples could be subjected to electrophoresis simultaneously. Malate dehydrogenase was detected using the developing solution of Atzpodien *et al.* (1968). Cell fractions were sonicated briefly before application to the gel to release any membrane-bound enzyme. Cell fractions were prepared from cells grown on lactate as carbon source; growth on glucose results in repression of the cytoplasmic (lower mobility) enzyme, while growth under anaerobic conditions results in a preferential loss of the mitochondrial (higher mobility) enzyme (M. J. Lowden and P. R. Stewart, unpublished).

compound, which deposits at the site of the enzyme in the gel (Fig. 1). Enzymatic reactions that can be coupled to oxidation–reduction reactions through a common product or substrate can also be detected in this way. Fumarase, for example, can be detected in yeast homogenates by using a mixture of fumarate, malate dehydrogenase, NAD, and nitro blue tetrazolium to stain the gel. Where fumarase is present, fumarate is converted to malate, and reduced dye is deposited by subsequent oxidation of malate catalyzed by the malate dehydrogenase.

Esterases, including phosphatases, can be located by using as substrates naphthyl esters and coupling the naphthol produced by the enzymatic reaction to diazotized aromatic amines to yield intensely colored products. Campbell *et al.* (1972) have described esterase "zymograms" as an aid to the taxonomic differentiation of yeast species.

Wilkinson (1970) and Latner and Skillen (1968) review general methods of isoenzyme analysis.

b. *Detection of Enzymes and Noncatalytic Proteins by Immunochemical Reaction.* In cases in which a specific protein can be purified to homogeneity, preparation of antisera offers a rapid and very specific means to detect and to isolate single proteins or polypeptides in complex cellular or subcellular mixtures of proteins. This approach has been used by Schatz and co-workers (Mason *et al.*, 1973; Ebner *et al.*, 1973) in their examination of the site of synthesis of subunit proteins of the cytochrome-c oxidase complex of yeast mitochondria.

Highly purified protein (multimer or subunits) is necessary as antigen, since minor but very antigenic contaminants are capable of eliciting an antibody response that may obscure antiserum interaction with the protein(s) of interest. Mason *et al.* (1973) purified subunits of cytochrome oxidase by $(NH_4)_2SO_4$ precipitation and ion-exchange chromatography, followed by fractionation on Sephadex G-100 equilibrated with buffered sodium dodecyl sulfate (SDS). The column fractions are concentrated by ultrafiltration, and the SDS removed by treatment with an anion-exchange resin in the presence of urea. The polypeptides are renatured by careful dilution of the urea solution, and concentrated and dialyzed. Up to 2 mg of protein in 1.5 ml of buffer is then homogenized by sonication with 1.5 ml of Freund's adjuvant. The milky suspension is injected subcutaneously into several sites on the back and into the footpads of a rabbit (Nelson *et al.*, 1973). After 4 weeks the animal is given an intravenous booster injection of up to 1 mg of protein in solution, via an ear vein. On the sixth, eight, and tenth days after the booster injection, blood (20–40 ml on each day) is collected and allowed to clot. The serum is decanted or collected by centrifugation, filtered, and stored at $-20°C$. A second booster injection can be given to the rabbit 30 days after the first, and bleeding carried out subsequently as described. Control serum (i.e., without antibodies, to test for nonspecific precipitation) is obtained from the animal just prior to the initial injection of antigen (Mason *et al.*, 1973; Nelson *et al.*, 1973).

Enzyme present in crude mitochondrial lysates is precipitated as follows, using the antiserum. Submitochondrial particles, prepared by sonicating mitochondria and centrifuging at high speed, are dispersed in potassium cholate, and cytochrome oxidase is precipitated with $(NH_4)_2SO_4$. The precipitated crude enzyme preparation is redissolved in buffered Triton X-100, the cholate and $(NH_4)_2SO_4$ are removed by dialysis and, after clarification by centrifugation, cytochrome oxidase is precipitated by adding antiserum. After incubation at 4°C overnight, the antigen–antibody complex is recovered by centrifugation, washed, and dissociated with SDS. If the enzyme has been labeled prior to its isolation from the yeast, the labeled material may be analyzed by gel electrophoresis in SDS and shown to be the same as the purified enzyme against which the antiserum was originally prepared (Fig. 2.)

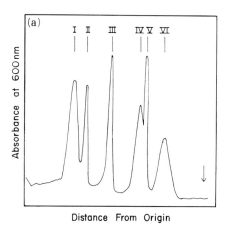 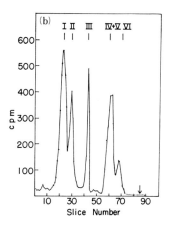

Fig. 2. Immunoprecipitation of cytochrome-c oxidase from crude mitochondrial extracts. (a) Subunits of purified cytochrome-c oxidase from *S. cerevisiae* were separated by electrophoresis on SDS–polyacrylamide gel. After electrophoresis, gels were stained with Coomassie brilliant blue, destained, and scanned at 600 nm. (b) Cells were grown in the presence of leucine-^{3}H, and mitochondria subsequently isolated. Rabbit antiserum to the preparation shown in (a) was used to precipitate material from a cholate-$(NH_4)_2SO_4$ precipitate prepared from the labeled mitochondria. The antigen–antibody complex was recovered by centrifugation, dissociated with SDS, and subjected to electrophoresis as in (a). Gels were sliced and assayed for radioactivity. The arrow in each case indicates the extent of migration of the tracking dye, bromphenol blue. (From Mason *et al.*, 1973.)

B. RNA

1. Quantitative Assay of RNA in Whole Cells and Subcellular Fractions

In general, the most useful methods for extracting nucleic acids from microorganisms involve $HClO_4$, at either 70°C (RNA plus DNA) or 37°C (RNA only). A detailed discussion of the principles and methods involved and their application to different types of microorganisms is given by Herbert et al. (1971). The method described below is that devised for yeasts by Trevelyan and Harrison (1956). Extraction of carbohydrate occurs when $HClO_4$ is used at 70°C with yeast cells, and this interferes with the subsequent estimation of RNA; extraction is therefore carried out in $HClO_4$ at 37°C. RNA (hydrolyzed to nucleotides) is estimated in the extract by the orcinol method for ribose. Ultraviolet spectrophotometric assay of the nucleotides can also be used as a means of assay (but see Herbert et al., 1971). RNA accounts for approximately 2–10% of yeast cell dry weight. As in bacteria, this proportion is almost certainly a function of growth rate of the cells.

In some cases lipid extraction is carried out prior to RNA (or DNA) extraction to prevent possible interference from this source. Lipid extraction of yeast cells is generally unnecessary, since assay methods based on ribose or bases are not likely to be interfered with by the small amount of lipid solubilized from normal cells with $HClO_4$. If lipid extraction is necessary, e.g., during assay of nuclcic acids prcscnt in ycasts grown on alkanes or under other conditions that may lead to a high lipid content of the cells, it is preferable that this be done before the cold-acid extraction (Herbert et al., 1971).

Reagents:

i. 0.5 *M* $HClO_4$

ii. 0.09% $FeCl_3 \cdot 6H_2O$ in concentrated HCl

iii. 1.0% orcinol solution; can be kept for 2 weeks at 2°C.

iv. Orcinol reagent; mix 1 vol of solution *iii* with 4 vol of solution *ii*, prepare fresh each day

Procedure: Add an equal volume (2–4 ml) of ice-cold 0.5 *M* $HClO_4$ to a chilled suspension of washed cells (20–50 mg dry weight). Incubate for 15 minutes on ice (shake occasionally to suspend cells), centrifuge at 0°C, and discard the supernatant. Resuspend the cells in ice-cold water and again add an equal volume of cold 0.5 *M* $HClO_4$. Incubate at 0°C as before for 15 minutes and recover the cells by centrifugation. These initial extractions serve to remove low-molecular-weight substances.

Resuspend the cells in 2–4 ml of $0.5M$ HClO$_4$ and incubate at 37°C for 2 hours; shake the tubes at intervals. Under these conditions RNA is hydrolyzed to nucleotides and is extracted from the cells. Pellet the cells by centrifugation and wash with a further 2–4 ml of 0.5 M HClO$_4$. Make the volume up to 5–10 ml and assay for ribose using the orcinol reagent as follows.

To a total volume of sample of up to 1.0 ml add 3.0 ml of orcinol reagent (*iv* above); prepare a blank and standards (50–400 μg RNA) containing 0.5 M HClO$_4$. Heat in a boiling water bath for 20 minutes, cool in tap water, and make up to 15 ml (calibrated tubes are useful) with *n*-butanol. Read the absorbance at 670 nm.

Two aspects of the method require checking for each organism used. First, the incubation time at 37°C in HClO$_4$ necessary to hydrolyze and extract RNA may vary with different organisms. Care should be taken that with longer extraction times carbohydrate does not contribute to color formation in the assay. This can be determined from the absorption spectrum of the color developed, and by comparison with that developed by hexoses (Herbert *et al.*, 1971).

Second, a heating time of 20 minutes with the orcinol reagent is near optimal for color development of ribose, and for purine ribosides of RNA which cleave rapidly under these conditions. Longer heating times result in further color development as a result of cleavage of pyrimidine ribosides, although contaminating hexoses also contribute increasingly to the reaction (Herbert *et al.*, 1971). If the absorption spectrum indicates that hexoses are interfering, shorter heating times should be adopted. Moreover, since the assay is sensitive to the relative proportion of purine and pyrimidine residues in the sample, the RNA used as a standard should, if possible, have a base composition similar to that of the RNA being examined.

2. IDENTIFICATION OF SPECIFIC RNA COMPONENTS IN YEAST CELLS

Quantitative isolation of all species of RNA in an intact state is difficult using techniques currently available. The most that can be hoped for with yeasts and other microbial cells is a representative sampling of different RNA types, the recovery of which is largely a function of cell lysis or breakage methods, and the extractive procedures used. The methods set out below were devised to isolate certain of the more structurally labile species of RNA from yeast cells and, although not quantitative, yield a representative selection of the metabolically stable RNA species normally found in the yeast cell. In combination with appropriate pulse-labeling procedures [including a rapid cell disruption (i.e., mechanical) technique], the procedures can also be used to study metabolically unstable species of RNA in yeasts (Yu *et al.*, 1972).

Reagents:

i. Solutions for preparation of cell-free homogenate, enzymatically or by mechanical cell disruption (see Chapter 8, Vol. XI and Chapter 6, Vol. XII).

ii. 0.1 M Sodium glycinate (pH 8.0), 0.1 M NaCl, 10 mM disodium EDTA, 1% bentonite, and 1% sodium deoxycholate (the method of preparation of bentonite is given by Poulson, 1973).

iii. Phenol saturated with aqueous 1 mM disodium EDTA.

iv. Diethyl ether (anhydrous).

v. 95% ethanol and 10 mM sodium acetate.

Procedure: Solutions should be ice-cold, and all operations should be carried out at $0°–4°$C. The cell homogenate is mixed with 3–5 vol of solutions *ii* and *iii* which have been mixed in a ratio 1:2 just prior to use. The mixture is shaken vigorously for 10 minutes, and then centrifuged at 10,000 g for 5 minutes to resolve the aqueous (upper) and phenol (lower) phases. The aqueous phase, containing extracted RNA, is carefully removed and extracted twice with 1 vol of ether to remove residual phenol. Two volumes of ethanol–sodium acetate are added to the aqueous extract, and the mixture is refrigerated at $-20°$C for 18 hours to precipitate RNA. RNA is recovered by centrifugation at 10,000 g for 30 minutes. The RNA can then be analyzed by acrylamide electrophoresis or by density gradient centrifugation using standard techniques (Poulson, 1973).

Particulate fractions from yeast cells (e.g., ribosomes, mitochondria, nuclei) can be extracted using the same procedures, except that the glycinate–sodium deoxycholate solution is mixed separately with the particulate fraction, and phenol is then added to complete the extraction mixture (Yu *et al.*, 1972). The range of species of RNA that can be extracted and fractionated from mitochondria and cytosol of *Candida parapsilosis* using these procedures is illustrated in Fig. 3.

A detailed description of the strategies and methods appropriate to the extraction of RNA from cells is given by Poulson (1973).

FIG. 3. Isolation of RNA species from cytoplasm and mitochondria of Candida parapsilosis. Nucleic acids were extracted with phenol from mitochondria or from the postmitochondrial supernatant, precipitated with ethanol overnight at $-20°$C, and then analyzed by electrophoresis on polyacrylamide gels using a Tris–acetate–EDTA–SDS buffer system as described in the text. Gels were prerun for 30 minutes, the sample applied, and electrophoresis continued for 2 hours at $22°$C and 5 mA/gel. After removal from tubes, the gels were soaked in water for 1 hour and then scanned at 265 nm. (a) and (b) Extracts from mitochondria prepared by two different methods. (c) Extract of postmitochondrial supernatant. (d) Mixture of RNA equivalent to that applied to gels in (b) and (c). C_1, M_1, C_2, and M_2 denote major cytoplasmic and mitochondrial rRNA species. 4 and 5 S RNA represent transfer and lowmolecular weight ribosomal species, respectively. The presence of cytoplasmic RNA species in mitochondrial extracts is probably a consequence of contamination occurring during cell fractionation (From Yu *et al.*, 1972), with permission of Springer-Verlag, Heidelberg.

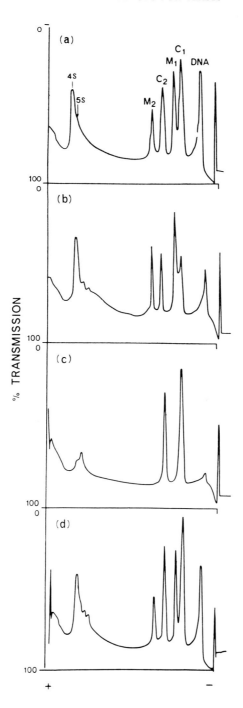

C. DNA

1. Quantitative Assay of DNA in Cells and Subcellular Fractions

DNA is most suitably determined in yeast cells by preparing hot $HClO_4$ extracts of cells and assaying these with the diphenylanine reagent, as described by Burton (1956) and modified by Richards (1974). The specificity of this assay for deoxyribose is such as to permit accurate assay of DNA even in the presence of a large excess of RNA and carbohydrate. An initial brief extraction with cold $HClO_4$ removes low-molecular-weight substances, primarily other deoxyribose compounds. Lipid extraction of yeasts, as mentioned earlier in relation to RNA extraction and assay, is generally unnecessary because of the specificity of the assay. However, if prior lipid extraction is thought to be desirable, it should be carried out before the cold-$HClO_4$ extraction. DNA accounts for less than 1% of the dry weight of most yeasts; 10–20% of this is found in the mitochondrial fraction.

Detailed surveys of methods of extraction and assay of DNA and RNA from microorganisms, and the difficulties that may arise, are given by Munro and Fleck (1966) and by Herbert *et al.* (1971). The method described below (Herbert *et al.*, 1971) is essentially that of Trevelyan and Harrison (1956).

Reagents:

i. 0.5 *M* $HClO_4$.

ii. 1.5% (w/v) diphenylamine in glacial acetic acid; add H_2SO_4 to a final concentration of 1.5% (v/v) and store in the dark.

iii. Diphenylamine reagent; add 0.5 ml of acetaldehyde solution (16 mg/ml in water) to 100 ml of solution *ii*; prepare fresh each day.

All reagents should be AR grade. The diphenylamine solution should remain colorless when acetaldehyde is added. If it does not, either different brands of AR diphenylamine and AR glacial acetic acid should be tried, or the diphenylamine should be steam-distilled and the glacial acetic acid redistilled to remove impurities (Herbert *et al.*, 1971).

Procedure:

i. Extraction. Add an equal volume (1–2 ml) of ice-cold 0.5 *M* $HClO_4$ to a chilled suspension (10–40 mg dry weight) of washed cells. Incubate for 15 minutes on ice and shake occasionally to resuspend the cells. Centrifuge, discard the supernatant, resuspend the cells in ice-cold water, and again add an equal volume of cold 0.5 *M* $HClO_4$. Incubate at 0°C as before and recover the cells by centrifugation. Resuspend the cells in 1–2 ml of 0.5 *M* $HClO_4$ and heat for 15 minutes at 70°C; shake the tubes at intervals. Centrifuge, repeat the extraction at 70°C two more times, and pool the supernatants. Both DNA and RNA are hydrolyzed and extracted under these conditions, but the diphenylamine assay is not interfered with significantly by concentrations of RNA generally found in yeasts. Make up the extract volume to 5–10 ml with 0.5 *M* $HClO_4$.

ii. Assay. Mix 1–2 ml of sample in 0.5 M HClO$_4$ with 2 ml of diphenyl-lamine reagent (reagent *iii* above) and incubate overnight at 30°C. Read the absorbance at 600 nm against a reagent blank and compare with standards containing up to 100 μg of DNA. Since deoxyribose residues set free from purine deoxyribonucleotides are primarily responsible for the color reaction, the standard DNA used and the DNA of the yeast should contain, if possible, a similar proportion of purine bases. A stable standard DNA solution can be prepared by dissolving purified DNA in 5 mM NaOH at 200 μg/ml and storing at 0°–4°C. At the time of assay, equal volumes of standard and 1.0 M HClO$_4$ are mixed and incubated for 15 minutes at 70°C. Aliquots of the standard sample are then assayed in parallel with supernatants of the cell extracts.

A useful modification of this procedure has been described by Richards (1974). Acetaldehyde is replaced by its much less reactive and volatile dimer, paraldehyde. Paraldehyde can also be included in the diphenylamine reagent with little or no loss of stability, and the mixture remains usable for up to 1 month. The method also becomes slightly more sensitive.

It should be noted that the quantity of cells used, and the initial steps in the extraction procedure, are similar for DNA and RNA extraction. More-over, the residue remaining after RNA hydrolysis and extraction as de-scribed earlier retains the DNA of the cell; this can thus be heated with HClO$_4$ to hydrolyze the DNA for extraction and assay. Thus a single sample of cells may be used to carry out quantitative analysis of both DNA and RNA.

2. Isolation of Species of DNA from Homogenates and Subcellular Fraction of Yeasts

Isolation and separation of linear, nicked circular, and covalently closed circular DNA species from cell-free homogenates or unpurified subcellular fractions of yeasts can most readily be achieved using ethidium bromide–cesium chloride gradients (Piko *et al.*, 1968; Clark-Walker, 1972; Clark-Walker and Miklos, 1974). Cells are harvested, and cell-free homogenates or subcellular fractions (nuclei, mitochondria) are prepared directly from disrupted cells, or from protoplasts, and processed as follows (Clark-Walker, 1972).

Resuspend the pellets (30–60 mg of protein) in 1–2 ml of 2% (w/v) Sarkosyl L (Geigy), or add concentrated Sarkosyl solution to homogenates to bring the detergent to this final concentration. DNA is released by gentle agitation for 10 minutes at 0°–4°C. Prepare ethidium bromide–cesium chloride gradients in polyallomer centrifuge tubes by mixing 4.3 gm cesium chloride with 0.3 ml of 0.1 M sodium EDTA (pH 7.0), 1.35 ml of ethidium bromide [1.75 mg/ml in 0.1 M sodium phosphate buffer (pH 7.0)], up to 2 ml of Sarkosyl-treated sample, and water to a total volume of 8.8 ml. Overlay the gradient mixture with liquid paraffin to fill the tube, and centrifuge for 48–72 hours at 10°C and 160,000 g (average).

Under these conditions of gradient centrifugation, covalently closed circular DNA binds less dye than do linear or nicked circular molecules, and consequently equilibrates at a higher density position in the gradient. The other two forms of DNA, banding at lower density, may separate depending on their relative intrinsic densities and the amount of dye bound by each (Fig. 4).

FIG. 4. Isolation of DNA Species from *S. cerevisiae*. A crude mitochondrial preparation from cells disrupted with glass beads was lysed in Sarkosyl, mixed with cesium chloride and ethidium bromide, and centrifuged for 50 hours as described in the text. The centrifuge tube and contents were then photographed under ultraviolet (350 nm) light. Three fluorescing bands can be seen: the upper one is open circles of mitochondrial DNA (density = 1.684), the middle is linear nuclear DNA (1.701), both of which bind more ethidium bromide than the lowest band, which is covalently closed circular DNA. The covalently closed circular DNA has a density similar to nuclear DNA when measured in the absence of ethidium bromide (Clark-Walker, 1972). (From R. O'Connor and G. D. Clark-Walker, unpublished).

If the DNA has been labeled by prior incubation of the cells in labeled precursor, the DNA may be fractionated by puncturing the bottom of the tube and collecting fractions either directly on filter paper discs (Bollum, 1966) or in tubes (Clark-Walker, 1972). After further processing, e.g., to remove contaminating RNA, fractions are assayed for radioactivity.

Alternatively, the bands of DNA (which can be visualized within the gradients by their fluorescence when exposed to 350-nm ultraviolet source) can be removed directly from the gradient by puncturing the side of the tube with a hypodermic needle and withdrawing the band with a syringe. Bound ethidium bromide is removed by extracting samples four times with isoamyl alcohol. Residual alcohol and cesium chloride are removed by dialysis overnight against 10 mM EDTA–15 mM NaCl–1.5 mM sodium citrate buffer (pH 7.0), and the DNA samples are then used for analytical ultracentrifugation or for other physical or chemical analysis.

By using this technique three or four species of DNA can be isolated from wild-type strains of *S. cerevisiae*: linear molecules of density approximately 1.701 corresponding to nuclear DNA; linear molecules of density 1.684 corresponding to broken circles of mitochondrial DNA; small circles of covalently closed DNA, of density very similar to that of the nuclear species; and linear species of density approximately 1.704, corresponding to γ-satellite DNA (Clark-Walker, 1972; Clark-Walker and Miklos, 1974).

D.   Polysaccharides and Glycoproteins

1.   SEPARATION AND DETERMINATION OF GLYCOGEN, MANNAN, AND GLUCAN

The major forms of polysaccharide in yeast cells are glycogen (storage carbohydrate), and glucan and mannan (cell wall materials). Together, they account for approximately one-half of the dry weight of the cells. The structure of these components was reviewed recently by Phaff (1971), Manners (1971), and Ballou and Raschke (1974). Glycogen exists primarily as finely dispersed granules in the cytoplasm of the cell (Matile *et al.*, 1969), although some may also be present in soluble form; glucan and mannan are integral structural components of the cell wall. A preliminary separation of glycogen from the structural carbohydrates can thus be achieved by disruption of the cell wall and sedimentation of wall fragments away from the glycogen at low centrifugal forces. A purely extractive procedure has been devised for *S. cerevisiae* (Trevelyan and Harrison, 1956) and found to be effective with other yeast species (Herbert *et al.*, 1971). At the same time trehalose, a major disaccharide component of yeasts, is also fractionated and may be assayed.

Chitin, a polymerized amino sugar found in certain yeast walls, is usually assayed separately. The extraction method is as follows.

Reagents:

i. 1.0 M trichloroacetic acid

ii. 0.25 M Na_2CO_3

iii. 0.1 M NaOH

iv. 0.28 M $CuSO_4$ in 5 mM H_2SO_4

v. 1.2 M sodium potassium tartrate in 3 M NaOH (filter if cloudy)

vi Fehling's solution; mix equal volumes of solutions *iv* and *v* on day of use.

vii. 1.0 M H_2SO_4

viii. 0.5 M $HClO_4$

ix. 2.0 M NaOH

Procedure:

i. *Extraction of trehalose.*   To a cold suspension (4–6 ml) of yeast cells (20–50 mg) in water add an equal volume of ice-cold 1.0 M trichloroacetic acid. Incubate the mixture at 0°C for 60 minutes with frequent shaking. After centrifuging, wash the cells once with 3 ml of water, combine the supernatants, and make up to 10 ml. Remove 1–2-ml aliquots for assay by the anthrone method described below.

ii. *Extraction of mannan and alkali-soluble glycogen.*   Resuspend the pelleted cells in 2–3 ml of 0.25 M Na_2CO_3, cover the tubes with glass stoppers, and heat for 45 minutes at 100°C. Cool the cell suspensions in tap water, centrifuge, and collect the supernatant; then wash the pellets with 2 ml of water. Make up the combined supernatants to 5.0 ml, and use 2.0 ml to assay glycogen (alkali-soluble) plus mannan by the anthrone method.

Mannan is precipitated from the mixture by treatment with an alkaline copper solution (Fehling's solution). Mix 2.0 ml of cell extract with 1.0 ml of Fehlings' solution in a centrifuge tube. Stopper the tubes and incubate overnight at room temperature. Recover the precipitated mannan–copper complex by centrifugation, wash with 2–3 ml of 0.1 M NaOH, dissolve in 0.5 ml of 1.0 M H_2SO_4, and make up to 10 ml. Use 1- to 2-ml aliquots for the anthrone assay. The glycogen (alkali-soluble) content of the cells can then be calculated from the difference between anthrone assays of mannan plus glycogen (alkali-soluble) and mannan.

iii. *Extraction of acid-soluble glycogen.*   Suspend the pelleted cell residue remaining from step *ii* in 2 ml of 0.5 M $HClO_4$ and heat for 30 minutes at 100°C. Cool the tubes, centrifuge, and wash the pellets once with 2 ml of water. Make up the pooled supernatants to 10 ml and use aliquots of 1–2 ml for the anthrone assay. Calculate the glycogen content as the sum present in the alkali- and acid-soluble extracts.

iv. *Glucan.*   Suspend the residue from the previous step in 1 ml of 2 M

NaOH, transfer quantitatively to a 25-ml stoppered flask or cylinder, and make up to volume with water. Shake the suspension vigorously to suspend the cells evenly; this suspension usually does not tend to settle. Take 1- to 2-ml aliquots for analysis by the anthrone method.

v. Anthrone assay of hexoses. The most suitable routine method is that described by Herbert *et al.* (1971); earlier methods, which relied on the heating caused by the dilution of the aqueous sample by concentrated H_2SO_4 in the anthrone solution, are generally less reproducible. The basis of the assay is the formation of a colored complex between the enol tautomer of anthrone and hydroxymethyl furfural formed by dehydration of the sugar in hot acid.

Reagents:

Standard glucose; stock solution 1.0 mg/ml in 0.15% benzoic acid, diluted on day of use to 0.1 mg/ml with water: H_2SO_4, 75% (v/v).

Anthrone reagent; 200 mg anthrone in 5 ml of ethanol made up to 100 ml with 75% H_2SO_4; avoid undissolved particles of anthrone, prepare fresh each day and keep on ice.

Procedure: Pipette 1.0-ml samples into thin-walled boiling tubes and chill on ice. Add 5.0 ml of chilled anthrone reagent with rapid stirring, and return the tubes to the ice as quickly as possible. After chilling is complete, transfer the tubes simultaneously to a 100°C water bath, heat for exactly 10 minutes, and return them to the ice. After cooling, record the absorbance of samples at 625 nm using a reagent blank, and compare with appropriate standards.

The anthrone assay can also be carried out directly with intact cells to give values for total hexose (including 6-deoxyhexose; hexosamines and pentose do not interfere significantly). Although the color yield in the anthrone reaction differs among hexoses, the relative constancy of proportion of mannan to glucose polymers, which together constitute most of the polysaccharide fraction (Trevelyan and Harrison, 1956), means that the anthrone method may with care be used directly to compare the total carbohydrate content of different strains, or of the same strain grown under different conditions. Alternatively, phenol can be used in place of anthrone (Herbert *et al.*, 1971), in which case total pentose and hexose is estimated with little difference in the color contributed by different sugars.

2. Minor Polysaccharide Components of Yeasts

a. Chitin. Polymers of 1,4-linked *N*-acetylglucosamine may remain as insoluble residue after extraction of mannan, glucan, protein, and lipid from cell walls (Eddy, 1958; Phaff, 1971). The total amount found in the cell walls of most yeasts is small (Phaff, 1971). Moreover, much of the *N*–acetylglucosamine content of yeast cell walls may not be present as chitin, but may occur in combination with mannan and protein (Sentandreu and Northcote, 1969). The true chitin in the cell walls of yeasts may occur predominantly in bud scars, as reported in Korn and Northcote (1960) and Bacon *et al.* (1966). Methods for the extraction and assay of chitin are given in these references.

b. Polysaccharide–Protein Complexes. The presence of a phosphoglycopeptide fraction in the cell walls of yeasts has been known for some time (Phaff, 1971). Chemical and enzymatic methods for extraction and partial characterization of a phosphomannan–protein complex are given by Sentandreu and Northcote (1969), and Eddy and Longton (1969). Ballou and Raschke (1974) have surveyed recent progress in this field, and Taylor and Cameron (1973) discuss the strategy and tactics of yeast and fungal cell wall analysis.

c. Capsular Polysaccharides. Since the formation of capsules by yeasts is by no means a universal phenomenon, analysis of their composition and structure is not considered here. A useful review of this subject is given by Phaff (1971).

II.   Small-Molecule Pools in Yeasts

Yeast cells contain pools of low-molecular-weight substances representing macromolecule precursors, intermediary metabolites, secondary metabolites awaiting release from the cell, coenzymes, vitamins, and a range of other substances. The amount of a given compound is dependent on a balance existing between uptake, synthesis, and salvage of the compound on the one hand, and its polymerization, excretion, and further metabolism on the other. The amount of material existing in a given pool is thus closely related to the metabolic and physiological state of the cell, and the physical and chemical nature of the cell's environment (Conway and Armstrong, 1961; Stebbing, 1974).

Because of this dependence on the physiological status of the cell and on culture conditions, a crucial factor in the reproducibility of the assay of pool constituents in yeast cells is the consistency with which extraction procedures can be carried out from time to time. Moreover, changes in osmolarity or ionic strength, pH, temperature, and so on, during extraction may result in leakage of low-molecular-weight substances from the cells. During harvesting and washing these parameters should be kept as close as possible to those existing in the culture from which the cells have been taken. To be balanced against this, however, is the continuing metabolism of the cell, particularly at or near normal growth temperatures. Hence the possibility of untoward physiological effects resulting from nutrient depletion and end product buildup in concentrated suspensions and pellets of cells may sometimes compromise the conflicting requirements of minimum change between culture and extraction conditions.

In general terms the techniques found most effective in releasing low-molecular-weight substances from yeast cells involve acids or alcohols at low temperature, water or alcohols at high temperature, or cycles of freezing and thawing. The aim in each case is rapidly and effectively to destroy permeability barriers of and within the cell, and to inhibit chemical or enzymatic degradation of preexisting compounds.

Of the extraction solvents used, cold aqueous $HClO_4$ meets most of the above requirements, except that certain nucleotides and polynucleotides are sufficiently labile under these conditions to undergo significant hydrolysis even during short extraction times. $HClO_4$ also offers the advantage, in this and other extractive procedures, that it can be removed by centrifugation as insoluble $KClO_4$ after neutralizing extracts with KOH or K_2CO_3. Hot-water has also been found useful as a small-molecule extractant, although it seems that some substances insoluble in $HClO_4$ dissolve in hot water. The methods set out below offer by no means a complete account of the methods available. They have been found suitable in the particular cases concerned. As before, when an organism not previously investigated is to be examined, trial experiments with different methods of extraction and assay are recommended.

A. Amino Acids

The free amino acid pool in yeasts accounts for the greatest proportion of low-molecular-weight material extractable from most yeasts. In *Candida* and *Saccharomyces* respectively, for example, approximately 90% of the carbon extractable with cold trichloroacetic acid and freezing and thawing is accounted for as amino acids (Cowie and Walton, 1956; Conway and Armstrong, 1961). In *Schizosaccharomyces pombe*, however, more carbohydrate than amino acid is released by freezing and thawing (Stebbing, 1971). A method with excellent reproducibility when applied to various species of *Saccharomyces* and *Candida* involves extraction with hot water. As in the case of fungi (Thornton and McEvoy, 1970), hot water extracts amino acids efficiently, while at the same time extracting less material interfering with the subsequent separation of individual amino acids.

Reagents:
Cation-exchange resin, e.g., Dowex 50–X8 or ZeoKarb 225, 20–50 mesh
5 M NH_4OH
0.05 M potassium phosphate buffer (pH 6.0).

Procedure: Mix an aliquot of culture containing 10–20 mg (dry weight) of cells with ice to chill the cells rapidly, and harvest by filtration or centrifugation at 2°C. Wash the cells three times with 4–5 ml of ice-cold 0.05 M KH_2PO_4 buffer, and once with ice-cold water. Resuspend the cells in 2–3 ml of water and heat in capped tubes in a boiling water bath for 15 minutes. Pellet the cells by centrifugation at room temperature, keep the supernatant, and wash twice with 1–2 ml of water. Pool the supernatants, and add ion-exchange resin until the pH of the supernatant is 2.5–3.0 (determined with small pieces of narrow-range indicator paper) and remains at this pH after 5 minutes of intermittent shaking. Allow the resin to settle, and aspirate and discard the supernatant. Wash the resin twice with 2–3 vol of water. This procedure disposes of protein, together with nonionic and anionic contaminants that do not bind to the resin. Elute the bound amino acids (and other substances that exist as cations at this pH) from the resin by three washes with two resin volumes of 5 M NH_4OH. Combine the eluates and evaporate to dryness under vacuum. Dissolve the residue in water or a suitable buffer. The amino acids may now be assayed and separated by standard methods.

Ethanol, methanol, hot water, and cold tricholoacetic acid give identical results when used to extract amino acids from *C. utilis* (Nurse and Wiemkin, 1974). The cleanup procedure using cation-exchange resin should also be applied to these types of extracts before submitting them to quantitative or qualitative analysis.

B. Sugars

Sugars can be measured in cell or culture extracts prepared using $HClO_4$, water, or ethanol as described for amino acid extraction. The method of assay of sugars in these extracts may be specific (e.g. glucose, galactose) where appropriate enzymatic or colorimetric methods are available; it may be a total assay, for example, using the anthrone method described earlier; or it may require a separation by chromatographic procedures for precise measurement of minor components (Dutton, 1973). Chemical methods are available for the determination of hexosamines, pentoses (the orcinol reaction described earlier), heptoses, 6- and 2-deoxy sugars, and total reducing sugars, and for discriminating between aldoses and ketoses. These methods as applied to microbial cells in general are comprehensively discussed and summarized by Herbert *et al.* (1971). They appear to be applicable, with usual precautions, to the analysis of yeast cells and culture fluids.

C. Bases, Nucleosides, and Nucleotides

The estimation of these substances is rather more complex than procedures for amino acids and sugars, because (1) they frequently occur at very low concentration in the cells, (2) a wide diversity of molecular species is present, and (3) the lability of the different molecular species varies under different conditions of extraction. The general topic of extraction, separation, and quantitative determination of bases and nucleotides in cell extracts is treated in detail elsewhere (Henderson and Paterson, 1973).

In the specific case of yeasts, the need frequently arises to assay individual species of adenine nucleotides, e.g., to determine the energy (adenylate) charge of the cells, or to determine variations in the level of cyclic nucleotides in response to, or as a consequence of, altered regulatory status within the cells.

1. Measurement of Adenine Nucleotides (Energy Charge)

The adenylate energy charge of a range of actively metabolizing cells has constancy (Chapman *et al.*, 1971) which suggests that this parameter may describe a fundamental and universal property of living cells. The value for energy charge (*EC*) is given by the relationship

$$EC = \frac{[ATP] + 0.5[ADP]}{[ATP] + [ADP] + [AMP]}$$

where [ATP], [ADP], and [AMP] denote the intracellular concentration of adenosine-5-tri-, di-, and monophosphate, respectively. The value of *EC* provides a single numerical measure of the amount of energy metabolically available to the cell in the adenylate system. Response curves of activity as a function of energy charge for a range of enzymes are such as to suggest that the energy charge is normally stabilized in a range near 0.85 (Atkinson, 1969). This has been amply confirmed, as mentioned above (Chapman *et al.*, 1971).

In the case of yeast cells, adenylate concentrations have been measured using several extractive and assay methods (Betz and Chance, 1965; Polakis and Bartley, 1966; Somlo, 1970; Weibel *et al.*, 1974). Somlo, and Weibel and co-workers, have described the lack of effectiveness of standard extractive techniques for the extraction of adenine nucleotides from yeast cells. The method finally adopted by Somlo (1970) involves incubation of cells in 5% (w/v) $HClO_4$ for 15 hours at $0°C$, followed by assay using standard enzyme-coupled fluorimetric techniques for the individual nucleotides (Lowry and Passoneau, 1972). Polakis and Bartley (1966) extracted cells with 30% (w/v) $HClO_4$ at $-10°C$ for an unspecified but presumably shorter time, followed by two cycles of freezing and thawing of the cells. Assay was also by standard fluorimetric procedures. However, the difference in energy change measured in these two studies (cf. Chapman *et. al.*, 1971) points to possible differences in the efficiency of the extraction methods used, and suggests that the extractability of cofactors of this type might need careful examination in future studies.

2. MEASUREMENT OF CYCLIC NUCLEOTIDES IN YEASTS

The role of cyclic AMP in the mechanism of catabolite repression in *Escherichia coli* is now well established (for review, see Polya, 1973). In *Saccharomyces carlsbergensis*, a similar regulatory role is suggested by the experiments of Van Wijk and Konijn (1971). During growth on galactose, a sugar that does not repress the activity of enzymes of alternative catabolic pathways (e.g. the tricarboxylic acid cycle), cyclic AMP concentrations were approximately six times higher than in cells grown on glucose. The values obtained under conditions of derepression were in the range 0.2–0.3 μM (intracellular concentration), which interestingly is about 1000 times lower than that found in derepressed prokaryotes (Makman and Sutherland, 1965) but similar to that found in animal cells and tissues (Robison *et al.*, 1971).

This nucleotide is stable in dilute acid and thus may be extracted either with cold 15% $HClO_4$ (Van Wijk and Konijn, 1971), or with 0.05 M HCl, as in the case of bacteria (Makman and Sutherland, 1965). In view of the report of Somlo (1970) mentioned earlier, concerning the difficulty of extracting nucleotides from yeast cells using $HClO_4$, the question of the extractability of cyclic nucleotides should also be examined more closely. The assay methods described by Makman and Sutherland (1965) and Van Wijk and Konijn (1971), involving activation of dog liver phosphorylase and migration of cultures of *Dictyostelium discoideum*, are perhaps unlikely to be generally available. At least two alternative procedures are available. Commercial kits, based either on radioimmunoassay or displacement from binding protein are now marketed (e.g., by Calbiochem and Amersham). A similar assay kit for cyclic GMP is also available. Brooker *et al.* (1968) describe an enzymatic radioisotope displacement method for assay of either cyclic nucleotide. A modification of this method, applied to cyclic AMP, is described by Jost *et al.* (1970).

D. Metabolic Intermediates

Generally, the methods that have been devised and refined for analysis of intermediary metabolites in animal tissues can be adapted with little difficulty for use with yeasts and other microorganisms. Two excellent treatises on this subject are available: Bergmeyer (1974) and Lowry and Passonneau (1972). The reader is referred to these for specific details of assay of individual metabolites and enzymes. These investigators also describe "enzymatic cycling," a means for chemically amplifying very small quantities of intermediates and cofactors in cell extracts.

For realistic measurement of metabolite levels in cells, metabolism must be arrested as quickly and as effectively as possible. The rate at which yeasts or other rapidly growing microorganisms can exhaust the supply of internal or local external nutrients, even at 0°C, should not be underestimated. Weibel *et al.* (1974), for example have estimated that the time required for exhaustion of oxygen and glucose in a glucose-limited chemostat culture of yeasts is 8.4 and 32 seconds, respectively, when supply of these nutrients to the cells is cut off. Since the turnover time for the ATP pool in yeast cells is only 15 seconds, substantial changes in the concentration of other metabolites in the cell might be expected. For this reason centrifugation and similar harvesting techniques cannot be considered satisfactory. However, quenching methods involving very low or high temperatures are unnecessary with yeast cells, since acid, alkali, or organic solvents (which also act as extractants) penetrate and quench metabolism in the individual cells of a culture very rapidly. Nevertheless, the denaturing effect of chemical quenchers on enzymes in the cell should not necessarily be assumed to be complete, and the extraction of metabolites from the cell should proceed as rapidly as possible. In turn, extracts should be analyzed as soon as possible, and in any case they should not be stored in acid or alkali unless it is known that the intermediates concerned are stable at low or high pH. If storage is necessary, very low temperatures (at least $-50°C$) should be used. Storage times should be constant where different preparations are being compared.

A general procedure for the extraction of intermediary metabolites which might be adopted is as follows, Transfer samples of culture or cell suspensions with rapid mixing to 2 vol of ice-cold $1.2 M$ $HClO_4$–1 mM EDTA held in an alcohol or ice–salt water bath at $-5°C$. Continue mixing to bring the temperature of the quenched cells down quickly. Remove the cells by brief centrifugation at $-5°$ to $0°C$ and wash twice with ice-cold $0.6 M$ $HClO_4$. Neutralize the pooled extracts with cold K_2CO_3 or $KHCO_3$ and centrifuge to remove the precipitated $KClO_4$. The clarified extract is used for assay, or is stored at $-50°C$ or lower. It should be pointed out that, if cultures have been used at starting material, this procedure is in fact an extraction of both cells and culture medium. This is deliberately so since, as mentioned earlier, even the most rapid harvesting procedures result in some perturbation of cellular metabolism. The contribution of substances in the culture fluid to the assay can be determined by separate treatment of a sample of the culture fluid from which cells have been removed. If this contribution is large and interferes with the assay, or in some way complicates interpretation of the results obtained with cells plus culture fluid, resort must be made to some rapid harvesting method (e.g., pressure filtration through membrane or glass-fiber filters) before processing the cells with $HClO_4$.

The analysis of pyridine nucleotides presents particular problems because of the instability of reduced forms in acid, and of oxidized forms in alkali. Polakis and Bartley (1966) have carefully evaluated this problem as it relates to extraction and assay of these metabolic cofactors in yeasts.

E. Hemes, Porphyrins, and Cytochromes

Porphyrins and metalloporphyrins (primarily heme) are present as prosthetic groups of cytochrome and hemoproteins, or as intermediates in the synthesis of these substances in yeast cells. As a group their chemistry and pathways of synthesis have been extensively explored, and this subject has been comprehensively reviewed by Falk *et al.* (1961), Eddy (1958), and Falk (1964). Interest more recently has centered on mechanisms of regulation of synthesis of the heme entity (Lascelles, 1964; Poulson and Polglase, 1974), and of the interaction of intermediates and products of this pathway with the synthesis of apoproteins of the cytochromes (Charalampous, 1974).

Detailed methods of extraction, purification, and characterization of an array of porphyrins and metalloporphyrins from a wide range of different tissues and cells are given by Falk (1964). A particular application to yeast cells is given in Section III,E,1. Following this, methods for assaying specifically the major cytochrome species of yeast cells are described. Usually, most of the heme of the yeast cell is associated with cytochrome apoprotein, so values obtained for heme and cytochrome concentration should be approximately the same. Important exceptions to this general case exist, and these will be referred to subsequently.

1. EXTRACTION AND ESTIMATION OF HEME

Protoheme (the prosthetic group of cytochrome b), heme a, and other hemes that are not covalently bound to protein may be extracted from lyophilized preparations of cell-free homogenates or subcellular fractions with acidified organic solvents. The hemes are then converted to the corresponding pyridine hemochromogens by reaction with pyridine under alkaline conditions (Falk, 1964). The absorption spectrum of the derived pyridine hemochromogens provides a means to assess qualitatively and quantitatively the heme moieties present in cells or organelles (Falk, 1964). The method set out below is essentially that described by Jacobs and Wolin (1963).

Reagents:
i. Acetone
ii. Chloroform–methanol (2:1 by volume)
iii. Acetone–HCl (acetone containing 1% 2.4 M HCl by volume)
iv. Ether
v. 1.5 M HCl

vi. 0.5 *M* HCl
vii. NaCl, 3% (w/v)
viii. Pyridine
ix. 0.2 *M* KOH

Procedure: Mix samples (containing approximately 100 mg of protein) of lyophilized cell-free homogenate or cell fraction with 40 ml of ice-cold acetone. Centrifuge, discard the supernatants, and wash the pellet with 40 ml of cold chloroform–methanol. Centrifuge and wash the pellet once more with cold acetone. Extract noncovalently bound heme from the pellet with two lots of 40 ml of acidified acetone (reagent *iii*). Combine the supernatants and reduce the volume under vacuum to about 20 ml. Extract the concentrate by shaking with 20 ml of ether plus 5 ml of 1.5 *M* HCl. Collect the ether layer and reextract the aqueous acetone layer with 8 ml of ether plus 2 ml of 1.5 *M* HCl. Combine the ether extracts, wash with 30 ml, then with 20 ml of 0.5*M* HCl, and finally with 20 ml of 3% NaCl. Remove the ether under vacuum and resuspend the residue in 3.5 ml of pyridine and 3.5 ml of 0.2 *M* KOH. If necessary, centrifuge to remove any precipitate that may be present, and then divide the samples between two cuvettes for difference spectroscopy. Add and mix a few grains of potassium ferricyanide to one, and the same amount of sodium dithionite to the other; record the spectrum in the region 500–600 nm. Check that oxidation and reduction are complete by adding a few more grains each of ferricyanide and dithionite, and rerecord the spectrum. Absorption band positions and extinction coefficients characteristic of the different pyridine hemochromogens are given by Falk (1964). Heme a is unstable and moreover associates tenaciously with lipid (Williams, 1964; Falk, 1964). Special methods are thus necessary to determine its presence quantitatively (Falk, 1964). In the method described above, the amount of heme a extracted and converted to the pyridine hemochromogen will almost certainly be a underestimate of the amount originally present in the cells.

The covalently bound hemes of cytochromes c and c_1 may be estimated in the residue remaining after extraction with acidified acetone. The residue is suspended in 10–15 ml of a mixture of equal amounts of pyridine and 0.2 *M* KOH, using a tissue homogenizer (Jacobs and Wolin, 1963; Williams, 1964). The sample is placed in two cuvettes, treated with ferricyanide and dithionite as described above, and a difference spectrum is recorded using a spectrophotometer capable of analyzing turbid samples. In cytoplasmic and nuclear petite yeast mutants, in anaerobically cultured cells, and in cells subjected to glucose repression, relatively large amounts of free porphyrin and zinc porphyrin are formed, as reported by Pretlow and Sherman (1967) and Poulson and Polglase (1974). These investigators describe methods for the separation of porphyrins and metalloporphyrins and for the identification of individual molecular species in these mixtures. Barrett (1961) describes the isolation of cryptohematin a, a particular form of heme a, from yeasts grown under different conditions of respiratory development.

2. ANALYSIS OF CYTOCHROMES IN CELL HOMOGENATES AND MITOCHONDRIA

Analysis of the cytochrome content of cell suspensions or of mitochondria or other particulate subcellular fractions requires a spectrophotometer capable of analyzing small absorbance differences at small bandwidths, using turbid solutions. A split-beam instrument is essential for such a study; Chance (1957) gives a detailed account of the critical aspects of spectrophotometric analysis of turbid samples.

Absolute absorption spectra of whole yeast cells may be obtained by comparing the spectrum of a cell suspension [15–25 mg (dry weight)/ml in 50% (v/v) glycerol] reduced either by inducing anaerobiosis or by adding a strong reducing agent such as dithionite to the cell suspension. A blank which generates comparable scatter, but without significant specific absorption in the region 500–650 nm, is prepared as follows (Linnane, 1965). Mix soluble starch (80 mg) and plain flour (800 mg) vigorously while adding 15 ml of water. Heat the slurry carefully over a flame until it just begins to boil. Pour the thickened slurry into spectrophotometer cuvettes (5- or 10-mm light path) while still hot; avoid trapping air bubbles. After the mixture sets at room temperature, seal the cuvettes with caps or Parafilm. Pairs of blank cuvettes prepared in this way should be used to establish or check the zero absorbance trace (baseline) of the instrument. A yeast suspension is then substituted for the appropriate blank cuvette, and the absorbance spectrum is determined. Adjustment of the yeast cell density in the suspension is indicated if the baseline trend of the absorption spectrum is upward as the wavelength decreases (i.e., scatter by the cells is greater than scatter in the blank). If the baseline trend is downward, cell density in the suspension should be increased.

Quantitative estimations of cytochromes are sometimes difficult to reproduce using absolute absorption spectra of whole-cell suspensions. A useful method of quantitating the spectral assay of cytochromes is to prepare either cell-free homogenates or suspensions of subcellular fractions, and to examine the spectra of oxidized versus reduced samples of the same material. A detailed account of the information that can be obtained from such preparations is given by Chance (1957), but a simple procedure for determining the content of total and individual cytochromes of yeast mitochondria, for example, is as follows.

Reagents:

   i. Mitochondrial suspension (5–10 mg protein/ml)

ii. Glycerol, 50% (v/v)

iii. Sodium deoxycholate, 20% (w/v)

iv. Potassium ferricyanide or ammonium persulphate

v. Sodium dithionite

Procedure: The concentration of mitochondria required in suspension to obtain useful spectra is in general dependent on the sensitivity of the spectro-photometer used. The concentration indicated above (5–10 mg protein/ml) is satisfactory with a relatively sensitive instrument, such as the Cary 14 fitted with a scattered-transmission accessory. Moreover, with this instrument dispersion and partial clarification of the sample with deoxycholate, as described below, is not essential [e.g., see Clark-Walker and Linnane (1967), but also note that the spectra described were obtained with an instrument especially adapted for the examination of samples at 77°K to improve the resolution of absorption bands]. The addition of deoxycholate to a final concentration of approximately 2% (w/v) reduces scatter by the sample and permits the use of instruments with lower signal-to-noise characteristics.

Deoxycholate is added to the mitochondrial suspension to a final concentration of 2%. The sample is divided between two cuvettes (light path not greater than 1 cm), and a difference trace recorded between 500 and 650 nm. This trace with both samples untreated provides a reference baseline for the difference spectrum. A few grains of ferricyanide or persulfate are mixed with the contents of the blank-position cuvette to oxidize the cytochromes, and a few grains of dithionite are added to reduce the cytochromes in the test cuvette. The difference spectrum is then recorded. It is useful after recording the spectrum to add a few more grains of oxidizing and reducing agent to the respective cuvettes and rerecord the spectrum to ensure that oxidation and reduction are complete (i.e., there should be no further change in the spectrum). After correction of the difference spectrum for variation in the baseline originally recorded, the concentrations of cytochromes a, b, c, and c_1 are calculated using either the extinction coefficients and the wavelength pairs described by Estabrook and Holowinsky (1961), or the simultaneous equations (also based on extinction coefficients of absorbance differences between specific wavelengths) derived by Williams (1964). Examples of traces obtained with crude mitochondrial pellets prepared from *S. cerevisiae* and *C. parapsilosis* are shown in Fig. 5. Cytochrome a_3 can be determined by scanning in the Soret region (400–460 nm) and obtaining the absorbance difference between 444 and 455 mn (Estabrook and Holowinsky, 1961). This may require dilution of the sample or a decrease in the sensitivity of the instrument, since the extinction coefficient in this region is greater than at longer wavelengths. Ferricyanide should not be used as oxidant when recording in the Soret region, since it has significant absorbance in this region.

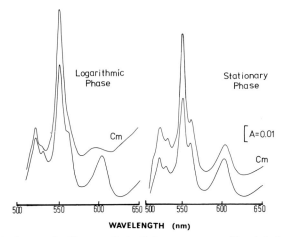

FIG. 5. Cytochromes in *C. parapsilosis*. Cells were grown either into log or stationary phase in the presence or absence of 3 mg/ml of chloramphenicol (Cm). After harvesting and washing, the cells were resuspended at approximately 15 mg/ml in 50% (v/v) glycerol and reduced with dithionite; a spectrum was recorded in a Cary 14 with scattered transmission accessory versus a starch-flour blank prepared as described in the text. Spectra show that synthesis of cytochromes aa₃ and *b* is inhibited by chloramphenicol only in rapidly growing cells. (From Yu and Stewart, 1974).

3. Extraction of Cytochrome c from Cells and Mitochondria

Cytochrome c is loosely associated with the mitochondrial inner membrane and can be selectively extracted with salt or with low concentrations of detergent. A procedure for the extraction of cytochrome c from cells, and its assay, is as follows (Sels *et al.*, 1965).

Reagents:

i. Glycerol, 8.5% (v/v)
ii. Sodium dithionite
iii. Ethyl acetate
iv. 0.5 *M* NaCl

Procedure: Cells (approximately 1 gm dry weight) are harvested, washed, and resuspended in 2 ml of 8.5% glycerol. Add 170 mg of fresh sodium dithionite to the suspension, followed by 1 ml of ethyl acetate. Flush the tube and contents with nitrogen and seal. Agitate gently for 16 hours at 25°C. Dilute the autolyzed cell suspension with 5 ml of 0.5 *M* NaCl and centrifuge for 10 minutes at 3000 *g*. Retain the supernatant and extract the residue with a further 8 ml of 0.5 *M* NaCl. Centrifuge again, combine the supernatants, and estimate the cytochrome-c content from absorbance at 550 nm (molar extinction coefficient = 28) for optically clear solutions. If the extract is turbid, a difference spectrum should be recorded by reading an untreated (reduced) sample against a second sample oxidized with a few grains of potassium ferricyanide. The absorbance difference $A_{550} - A_{535}$ is determined, and the cytochrome-c concentration may then be calculated using a molar extinction coefficient ($A_{550} - A_{535}$) of 21.0 (Williams, 1964).

The extraction of cytochrome c from mitochondria is necessary where estimation of the two c-type cytochromes in a preparation is sought. The overlap of spectra of these two cytochromes at normal temperatures is such as to preclude other than an estimate of the total cytochrome c plus c_1. Cytochrome c_1 remains bound to the mitochondrial inner membrane after the extraction of cytochrome c with low concentrations of detergent. Cytochrome c may then be estimated in the extract, and correction made to the c plus c_1 value of the mitochondria or alternatively, c_1 may be estimated directly by difference spectroscopy of the mitochondria as described earlier. This method could also be adopted for other subcellular fractions, provided other oxidation–reduction compounds with spectral properties similar to those of cytochrome c are not also detached and extracted. The method is as follows (Clark-Walker and Linnane, 1967).

Reagents:

i. 0.6 *M* potassium phosphate buffer (pH 7.4)

ii. Sodium deoxycholate

iii. Potassium ferricyanide

iv. Sodium dithionite

Procedure. Add an equal volume of potassium phosphate buffer to 3–4 ml of a suspension of mitochondria or submitochondrial particles containing approximately 20 mg protein/ml. Add solid sodium deoxycholate to a final concentration of 0.3 mg/mg of protein, homogenize briefly in a Potter-Elvejhem homogenizer, and leave on ice for 30 minutes. Centrifuge for 60 minutes at 105,000 *g* to sediment particles, and wash with 2–3 ml of phosphate buffer by homogenizing and centrifuging as described previously. Pool the supernatants and estimate the cytochrome-c content in oxidized versus reduced difference spectra as described earlier for cell homogenates or mitochondrial suspensions. Extraction efficiency, and independent assay of cytochrome c_1, may be obtained by recording the difference spectrum of the particle fraction.

Cytochrome c may also be extracted from mitochondria with a concentrated salt solution (Williams, 1964).

IV. Lipids

The major proportion of the lipid in yeast cells is associated with the plasma membrane or membranes of the cellular organelles. A minor amount of lipid is found in cell wall preparations, and some yeasts may store lipid in their cytoplasm. Most yeasts, grown on a balanced medium, contain lipid to the extent of 5–15% of dry weight, although certain species under appropriate conditions accumulate as much as 60% of their weight as lipid (Hunter and Rose, 1971).

A diverse range of individual solvents and solvent mixtures, of cell conditions (wet, freeze-dried, disrupted), and of extraction conditions (temperature, pH, ionic strength) has been tested for their effects on the extraction of lipids from yeasts. Ethanol, methanol, and acetone are almost universally included as extractants, since they rupture the more polar interactions between lipids and proteins and also serve to remove water from samples so that the lipophilic or nonpolar component (chloroform, diethyl ether, petroleum ether) of the extraction mixture can more effectively permeate the structure of the membranes, hence dissolve the lipid present. To avoid peroxidative reactions of unsaturated lipids, solvents should be freshly distilled and peroxide-free before use and, if possible, extraction should be carried out in an inert atmosphere.

The use of polar solvents, especially the lower alcohols, in extraction mixtures unavoidably leads to coextraction of low-molecular-weight, non-lipid substances such as sugars and amino acids. It is therefore essential that the crude lipid extract be partitioned against (i.e., washed with) a more polar solvent (water or dilute salt solution) to remove these contaminants. The methods described by Folch *et al.* (1957) and Bligh and Dyer (1959) specifically incorporate such a procedure.

Once extracted, lipids should not be stored dry, but should be redissolved in a suitable solvent and stored at a low temperature ($-20°C$ or lower) to retard oxidation. If storage is to be prolonged (longer than several days), an antioxidant such as 2,6-di-*tert*-butyl-4-methoxyphenol (BHT) should be added to a final concentration of 0.02–0.05%.

Because most of the water is removed, the lipids from freeze-dried cells generally are more readily extractable than those from wet cells. Nyns *et al.* (1968) carried out systematic experiments aimed at establishing the most efficient method of extracting lipid from *Candida lipolytica*. Optimum results were obtained by extracting freeze-dried cells with chloroform–methanol (1:1 by volume) at room temperature. Other solvents systems, including ethanol–ether, ethanol–benzene, petroleum ether–diethyl ether, and chloroform–methanol (2:1 by volume) after ethanol, were less effective when wet cell pastes were used as starting material.

The yeast cell wall frequently presents a significant diffusion barrier even to small lipid molecules, and for this reason extraction of disrupted yeast cells yields greater amounts of lipid than does extraction of intact cells with neutral solvents (Letters, 1966). On a dry-weight basis, reproducibility may be poorer with cell homogenates, which may result from several possible causes: variable contamination of homogenates by unbroken cells, variable size of large particulate material formed during disruption, losses of large membrane fragments when unbroken cells are removed by centrifugation prior to extraction, or release and/or activation of lipases in cells prior to contact with the extracting solvent. Letters (1968) attempted to overcome the problems of enzymatic degradation of lipids during extraction by incubating intact cells initially with ethanol at 70°C to inactivate enzymes, and by using acidified chloroform–methanol mixtures, which are more effective than neutral mixtures, particularly for the extraction of tightly bound phospholipid. The following method is based on that described by Letters (1968) and includes the washing procedure of Folch et al. (1957). Cell disruption is included in this technique, and it is important that the method of disruption used be rapid and achieve at least 98% breakage of cells. Also described below is a saponification procedure which, while not yielding lipids intact, is the only effective means to bring about a quantitative extraction of lipid from yeast cells [see Ratledge and Saxton (1968) for further comments on this matter].

Reagents:
i. 0.05 M potassium phosphate buffer (pH 6.0)–1 mM disodium EDTA
ii. Chloroform–methanol (2:1 by volume)
iii. Chloroform–methanol–concentrated HCl (124:61:1 by volume).

Procedure: Cells are harvested and washed twice with phosphate–EDTA. The cells are then disrupted in a Braun MSK homogenizer or a French press, and the homogenate is extracted as follows. Add 4 vol of absolute ethanol and heat the mixture at 80°C for 15 minutes. Transfer to sealed tubes, flush with nitrogen, and shake overnight at room temperature. Centrifuge, decant, and retain the extract at 0°C. Add 4 vol of chloroform–methanol to the pellet and shake for 2 hours at room temperature. Centrifuge, and then reextract the pellet with chloroform–methanol in exactly the same way. Combine these two extracts with the first ethanol extract. Finally, extract the pellet by shaking overnight at room temperature under nitrogen with 4 vol of chloroform–methanol–HCl. After centrifuging neutralize this extract with dilute KOH before combining with the other extracts.

Dry the lipid extract under vacuum at 30°C, and take up the residue in about 1 vol of chloroform–methanol. Wash by shaking with $\frac{1}{5}$ vol of dilute salt solution (0.05% CaCl$_2$; 0.04% MgCl$_2$; 0.73% NaCl; or 0.88% KCl, as preferred), and resolve the two phases by centrifugation (Folch et al., 1957). Remove as much of the upper phase as possible, and complete the extraction of water-soluble material from the lipid extract by mixing the interface three times with small amounts of chloroform–methanol containing water and salt solution (0.04% CaCl$_2$; 0.034% MgCl$_2$; 0.58% NaCl; or 0.74% KCl, as appropriate) mixed in the ratio 3:48:47 (chloroform–methanol : water:salt solution) by volume. Disperse the remaining rinse fluid into the lower phase by adding a small amount of methanol; make the extract up to known volume with chloroform–methanol (2:1 by volume). A single wash with an aqueous phase, as described above, usually suffices to remove nonlipid material from the extract. However, when radioactively labeled material is present, the number of washes necessary is determined by the specific radioactivity of the contaminating nonlipid.

An alternative and frequently used lipid extraction procedure, particularly useful if information on fatty acid or sterol composition is being sought, is saponification, followed by extraction of neutral (uncharged) and acidic lipids (primarily fatty acids). The saponification conditions given below are suitable for whole cells and subcellular fractions of *S. cerevisiae* (Jollow *et al.*, 1969; Gordon and Stewart, 1971), but extraction from *Candida* (Ratledge and Saxton, 1968) and *Mucor* (Gordon *et al.*, 1971) is incomplete, and these organisms require a preliminary acid hydrolysis step for effective total extraction of lipid.

Reagents:

i. 7 *M* KOH

ii. Pentadecanoic acid; 10 mg/ml in $CHCl_3$, as internal fatty acid standard

iii. Cholesterol (or similar sterol, not found in the yeast under examination); 5 mg/ml in $CHCl_3$, as internal sterol standard

iv. Methanolic HCl; bubble 2.5 gm of HCl gas into 100 ml of anhydrous methanol; alternatively, add 1.0 ml of concentrated HCl to 50 ml of anhydrous methanol and shake several times with 2–3 gm of anhydrous Na_2SO_4; store over Na_2SO_4

Procedure: After harvesting and washing the cells (approximately 500 mg dry weight), add 0.5 ml each of pentadecanoic acid and sterol standard solutions to the pellet. Remove the chloroform by carefully heating the tube in a boiling water bath. Add 15 ml of 7 *M* KOH, stir well, and heat the mixture in a boiling water bath for 60 minutes; resuspend the settled material at intervals. If subcellular fractions rather than whole cells are being extracted, a suspension containing about 100 mg of protein should be used as starting material; add 0.5 ml of each standard solution and remove the chloroform by careful heating as described previously. Add an equal volume of 3.5 *M* KOH and carry out the saponification as described for whole cells. The extraction of neutral and acidic lipids is then the same for cells and subcellular fractions, and is as follows.

Cool the saponification mixture and extract three times by shaking with approximately 25 ml of diethyl ether. Combine the ether extracts and wash twice with 25 ml of water. Dry the ether extract by shaking with 2–3 gm of anhydrous Na_2SO_4, and concentrate by evaporating under vacuum. The residue, after dissolving in the solvent, can be used directly for analysis (more labile neutral lipids, such as quinones, require special protection against oxidation, as will be detailed subsequently.)

Fatty acids are then extracted as follows. To the chilled saponification mixture on ice carefully add concentrated HCl to bring the pH to 1.5–2. Extract the fatty acids with three 25-ml volumes of diethyl ether. Wash the ether extract twice by shaking with 25 ml of water, and then dry over 2–3 gm of anhydrous Na_2SO_4. Remove the ether by evaporating *in vacuo*, and take up the residue in 2 ml of methanolic HCl. Transfer to screw-top tubes, flush the tube and contents with nitrogen, seal, and heat at 60°C for 2 hours. After methylation remove excess methanol and HCl with a stream of nitrogen, and extract the methyl esters of the fatty acids into 1–2 ml of hexane. Reduce the volume of the hexane with a steam of nitrogen. The hexane extract can then be examined by gas–liquid (James, 1960) or thin-layer chromatography (Gordon *et al.*, 1971). The internal standards provide a means to quantitate amounts of components derived from the yeast cells or subfractions.

The general extraction procedure above yields extracts suitable for detailed analysis. The various analytical procedures that may be used in such an examination are described in detail in several excellent texts and reviews (e.g., Johnson and Davenport, 1971; Kates, 1972). It should be noted that the addition of pentadecanoic acid and cholesterol (or another sterol) as internal standards in the saponification method described above is only possible if *separation* of components in the mixtures is subsequently carried out. The amounts of internal standards used may need to be varied from those given above to correspond approximately to the amounts of major fatty acids or sterols present in the cell samples.

Among neutral lipids the quinones and carotenoids present special difficulties in terms of their stability during extraction. The extraction and fractionation of carotenoids from yeast cells is surveyed by Simpson *et al.* (1971). In the case of quinones saponification conditions must be milder than those described above, and reducing conditions must be maintained during extraction otherwise these compounds will be recovered in low yield. Extraction procedures that do not involve saponification, e.g., extraction with chloroform–methanol or acetone, result in low, irregular values for yeast cells. The following method of extraction of ubiquinone has been found to be effective with *S. cerevisiae* (Gordon and Stewart, 1969).

Reagents:

i. 15% (w/v) KOH in 95% (v/v) ethanol
ii. Pyrogallol
iii. *n*-Heptane

iv. Silica gel G TLC plates, activated by heating at 100°C for 30 minutes

v. Benzene–chloroform, 1:1 by volume

vi. Potassium borohydride

Procedure: Suspend washed cells (approximately 0.5 gm dry weight) in water (total volume 5–10 ml). Add an equal volume of alcoholic KOH and 0.5 gm pyrogallol/gm of cells, and boil gently under reflux for 30 minutes. If subcellular fractions are being extracted, samples should be diluted to 5 mg protein/ml before ethanolic KOH and pyrogallol are added, and the mixtures heated for 10 minutes at 80°C. Cool and extract the saponification mixture twice with 100 ml of *n*-heptane. Pool the heptane extracts, wash twice with 100 ml of water, and recover the residue from heptane by evaporating at 40°C under vacuum. The residue is taken up in a small volume of ethanol, concentrated under a stream of nitrogen, and chromatographed on silica gel *G* using benzene–chloroform to develop the plates. Authentic ubiquinone-6 is chromatographed in parallel to identify the position of ubiquinone in the sample. This region is scraped off the plate, and ubiquinone is eluted from the gel with absolute ethanol. After washing the gel twice with ethanol, a quantitative determination of ubiquinone in the pooled supernatants is given by the extent of absorbance change at 275 nm when a sample of the ubiquinone in ethanol is reduced with a few grains of borohydride in a spectrophotometer cuvette. The extinction coefficient used is $E_{1\,cm}^{1\%} = 206$ at 275 nm.

V. Concluding Remarks

The methods of extraction and analysis outlined or referred to in this chapter should provide a starting point at least for a quantitative description of the major components of yeast cells. Emphasis has been given to the importance of testing and modifying extraction procedures, in particular when new strains or species are examined for the first time.

One other consideration should also be restated, especially in view of recent data on the permeability and porosity of the yeast cell wall. Measurements with a series of polyols indicate an effective cutoff limit by the cell wall of *S. cerevisiae* for molecules larger than 600–800 daltons (Scherrer *et al.*, 1974). It seems unlikely that this limit is lower for charged or hydrophobic molecules and, although extraction solvents undoubtedly alter the permeability properties of the wall, extractive and analytical procedures applied to intact cells should be used with this consideration always in mind.

REFERENCES

Atkinson, D. E. (1969). *Annu. Rev. Microbiol.* **23**, 47–68.

Atzpodien, W., Gancedo, J. M., Duntze, W., and Holzer, H. (1968). *Eur. J. Biochem.* **7**, 58–62.

Bacon, J. S. D., Davidson, E. D., Jones, D., and Taylor, I. F. (1966). *Biochem. J.* **101**, 36C–38C.

Ballou, C. E., and Raschke, W. C. (1974). *Science* **184**, 127–134.

Barrett, J. (1961). *Biochim. Biophys. Acta* **54**, 580–582.

Bergmeyer, H. U., (1974). "Methods of Enzymatic Analysis." Academic Press, New York.

Betz, A., and Chance, B. (1965). *Arch. Biochem. Biophys.* **109**, 585–594.

Bligh, E. G., and Dyer, W. J. (1959). *Can. J. Biochem. Physiol.* **37**, 911–917.

Bollum, F. J. (1966). *In* "Procedures in Nucleic Acid Research" (G. L. Cantoni and D. R. Davies, eds.), pp. 296–300. Harper, New York.

Brooker, G., Thomas, L. J., and Appleman, M. M. (1968). *Biochemistry* **7**, 4177–4181.

Burton, K. (1956). *Biochem. J.* **62**, 315–323.

Campbell, I., Gilmour, R. H., and Rous, P. R. (1972). *J. Inst. Brew.* **78**, 491–499.

Chance, B. (1957). "Methods in Enzymology" (S. P. Colowick and N. O. Kaplan, eds.), Vol. 4, pp. 273–329. Academic Press, New York.

Chapman, A. G., Fall, L., and Atkinson, D. E. (1971). *J. Bacteriol.* **108**, 1072–1086.

Charalampous, F. C. (1974). *J. Biol. Chem.* **249**, 1014–1021.

Clark-Walker, G. D. (1972). *Proc. Nat. Acad. Sci. U.S.* **69**, 388–392.

Clark-Walker, G. D., and Linnane, A. W. (1967). *J. Cell Biol.* **34**, 1–14.

Clark-Walker, G. D., and Miklos, G. L. (1974). *Eur. J. Biochem.* **41**, 359–365.

Conway, E. J., and Armstrong, W. McD. (1961). *Biochem. J.* **81**, 631–639.

Cowie, D. B., and Walton, B. P. (1956). *Biochim. Biophys. Acta* **21**, 211–226.

Dutton, G. G. S. (1973). *Advan. Carbohyd. Chem. Biochem.* **28**, 11–160.

Ebner, E., Mason, T. L., and Schatz, (1973). *J. Biol. Chem.* **248**, 5369–5378.

Eddy, A. A. (1958). *In* "The Chemistry and Biology of Yeasts" (A. H. Cook, ed.), pp. 157–250. Academic Press, New York.

Eddy, A. A., and Longton, J. (1969). *J. Inst. Brew.* **75**, 7–9.

Estabrook, R. W., and Holowinsky, A. (1961). *J. Biophys. Biochem. Cytol.* **9**, 19–28.

Falk, J. E. (1964). "Porphyrins and Metalloporphyrins." Elsevier, Amsterdam.

Falk, J. E., Lemberg, R., and Morton, R. K., eds. (1961). "Hematin Enzymes." Pergamon, Oxford.

Folch, J., Lees, M., and Sloane-Stanley, G. (1957). *J. Biol Chem.* **226**, 497–509.

Gordon, P. A., and Stewart, P. R. (1969). *Biochim. Biophys. Acta* **177**, 358–360.

Gordon, P. A., and Stewart, P. R. (1971). *Microbios* **4**, 115–132.

Gordon, P. A., Stewart, P. R., and Clark-Walker, G. D. (1971). *J. Bacteriol.* **107**, 114–120.

Henderson, J. F., and Paterson, A. R. P. (1973). "Nucleotide Metabolism." Academic Press, New York.

Herbert, D., Phipps, P. J., and Strange, R. E. (1971). *Methods Microbiol.* **5B**, 209–344.

Hunter, K., and Rose, A. H. (1971). *In* "The Yeasts" (A. H. Rose and J. S. Harrison, eds.), Vol. 2, pp. 211–270. Academic Press, New York.

Jacobs, N. J., and Wolin, M. J. (1963). *Biochim. Biophys. Acta* **69**, 18–28.

James, A. T. (1960). *Methods Biochem. Anal.* **8**, 1–59.

Johnson, A. R., and Davenport, J. B. eds. (1971). "Biochemistry and Methodology of Lipids." Wiley (Interscience), New York.

Jollow, D., Kellerman, G. M., and Linnane, A. W. (1969). *J. Cell Biol.* **37**, 221–230.

Jost, J. P., Hsie, A., Hughes, S. D., and Ryan, L. J. (1970). *J. Biol. Chem.* **245**, 351–357.

Kates, M. (1972). *In* "Laboratory Techniques in Biochemistry and Molecular Biology" (T. S. Work and E. Work, eds.), pp. 267–610. North-Holland Publ., Amsterdam.

Korn, E. D., and Northcote, D. H. (1960). *Biochem. J.* **75**, 12–17.

Lascelles, J. (1964). "Tetraphyrrole Biosynthesis and its Regulation." Benjamin, New York.

Latner, A., and Skillen, A. W. (1968). "Isoenzymes in Biology and Medicine." Academic Press, New York.

Letters, R. (1966). *Biochim. Biophys. Acta* **116**, 489–499.

Letters, R. (1968). *In* "Aspects of Yeast Metabolism" (R. D. Mills, ed.), pp. 303–319. Blackwell, Oxford.

Linnane, A. W. (1965). *In* "Oxidases and Related Redox Systems" (T. E. King, H. Mason, and M. Morrison, eds.), Vol. 2, pp. 1102–1128. Wiley, New York.

Lowry, O. H., and Passonneau, J. V. (1972). "A Flexible System of Enzymatic Analysis." Academic Press, New York.

Lowry, O. H., Rosebrough, W. J., Farr, A. L., and Randall, R. J. (1951). *J. Biol. Chem.* **193**, 265–275.

Makman, R. S., and Sutherland, E. W. (1965). *J. Biol. Chem.* **240**, 1309–1314.

Manners, D. J. (1971). *In* "The Yeasts" (A. H. Rose and J. S. Harrison, eds.), Vol. 2, pp. 419–440. Academic Press, New York.

Mason, T. L., Poyton, R. O., Wharton, D. C., and Schatz, G. (1973). *J. Biol. Chem.* **248**, 1346–1354.

Matile, P., Moor, H., and Robinow, C. F. (1969). *In* "The Yeasts" (A. H. Rose and J. S. Harrison, eds.), Vol. 1, pp. 219–302. Academic Press, New York.

Munro, H. N., and Fleck, A. (1966). *Methods Biochem Anal.* **14**, 113–176.

Nelson, N., Deters, D. W., Nelson, H., and Racker, E. (1973). *J. Biol. Chem.* **248**, 2049–2055.

Nurse, P., and Wiemkin, A. (1974). *J. Bacteriol.* **117**, 1108–1116.

Nyns, E. J., Chiang, N., and Wiaux, A. L. (1968). *Antonie van Leeuwenhoek; J. Microbiol. Serol.* **34**, 197–204.

Phaff, H. J. (1971). *In* "The Yeasts" (A. H. Rose and J. S. Harrison, eds.), Vol. 2, pp. 135–210. Academic Press, New York.

Piko, L., Blair, D. G., Tyler, A., and Vinograd, J. (1968). *Proc. Nat. Acad. Sci. U.S.* **59**, 838–845.

Polakis, E. S., and Bartley, W. (1966). *Biochem. J.* **99**, 521–533.

Polya, G. (1973). *In* "The Ribonucleic Acids" (P. R. Stewart and D. S. Letham, eds.), pp. 7–36. Springer-Verlag, Berlin and New York.

Poulson, R. (1973). *In* "The Ribonucleic Acids" (P. R. Stewart and D. S. Letham, eds.,), pp. 243–261. Springer-Verlag, Berlin and New York.

Poulson, R., and Polglase, W. J. (1974). *FEBS. (Fed. Eur. Biochem. Soc.) Lett.* **40**, 258–260.

Pretlow, T. P., and Sherman, F. (1967). *Biochim. Biophys Acta* **148**, 629–644.

Ratledge, C., and Saxton, R. K. (1968). *Anal. Biochem.* **26**, 288–294.

Richards, G. M. (1974). *Anal. Biochem.* **57**, 369–376.

Robison, G. A., Butcher, R. W., and Sutherland, E. W. (1971). "Cyclic AMP." Academic Press, New York.

Scherrer, R., Louden, L., and Gerhardt, P. (1974). *J. Bacteriol.* **118**, 534–540.

Sels, A. A., Fukuhara, H., Pere, G., and Slonimski, P. (1965). *Biochim. Biophys. Acta* **95**, 486–502.

Sentandreu, R., and Northcote, D. H. (1969). *Carbohyd. Res.* **10**, 584–585.

Simpson, K. L., Chichester, C. O., and Phaff, H. J. (1971). *In* "The Yeasts" (A. H. Rose and J. S. Harrison eds.), Vol. 2, pp. 493–515. Academic Press, New York.

Somlo, M. (1970). *Arch. Biochem. Biophys.* **136**, 122–133.
Stebbing, N. (1971). *J. Cell Sci.* **9**, 701-717.
Stebbing, N. (1974). *Bacteriol. Rev.* **38**, 1–28.
Taylor, I. E. P., and Cameron, D. S. (1973). *Annu. Rev. Microbiol.* **27**, 243–259.
Thornton, D. R., and McEvoy, J. (1970). *Experientia* **26**, 24–26.
Trevelyan, W. E., and Harrison, J. S. (1956). *Biochem. J.* **63**, 23–33.
Van Wijk, R., and Konijn, T. M. (1971). *FEBS (Fed. Eur. Biochem. Soc.) Lett.* **13**, 184–186.
Weibel, K. E., Mor, J. R., and Fiechter, A. (1974). *Anal. Biochem.* **58**, 208–216.
Wieme, R. J. (1965). Agar Gel Electrophoresis." Elsevier, Amsterdam.
Wilkinson, J. H. (1970). "Isoenzymes," 2nd ed. Spon, London.
Williams, J. N. (1964). *Arch. Biochem. Biophys.* **107**, 537–543.
Yu, R. S. T., and Stewart, P. R. (1974). *Cytobios* **9**, 175–192.
Yu, R. S. T., Poulson, R., and Stewart, P. R. (1972). *Mol. Gen. Genet.* **114**, 325–338.

Chapter 9

Methods for Avoiding Proteolytic Artefacts in Studies of Enzymes and Other Proteins from Yeasts[1]

JOHN R. PRINGLE

Institute of Microbiology,
Swiss Federal Institute of Technology,
Zürich, Switzerland

I. Introduction

Yeast cells contain a variety of different proteolytic enzymes. During the past several years, it has become increasingly apparent that these proteolytic enzymes pose a very serious problem for attempts to study other yeast proteins. This conclusion is based on the recognition of proteolytic artefacts in

[1] Portions of this work were supported by Swiss National Science Foundation Grant No. 3.628.71 (to A. Fiechter); by U.S. Public Health Service Grant HE-08893 and by U.S. National Science Foundation Grant GB-6360 (both to G. Guidotti); and by U.S. Public Health Service postdoctoral fellowship FO2-GM41910.

a remarkable number of studies of a remarkable variety of yeast proteins. However, some of the relevant observations have not been published at all, many others have been mentioned only briefly in articles dealing primarily with other topics, and no review of these studies (or even a comprehensive bibliography) has appeared. Probably as a result of this lack of publicity, careful attention to the protease problem is still far from being a universal feature of studies of yeast enzymes and other proteins.

Although much of the discussion is in fact more generally applicable, this chapter refers explicitly only to *Saccharomyces cerevisiae* and closely related yeasts. The chapter consists of three sections. In the first section, we have attempted a very brief summary of the many studies known to us in which the activity of yeast proteases *in vitro* has presented a technical problem. Special attention is drawn to several cases in which proteolytic artefacts are known to have been particularly persistent, even after the existence of the problem was clearly recognized. By presenting this summary, we hope to make the reader aware that a great variety of types of studies, of a great variety of types of proteins, has been subject to proteolytic artefacts, and that these proteolytic artefacts can be quite difficult to eliminate. In particular, we hope to convince each reader of the validity of the conclusions we have reached:

1. No study of a yeast enzyme or structural protein that does not attend to the protease problem can be fully relied on for information regarding the properties of that protein in the living cell.

2. Anything less than *careful* (and thoughtful) attention to the protease problem is inadequate to guarantee that an artefact-free study has been conducted. In particular, several studies have clearly shown that simply treating crude and partly purified extracts with phenylmethanesulfonyl fluoride (PMSF) is not sufficient to eliminate artefacts, even in cases in which a PMSF-sensitive protease is known to be the major culprit.

In the second section of this chapter, we briefly summarize the most important properties of known yeast proteases. This information is important, because a careful, thoughtful approach to the artefact problem must be conducted with the properties of the proteases in mind.

Finally, in the third section, we consider the virtues and limitations of the various methods that have been used, or that might be used, in attempts to cope with the protease problem. Although we return to this point, we emphasize at the outset that it is not possible to give a magic antiprotease formula guaranteed to eliminate all artefacts in every case in which it is applied. Effective methods are available, but they must be applied in individual cases with the investigator aware of the possibility that their effectiveness is not complete.

This chapter is concerned solely with a technical problem: How can one avoid proteolytic degradation of the protein of interest between the time that harvesting of cells is begun and the time that a conclusion, based on the properties of the isolated protein, is drawn? Thus we do not address directly the very interesting question of the functions of the proteases within the living cell. We do, however, note that studies of these questions must be especially attentive to the proteolytic artefact problem. In particular, attempts to identify physiologically significant "maturation" or "processing" of proteins (see, e.g., Shaffer *et al.*, 1972) must take special care to ensure that any modifications observed have not actually occurred *in vitro*; as the examples cited in Sections II and V make clear (the case of RNA polymerase from sporulating *Bacillus subtilis* is perhaps the most instructive), it can be rather difficult to eliminate this possibility. A further point is that the interesting attempts to infer specific functions for the proteases *in vivo* from their various particular actions *in vitro* (Afting *et al.*, 1972; Katsunuma *et al.*, 1972; Holzer *et al.*, 1973, 1974; Cabib and Ulane, 1973; Cabib *et al.*, 1973, 1974; Hasilik and Holzer, 1973; Molano and Gancedo, 1974; Holzer, 1974; Hasilik, 1974a,b) should be approached with some caution. This follows from the facts that (1) the overwhelming impression of protease action *in vitro* is one of versatility, not one of selectivity (see Section II); and (2) much of the *in vivo* control over protease action (particularly compartmentation and its consequences) is undoubtedly broken down when extracts are made. Neeff *et al.* (1974) have recently had some success in attempting to relate the proteolytic attack on malate dehydrogenase that occurs *in vitro* (Table I and Section II,A,1) to the rapid disappearances of malate dehydrogenase activity that sometimes occur *in vivo* (Witt *et al.*, 1966; Ferguson *et al.*, 1967; Duntze *et al.*, 1968; Chapman and Bartley, 1968; van Rijn and van Wijk, 1972), but this is clearly not a simple problem. For purposes of this chapter, we simply regard all proteolyses observed *in vitro* as artefacts, without passing judgment on their possible physiological significance.

II. Known Cases of Proteolytic Artefacts in Yeast Studies

In Table I is given a compact summary of the cases known to us in which proteolytic artefacts have been recognized in studies of yeast proteins. Several especially instructive cases are then discussed briefly in Section II,A. We then attempt, in Section II,B, to define the scope of the protease artefact problem as encountered in studies of yeast.

TABLE I

KNOWN CASES OF PROTEOLYTIC ARTEFACTS IN YEAST STUDIES[a]

Protein	Protease effects on native protein		Protease effect at denaturation[c]	Protease contamination of highly purified preparations[d]	Reference[e]
	Inactivation[b]	Modification[b]			
Alcohol dehydrogenase	+(P)	+?(P)	+(P)	+(P)	Bühner and Sund (1969); Pringle (1970b)
Aldehyde dehydrogenase	+(P)	+(P)	—	+(P?)	Clark and Jakoby (1970)
Aldolase	+(P)	—	—	+(P)	Harris et al. (1969); Collins (1974)
Catalase[f]	+?	—	—	—	Molano and Gancedo (1974)
Chitin synthetase	+(B)[g]	+(B)[g]	—	—	Cabib et al. (1973, 1974); Hasilik and Holzer (1973); Hasilik (1974a,b)
Coupling factor YF$_6$	—	+(P)	—	—	Fessenden-Raden and Hack (1972); J. M. Fessenden-Raden (personal communication, 1974)
Cytochrome b$_2$	+?	+(P)	+?	+(P)	Jacq and Lederer (1974)
Cytochrome c	—	—	+[h]	+[h]	J. W. Stewart and F. Sherman (personal communication, 1970)
Endonuclease	+(P)	—	—	—	Piñon (1970; also personal communication)
F$_1$ = ATPase inhibitor[f]	+(A,B)	—	—	—	E. Ebner and K. Maier (personal communication, 1974); Holzer et al. (1974)
Fructose-1, 6-DiP phosphatase	+ (B)	—	—	—	Molano and Gancedo (1974)
Gluconate-6-P dehydrogenase	—	—	+(P)	+(P)	Pringle (1970b)
Glucose-6-P dehydrogenase	+(P)	—	+(P)	+(P)	Noltmann et al. (1961); Pringle (1970b)

Enzyme					Reference
Glutamate oxaloacetate transaminase[f]	+(B)	—	—	—	Katsunuma et al. (1972)
Glutathione reductase	—	+?[j]	+(P)[i]	+(P)[i]	Pringle (1970b)
Glyceraldehyde-3-P dehydrogenase	+?[j]	+(P)	+(P)	+(P)	Halsey and Neurath (1955); Pringle (1970b)
Glycogen debranching enzyme	+(P)	—	—	—	Lee et al. (1970)
Glycogen phosphorylase	+(P)	+(P?)	—	—	Fosset et al. (1971)
Glycogen synthetase	—	+(P, A?)	+(P)	+(P)	Huang and Cabib (1974)
Hexokinase	+(P)	—	—	+(P)	See text
Histidinol-P phosphatase	+?	—	—	—	Fink (1970; personal communication, 1974)
α-Isopropylmalate synthase	+(P)	—	—	—	Ulm et al. (1972)
Malate dehydrogenase	+(B)	+?(P)	+(P)	+(P)	Pringle (1970b); Neeff et al. (1974)
Mitochondrial membrane proteins[f]	—	—	+?[k]	—	Groot et al. (1972)
Ornithine transaminase[f]	+?	—	—	—	Afting et al. (1972)
Phosphofructokinase	—	+(P)	+(P)	+(P)	Wilgus et al. (1971); Diezel et al. (1973)
Phosphoglucose isomerase	—	—	+(P)[i]	+(P)[i]	Pringle (1970b); Kempe et al. (1974)
Phosphoglycerate kinase	—	—	+(P)	+(P)	Pringle (1970b)
Phosphoglyceromutase	+(P)	+(P)	—	—	Sasaki et al. (1966)
Phosphomannose isomerase	—	—	+(P)	+(P)	Pringle (1970b)
Phosphoribosyl-AMP cyclohydrolase	+(P)	—	—	—	Shaffer et al. (1972)
Phosphoribosyl-ATP pyrophosphohydrolase	+(P)	—	—	—	Shaffer et al. (1972)

(cont.)

TABLE I (cont.)

Protein	Protease effects on native protein		Protease effect at denaturation[c]	Protease contamination of highly purified preparations[d]	Reference[e]
	Inactivation[b]	Modification[b]			
Phosphoribosyl-ATP pyrophosphorylase	+?	—	—	—	Fink (1970; personal communication, 1974)
Protease B	+?(B)	+?	—	—	See Section III,A,2
Protease C	—	+(B?)	—	—	Aibara et al. (1971)
Protease-A inhibitor	+(B)	—	—	—	See Section III,C,1
Protease-B inhibitor	+(A,B,C)	—	—	—	See Section III,C,1
Protease-C inhibitor	+(A,B)	—	—	—	See Section III,C,1
Pyruvate decarboxylase	+(B,A?)	+(B)	—	+(A,B)	Juni and Heym (1968, 1970, also personal communication)
Pyruvate dehydrogenase	+?[j]	—[j]	+?[k]	+?[k]	Wais et al. (1973)
Ribosomal protein	—	+?[j]	—	—	A. Hopper (personal communication, 1972)
RNA polymerase	—	+(P)	—	—	P. T. Magee (personal communication, 1974); J. Sebastian (personal communication, 1974)
Threonine dehydratase[f]	+(A)	—	—	—	Katsunuma et al. (1972)
Tryptophan synthase	+(A,B)	—	—	—	Manney (1968); Saheki and Holzer (1974a); Schött and Holzer (1974)
UDP-Galactose 4-Epimerase	—	+?	+	+	Darrow and Rodstrom (1970); R. Weil (personal communication, 1970)

[a] Listed here are all the cases known to us in which proteolysis has appeared as a technical problem during an attempt to isolate or characterize a yeast protein. The list includes several cases in which a proteolytic effect seems almost certain but has not been positively established (see footnote b); it also includes several cases in which proteolysis has been observed under somewhat artificial conditions (see footnote f). The list does not include many other cases in which proteolysis seems likely but in which the available data are not convincing enough to justify inclusion. Several of the cases listed were recognized during a survey (Pringle, 1970b) of highly purified, commercially available, enzymes; we were specifically looking for contamination of these preparations by traces of the SDS-resistant protease already observed in malate dehydrogenase preparations.

[b] "Inactivation" means a significant loss of total activity; "modification" means production of a modified, but still active, form of the protein. +, Proteolytic effect is firmly established; +?, proteolytic effect is almost certain; A, B, C, P, proteolytic effect is due, at least in part, to protease A, protease B, protease C, or a PMSF-sensitive protease (not further characterized), respectively. The notation that an effect of one particular protease has been established does not, of course, necessarily imply that other proteases have no effects in that system.

[c] With three exceptions, the effects noted were observed during denaturation by SDS. See text (Section II,B,4) for further discussion. For explanation of symbols see footnote b.

[d] The description of preparations as "highly purified" is somewhat arbitrary, but conforms to common usage. For explanation of symbols see footnote b.

[e] The references given are not in all cases comprehensive, but are an adequate guide to the literature.

[f] The sensitivity of these proteins to attack by yeast proteases has so far been demonstrated only in experiments in which the test protein was purposefully mixed with protease. In at least one case (the F_1-ATPase inhibitor), however, this sensitivity to proteolysis is thought to have played a role in previous attempts to study the protein (E. Ebner and K. Maier, personal communication, 1974).

[g] Protease B catalyzes either activation or inactivation of chitin synthetase, depending on the degree of exposure. Both activation and inactivation have been suggested to be physiologically important.

[h] Preparations of cytochrome c from one yeast strain were contaminated with traces of a protease which altered the "fingerprint" pattern obtained after digestion with trypsin.

[i] Proteolysis was partly, but not completely, prevented by PMSF.

[j] Since attack on the native protein was monitored only by a change in polypeptide chain molecular weight, it is not clear whether inactivation or modification was involved.

[k] In view of the results obtained with so many other proteins, the low-molecular-weight polypeptide chains observed by SDS–polyacrylamide gel electrophoresis were almost certainly a result of proteolysis during denaturation.

155

A.   Particularly Persistent and Pernicious Proteolysis Problems

1.   MALATE DEHYDROGENASE

Our attempts to study the isoenzymes and subunit structure of malate dehydrogenase (Pringle, 1970b) were for almost 3 years completely frustrated by the variability and instability of the preparations. When we finally realized that these problems were due to proteolytic attack, and began to treat the preparations with the protease inhibitor PMSF (Gold, 1967), the situation improved dramatically. Malate dehydrogenase preparations that lost activity at rates ranging from 5 to 99% per day in the absence of PMSF were completely stable in the presence of this reagent. A single treatment with PMSF was not, however, sufficient to keep malate dehydrogenase stable throughout the course of extended purification; repeated treatments were necessary. In this way the enzyme could be purified to near homogeneity in reasonable yield. Surprisingly, attempts to study the subunit structure of this purified enzyme by sodium dodecyl sulfate (SDS)–polyacrylamide gel electrophoresis met with further difficulties; again, a PMSF-sensitive protease was responsible. Evidently, despite extensive purification and repeated treatments with a potent, covalently binding inhibitor, traces of protease still contaminated the malate dehydrogenase preparations. Although action of this contaminating protease on the native malate dehydrogenase was difficult to detect, very extensive degradation could occur during denaturing treatment with SDS.

2.   HEXOKINASE

Yeast hexokinase was discovered by Meyerhof in 1927, and has been much studied ever since (the recent review by Colowick, 1973, lists about 90 references for the period since a previous review in 1961). It is sobering to realize that most of this work has been devoted either to (inadvertent) study of forms of the enzyme not present in the cell, or to explicit attempts to eliminate the *in vitro* proteolysis that gives rise to these forms.

In 1961 it became clear that the multiply recrystallized hexokinase preparations then in use were highly heterogeneous, containing at least six electrophoretically separable active species (Trayser and Colowick, 1961; Kaji et al., 1961). Over a period of years, it then became clear that most, if not all, of these multiple active species were in fact proteolytically modified derivatives of the form or forms that had been present in the cell (Kaji et al., 1961; Kenkare et al., 1964; Ramel, 1964; Kaji, 1965; Lazarus et al., 1966; Derechin et al., 1966, Gazith et al., 1968; Schulze and Colowick, 1969; Kopperschläger and Hofmann, 1969; Easterby and Rosemeyer, 1969, 1972; Ramel et al., 1971). Even after elaborate antiproteolysis measures are taken, two distinct forms of hexokinase are observed in yeast extracts (Ramel et al.,

1971; Womack *et al.*, 1973); probably these represent true intracellular iso-enzymes. However, in view of the history of this case, and of the striking similarities between the two enzyme species (Lazarus *et al.*, 1968; Rustum *et al.*, 1971; Schmidt and Colowick, 1973b), we reserve judgment until appropriate mutants, and/or the results of definitive primary structure analysis, are available.

Concurrent with the struggle to ascertain which form(s) of hexokinase are actually present in the cell has been a struggle to determine the subunit structure of the enzyme. It now seems clear that the polypeptide "subunits" of approximately 25,000 daltons reported by Ramel *et al.* (1961) and Kenkare and Colowick (1965) were proteolysis products. However, evidence for the existence of such subunits continued to be reported (Lazarus *et al.*, 1968; Gazith *et al.*, 1968; Easterby and Rosemeyer, 1969; Rustum *et al.*, 1970) for quite some time after explicit antiproteolysis measures were introduced (Lazarus *et al.*, 1966; Schulze *et al.*, 1966). The recognition that the undegraded subunit is a polypeptide of approximately 52,000 daltons (Pringle, 1970a; Schmidt and Colowick, 1970, 1973a; Rustum *et al.*, 1971; Easterby and Rosemeyer, 1972) came only with even more scrupulous efforts to avoid proteolysis, particularly at the time of denaturation. As in the case of malate dehydrogenase, even highly purified preparations [and even when these have been treated repeatedly with diisopropyl fluorophosphate (DFP); Rustum *et al.*, 1971] contain traces of a DFP-sensitive protease whose action is greatly enhanced under some denaturing conditions.

3. PHOSPHOFRUCTOKINASE

Early reports indicated that the constituent polypeptide chain of phospho-fructokinase had a molecular weight of 44,000 (Lindell and Stellwagen, 1968) or of 60,000 (Liebe *et al.*, 1970). After taking steps (including treatment with PMSF) to prevent proteolysis both during purification and during subsequent denaturation, we observed only subunits with a molecular weight of approximately 100,000 (Wilgus *et al.*, 1971). More recent evidence, however, has indicated that even these 100,000-dalton polypeptides are degradation products of the actual 130,000-dalton subunits of this enzyme (Diezel *et al.*, 1973).

4. PYRUVATE DECARBOXYLASE

Juni and Heym (1968) observed that pyruvate decarboxylase suffered both proteolytic inactivation and proteolytic modification during the course of purification or storage. Different proteases were apparently responsible for the two effects. The modification led to an alteration in the relative magnitude of two catalytic activities of the enzyme. Using the ratio of these activities as a sensitive assay for modification, Juni and Heym (1970; also personal communication) showed that none of the many approaches they tried was completely successful in eliminating proteolytic attack.

B. Generalizations and Conclusions

1. INACTIVATION

Protease action on the protein of interest can lead to losses of activity, and such proteolytic inactivations have in fact been observed in many studies of yeast proteins (Table I). Inactivation may present relatively minor problems of reduced yields and poor storage properties. In other cases, however, the activity losses may be so massive that successful purification is virutally impossible in the absence of effective anti-proteolysis measures. Glucose-6-P dehydrogenase (Noltmann et al., 1961), malate dehydrogenase (Pringle, 1970b), and the his4A and his4B activities (Shaffer et al., 1972) are good examples of the latter situation.

When inactivation is simple and does not involve the production of modi-fied active forms (see Section II,B,2), it is a benign proteolytic artefact. With the use of the techniques described in Section IV, it should always be possible to avoid devastatingly large losses of activity. Even if some inactivation does occur, the danger of obtaining misleading results is small if the truly native (i.e., unmodified) form is really the only active species present.

2. MODIFICATION

A simple inactivation of the protein of interest is unfortunately not the only possible result of protease action. In many cases partially degraded, but still active, forms of the protein of interest can also be produced. In some cases the native and modified forms are seen as apparent isoenzymes, while in other cases the native form may entirely disappear (hexokinase, phos-phofructokinase, and cytochrome b_2 provide examples of the latter situa-tion; see Table I and Section II,A for references). Artefacts of the modi-fication type are more dangerous than simple inactivation, because one may study, without knowing it, a form of the protein not present in the cell. The results obtained may then of course be inaccurate guides to the biological functions of that protein.

In some cases the modified forms have significantly reduced activities, and the occurrence of modification is signaled by a parallel inactivation. It is important to note, however, that many proteins can suffer a significant degree of proteolytic degradation and still retain essentially full activity. Many of the modifications noted in Table I proceed with little or no loss of total activity (aldehyde dehydrogenase and glycogen synthetase are particularly good examples). The modifications of β-galactosidase by trypsin and papain (Givol et al., 1966), of milk galactosyl transferase (Magee et al., 1973) and of muscle phosphorylase kinase (Hayakawa et al., 1973) by trypsin, and of DNA polymerase by various proteases (Setlow et al., 1972) are good examples from other well-defined systems. Modifications that proceed with little or no diminution of activity are particularly dangerous as artefacts because they are relatively difficult to detect.

Adding to the hazards of proteolytic modification is the fact that the modification of even a relatively small proportion of the total protein can lead to quite confusing results (e.g., as to catalytic parameters or isoenzyme pattern).

3. WIDESPREAD CONTAMINATION OF HIGHLY PURIFIED PROTEINS

A remarkable feature of Table I is the frequency with which proteolytic enzymes have been found to contaminate highly purified preparations of other proteins. It should be emphasized that many of the preparations found to be contaminated were actually "homogeneous" by conventional criteria; hexokinase (Rustum et al., 1971), alcohol dehydrogenase (Bühner and Sund, 1969; Pringle 1970b), cytochrome c (J. W. Stewart and F. Sherman, personal communication, 1970), and phosphofructokinase (Wilgus et al., 1971; Diezel et al., 1973) are particularly good examples. Even the commercially available enzymes surveyed (Pringle, 1970b) were very highly purified with respect to each other, as well as by the criterion of SDS–polyacrylamide gel electrophoresis. It is also worth emphasizing that the proteins found to be contaminated are very diverse in their molecular properties (compare, for example, phosphofructokinase and cytochrome c). Perhaps the most striking fact is that in at least two well-documented cases [malate dehydrogenase (Pringle, 1970b) and hexokinase (Rustum et al., 1971)] repeated treatment, during purification, with PMSF or DFP did not prevent the subsequent appearance of PMSF- or DFP-sensitive proteolytic contaminants. Presumably this effect is due to copurification, with the protein of interest, of an inactive (and thus PMSF- or DFP-resistant) protease–polypeptide inhibitor complex (see Section III) which subsequently becomes activated.

The explanation for this widespread proteolytic contamination is not fully clear, but probably involves the following three factors:

1. The yeast cell contains *at least* four or five different proteases (Section III). These, and their naturally occurring inactive complexes, can distribute themselves in different fractions during purification procedures, thus increasing the probability that at least one protease will copurify with any other particular protein.

2. The proteases may adsorb to other proteins and copurify with them for that reason (Rustum et al., 1971; Diezel et al., 1972).

3. The actual amounts of protease present as contaminants in highly purified preparations may be very small. For example, the rates of protease attack on highly purified native malate dehydrogenase (Pringle, 1970b) and on highly purified native hexokinase (Rustum et al., 1971) were orders of magnitude less than had been observed with less purified preparations of these enzymes. The rates of attack on other highly purified native proteins also tend to be low, although they may in some cases be significant [see Wilgus et al. (1971) and Diezel et al. (1973) on phosphofructokinase; Halsey and Neurath (1955) on glyceraldehyde-3-P dehydrogenase; and Lederer and Jacq (1971) on cytochrome b_2]. Even very small traces of contaminating protease can, however, produce major effects in experiments involving denaturation. In what is probably an extreme example, Pringle (1970b) found that 1 part of subtilisin mixed with 10^6 parts of serum albumin was sufficient to give an extensive digestion of the albumin during denaturation with 1% SDS and 1% 2-mercaptoethanol at 25°C. (SDS–polyacrylamide gel electrophoresis of the digested sample gave a pattern similar to that of gel g in Fig. 1; in the absence of proteolysis, a pattern like that of gel j was obtained.)

4. THE SPECIAL HAZARDS OF DENATURATION

It has been known for many years that proteolysis can often be greatly accelerated by exposing mixtures of protease and protein substrate to denaturing conditions (Anson and Mirsky, 1933; Northrup et al., 1948; Halsey and Neurath, 1955; Viswanatha and Liener, 1955). The occurrence of this acceleration can be explained by the following observations:

1. Many native proteins are resistant, in varying degrees, to attack by proteolytic enzymes. If the denaturing conditions employed lead to a loss of native structure, a very large increase in susceptibility to proteolytic attack may result. Several good examples of such large increases in susceptibility are provided by Kunitz (1947), Halsey and Neurath (1955), Cassman and Schachman (1971), and Hayashi et al., (1973a).

2. Many proteases are resistant, in varying degrees, to inactivation by one or more of the commonly employed denaturing conditions. If a protease remains active, even temporarily, under conditions in which substrate proteins are denatured, very high rates of proteolysis may result. Striking examples of resistance to the effects of urea (Northrup et al., 1948; Halsey and Neurath, 1955; Perlman, 1956), guanidinium chloride (Cassman and Schachman, 1971; Awad et al., 1972), and SDS (Guidotti, 1960; Nelson, 1971; Diezel et al., 1972) have been reported among proteases from various biological sources.

3. In some cases (including at least two proteases from yeast; see Lenney, 1956; Hayashi et al., 1968a,b, 1972; Hata et al., 1972; and Section III,C,1),

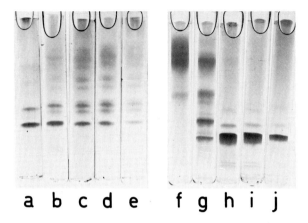

a   b   c   d   e   f   g   h   i   j

FIG. 1. Results of SDS–polyacrylamide gel electrophoresis of protein samples that have experienced various degrees of proteolytic attack during denaturation. Electrophoresis was performed as described by Weber *et al.* (1972). The gels are shown with their electrophoretic origins at the bottom of the figure; thus the higher a particular polypeptide is in the figure, the lower is its molecular weight. (a to e) Progressive digestion of a malate dehydrogenase sample following the addition of SDS and 2-mercaptoethanol (to 1% each) at 24°C (pH 7.6). At 0 minutes (a), 1 minute (b), 5 minutes (c), 30 minutes (d), and 310 minutes (e) after the addition of SDS, the samples were treated at 100°C for 5 minutes to terminate proteolysis. Extended incubation at 24°C *after* the 100°C treatment led to no further changes in the band patterns. Incubation at 24°C (without a 100°C treatment) for several days gave a pattern indistinguishable from that of (e); evidently, the protease in this sample was itself fully inactivated within 310 minutes in 1% SDS and 1% 2-mercaptoethanol at 24°C. (f to j) Digestion of a sample of phosphoglycerate kinase (Sigma) by contaminating protease. (g) SDS and 2-mercaptoethanol were added (to 1% each) at 24°C (pH 7.6), and the sample was incubated overnight before electrophoresis. (f) Sample like that of (g), except that ethanol was added to 4.5%, shortly before the addition of SDS; for unknown reasons, the ethanol accentuates the proteolytic breakdown. (h to j) Proteolysis during denaturation was prevented by (h) adding PMSF a few minutes before the addition of SDS [note that, since the PMSF was initially dissolved in ethanol, this sample had the same ethanol concentration as the sample in (f)]; (i) adding SDS and 2-mercaptoethanol with an *immediate* treatment at 100°C for 4 minutes; (j) denaturing initially with 6 *M* guanidinium chloride at 100°C (see Weber *et al.*, 1972, for the details of this procedure).

exposure to denaturing conditions leads to actual activation of the protease itself as a result of the destruction of protease–polypeptide inhibitor complexes.

The practical consequence of the acceleration of proteolysis is that amounts of proteolytic activity too small to be readily detected by conventional tests of protein homogeneity, by conventional protease assays, or by effects on the native protein of interest, can have profound effects when denaturing conditions are applied. The extensive degradations of hexokinase (Pringle, 1970a; Rustum *et al.*, 1971) and of malate dehydrogenase (Pringle, 1970b) that occur on the addition of SDS provide good examples among yeast proteins (see also the previous comments on these cases in Sections II,A,1; II,A,2; and II,B,3). Although an incubation of 24 hours or more at 25°C was necessary to detect either proteolytic inactivation of malate dehydrogenase or casein hydrolysis by malate dehydrogenase preparations (using the assay of Kunitz, 1947, with malate dehydrogenase at about 0.1 mg/ml), degradation in the presence of SDS was detectable within 1 minute and was extensive within 5 minutes (Fig. 1, gels a to e).

Similar proteolyses in SDS were observed with many other highly purified yeast proteins (Table I; Fig. 1, gels f to j), most of which seemed to have survived, in more-or-less intact form, their purification and handling as native proteins. The following comments should be made:

1. The rapid production of low-molecular-weight polypeptides suggests that endopeptidase action is involved. It should be noted that exopeptidase action is also frequently accelerated by denaturing treatment, but such effects are more difficult to detect unless one actually examines the chain-terminal amino acids.

2. In most cases, but not in all, the digestion was completely prevented by treatment with PMSF immediately before the SDS was added. Thus in most cases, but perhaps not in all, protease B seems to be responsible for the effects observed.

3. Most of the known denaturation-related artefacts in yeast studies involve denaturation with SDS. This situation is probably due in part to the relatively large number of proteins tested for proteolysis in SDS, and to the ease of detection of proteolytic effects (by gel electrophoresis). Note that accelerated degradations have also been observed at pH 10.5 (Kenkare and Colowick, 1965), after acylation (Easterby and Rosemeyer, 1972), and in the course of tryptic digestion (J. W. Stewart and F. Sherman, personal communication, 1970). Also, the resistance of various yeast proteases (see Section III) to denaturation by high concentrations of urea, or by low pH, virtually guarantees that accelerated degradations will occur in some studies employing these denaturing conditions.

5. THE PREVALENCE OF ARTEFACTS

Table I includes about 45 cases in which *in vitro* proteolytic effects were recognized. There is no doubt that other cases which might have been included have escaped our attention, and that in still other cases proteolytic effects occurred without being recognized by the investigators. (In particular, there are numerous reports of enzyme instability in which no tests for protease involvement were made, or in which the use of PMSF was the only control that was tried.) The cases listed in Table I involve proteins as diverse as the protease-B inhibitor, cytochrome c, phosphofructokinase, the pyruvate dehydrogenase complex, and mitochondrial membrane proteins. Studies of isoenzymes, of catalytic mechanisms, of subunit structure, and of amino acid sequence have been affected. Thus it seems clear that protease effects are the rule, rather than the exception, in studies of yeast proteins, and should always be taken into account.

It is worth emphasizing that most of the known proteolytic effects are due to one or more PMSF-sensitive proteases. This situation is almost certainly due in part to the greater ease of recognition of such effects. The number of effects attributed to protease A should increase as inhibitors of this protease come into use. Also, many exopeptidase effects have almost certainly escaped attention, both because of the lack of good inhibitors (at least for the aminopeptidases) and because such effects are frequently rather subtle.

6. THE PERSISTENCE OF ARTEFACTS

It is clear that the use of the specific protease inhibitors PMSF and DFP has in many cases been extremely useful in preventing proteolytic artefacts. It must be emphasized, however, that:

1. Since yeasts contain proteases that are not inactivated by PMSF and DFP, there are certainly protease effects that are not prevented by these reagents.

2. Even where PMSF- and DFP-sensitive proteases are involved, it cannot be assumed that casual use of these reagents will eliminate all problems. The examples in Section II,A make it clear that repeated treatments with the reagents are essential, and that even with such repeated treatments much caution is in order, particularly at the time of denaturation. It seems clear that the elimination of proteolytic artefacts can sometimes be quite difficult, even after the problem has been clearly recognized.

III. The Yeast Proteases

Yeast cells produce several different proteolytic enzymes. Although this protease complement has not been fully characterized, three of the enzymes (proteases A, B, and C) have been studied in detail, and useful information is available about several others. The short descriptions that follow focus on the properties relevant to the problem of preventing proteolytic artefacts. It should be noted that the enzymes described are predominantly *intracellular* enzymes. *Saccharomyces cerevisiae* produces little or no free extracellular proteolytic activity (Matile, 1969; Ahearn *et al.*, 1968), and *S. carlsbergensis* secretes no detectable protease, except during growth with proteins or peptides as nitrogen sources (Maddox and Hough, 1970). [Yeasts in some other genera produce larger amounts of extracellular protease activity (Ahearn *et al.*, 1968).]

A. Endopeptidases

1. PROTEASE A

Protease A is regarded as an acid protease. It digests hemoglobin and casein rapidly between pH 2 and 4, and is much less active against these substrates at higher pH values (Lenney, 1956; Lenney and Dalbec, 1967; Hata *et al.*, 1967a). It does, however, have some activity against casein at pH 6 (Hata *et al.*, 1967a), and it attacks and inactivates tryptophan synthase, and probably also threonine dehydrase, at pH 7 (Katsunuma *et al.*, 1972; Saheki and Holzer, 1974a). Protease A does not attack Azocoll (Juni and Heym, 1968; Saheki and Holzer, 1974a), or any synthetic peptide or ester substrate that has been tested (Lenny and Dalbec, 1967; Hata *et al.*, 1967a,b, 1972; Juni and Heym, 1968; Saheki and Holzer, 1974a). Protease A is not inhibited by EDTA, *p*-chloromercuribenzoate (PCMB), DFP, or PMSF (Lenney, 1956; Lenney and Dalbec, 1967; Hata *et al.*, 1967a; Juni and Heym, 1968; Saheki and Holzer, 1974a). It has recently been discovered (Saheki and Holzer, 1974b; Lenney, 1975) that pepstatin (Umezawa *et al.*, 1970a; Aoyagi *et al.*, 1971) is a highly effective inhibitor of protease A. These properties of protease A allow it to be readily differentiated from the other known yeast proteases (see Hata *et al.*, 1967b, 1972; Juni and Heym, 1968; Saheki and Holzer, 1974a).

Protease A is quite stable during incubation at pH 5–6 at temperatures up to 50°C, but it is much less stable at higher pH values; inactivation is rapid even at room temperature at pH 8 or above (Lenney, 1956; Hata *et al.*, 1967a; Saheki and Holzer, 1974a). Protease A may be stable in 6 *M* urea at some pH values (Hata *et al.*, 1967a), although it seems to be rapidly inactivated by urea at pH 2–4 (Lenney, 1956).

Protease A has a molecular weight of about 60,000 and an isoelectric point of about 3.7 (Hata *et al.*, 1967b). It seems to be effectively precipitated by $(NH_4)_2SO_4$ at 60% of saturation (Katsunuma *et al.*, 1972), although much is apparently soluble at 50% of saturation (Hayashi *et al.*, 1973a). In 0.01 *M* phosphate (pH 7), protease A is eluted from triethylaminoethyl (TEAE)-cellulose and DEAE-Sephadex columns at about 0.2 *M* NaCl (Hata *et al.*, 1967a,b, 1972; Hayashi *et al.*, 1973a; Saheki and Holzer, 1974a). Protease A is bound to hydroxyapatite in 1 m*M* phosphate (pH 7) and is eluted at slightly higher phosphate concentrations (Katsunuma *et al.*, 1972; Saheki and Holzer, 1974a).

2. PROTEASE B

Protease B hydrolyzes casein vigorously between pH 5.7 and 10, with an optimum at pH 9 (Hata *et al.*, 1967a). It attacks tryptophan synthase between pH 4.8 and 7, with an optimum at pH 5.5 (Schött and Holzer, 1974; Saheki and Holzer, 1974a), and attacks many other yeast enzymes in the neutral pH range (Table I). Protease B attacks Azocoll in the neutral pH range (Juni and Heym, 1968; Cabib and Ulane, 1973; Hasilik and Holzer, 1973; Saheki and Holzer, 1974a), has a weak activity toward α-N-benzoyl-L-arginine ethyl ester (BAEE) and α-N-acetyl-L-tyrosine ethyl ester (ATEE) (with a greater affinity for the former) between pH 6 and 10 (Hata *et al.*, 1967a), and does not attack carbobenzoxy (Cbz)-phe-leu, Cbz-glu-tyr, or Cbz-gly-leu (Lenney and Dalbec, 1967). Protease B is not affected by EDTA, but can be completely inhibited by DFP, PMSF, or PCMB (Lenney and Dalbec, 1967; Hata *et al.*, 1967a; Juni and Heym, 1968; Saheki and Holzer, 1974a), or by 0.5 mM Cu^{++} (A. Hasilik, personal communication, 1975). Lenney (1975) has reported that chymostatin (Umezawa *et al.*, 1970b) is a potent inhibitor of protease B.

At least some preparations of protease B are quite unstable, at all pH values, at 37°C (Hata *et al.*, 1967a; Saheki and Holzer, 1974a). However, protease B survives, and is in fact strongly activated during (see Section III,C) extended incubation of crude extracts at 25°C (Lenney and Dalbec, 1967; Hayashi *et al.*, 1967, 1968a; Juni and Heym, 1968), and even at 37°C (Cabib and Farkas, 1971). It also survived treatment at 60°C during purification to homogeneity by Schött and Holzer (1974), probably retains activity temporarily in 5.5 M urea (Lenney, 1956; Lenney and Dalbec, 1967), and is almost certainly responsible for most of the effects in SDS documented in Table I. [The activity in urea may have been due to contamination of the preparation by protease C, an enzyme known to be resistant to urea denaturation (Hayashi *et al.*, 1972, 1973a).] Thus the instability observed with protease B is very likely a self-digestion, and the enzyme (or in some cases its inhibited complex; see Section III,C) is clearly stable enough to survive extended manipulation and purification.

Protease B has an isoelectric point of about 5 and a molecular weight of about 82,000 (Schött and Holzer, 1974), although its behavior during gel filtration does not always seem to be in accord with this molecular weight (Lenney and Dalbec, 1969); possibly degraded, smaller, forms of protease B retain activity (Schött and Holzer, 1974), or the 82,000 molecular weight form may be a complex of some type. Protease B is precipitated by

$(NH_4)_2SO_4$ at 60% (Katsunuma et al., 1972; Saheki and Holzer, 1974a), or even 50% (Aibara et al., 1971; these data may actually relate to the inactive protease-polypeptide inhibitor complexes described in section III,C), of saturation. In 0.01 M phosphate (pH 7) protease B is only loosely bound to TEAE-cellulose, and is eluted with a slight increase in salt concentration (Hata et al., 1967a). At pH 7 protease B is eluted from hydroxyapatite at about 0.15 M phosphate (Saheki and Hozer, 1974a).

3. OTHER ENDOPEPTIDASES

Juni and Heym (1970; also personal communication) reported the tentative identification of a protease of distinctive molecular weight (175,000) and subcellular localization (see Section III,C). It was said to attack Cbz-dipeptide substrates, to be inhibited by PMSF but not by EDTA, and to retain activity when exposed to detergents, 4 M urea, or temperatures of 60°C. This enzyme might be an exopeptidase, and might even be an adsorbed or aggregated form of protease C.

Maddox and Hough (1970) described several apparent endopeptidases, isolated from *S. carlsbergensis*, which are difficult to relate to the known *S. cerevisiae* proteases. These enzymes have apparently not been studied further.

B. Exopeptidases

1. PROTEASE C

A carboxypeptidase from yeast, known as protease C or carboxypeptidase Y (it is probably also equivalent to the peptidase β described by Félix and Brouillet, 1966), has been characterized in considerable detail. The pure enzyme is devoid of endopeptidase activity and is a carboxypeptidase of broad specificity; it can remove carboxy terminal amide and ester groups, as well as carboxy terminal amino acids of all types, from a wide variety of peptide and protein substrates (Hayashi et al., 1970, 1973a; Hayashi and Hata, 1972a; Kuhn et al., 1974; Hermodson et al., 1972). It is active against various substrates in the pH range 4–9.5 (Doi et al., 1967; Hayashi and Hata, 1972a; Hermodson et al., 1972; Kuhn et al., 1974). Protease C is not affected by chelating agents, but can be completely inhibited by DFP, PMSF, or PCMB (Félix and Brouillet, 1966; Doi et al., 1967; Juni and Heym, 1968; Hayashi et al., 1972, 1973b; Kuhn et al., 1974; Hata et al., 1972; Saheki and Holzer, 1974a). Of particular usefulness in distinguishing protease C from other yeast proteases is its spectrum of activity against certain artificial substrates; it has a high activity against ATEE, Cbz-phe-leu, and Cbz-glu-tyr, a much lower activity against BAEE and Cbz-gly-leu, and is inactive against

Azocoll (Félix and Brouillet, 1966; Doi *et al.*, 1967; Lenney and Dalbec, 1967; Hata *et al.*, 1967b, 1972; Juni and Heym, 1968; Hayashi and Hata, 1972a; Hayashi *et al.*, 1973a; Saheki and Holzer, 1974a).

Protease C displays little self-digestion except after prolonged periods of storage (Hayashi *et al.*, 1973a), and loses little activity during a 30-minute incubation at 50°C at pH 6 (Doi *et al.*, 1967). It is, however, completely inactivated by a 5-minute incubation at 100°C (Hayashi *et al.*, 1973a,b). It survives incubation for 30 minutes at pH 3 (Hayashi *et al.*, 1972), and is quite resistant to denaturation by concentrated urea solutions (only a 20% loss of activity was observed during 1 hour in 6 M urea at 25°C; Hayashi *et al.*, 1972, 1973a; Kuhn *et al.*, 1974).

Protease C has an isoelectric point of pH 3.6, a molecular weight of about 60,000, and consists of a single polypeptide chain plus a carbohydrate moiety (Aibara *et al.*, 1971; Kuhn *et al.*, 1974; Hayashi *et al.*, 1973a). It is soluble in 50% saturated $(NH_4)_2SO_4$ solutions (Aibara *et al.*, 1971; Hayashi *et al.*, 1973a), but is at least partially precipitated at 60% of saturation (Katsunuma *et al.*, 1972; Saheki and Holzer, 1974a). (These data may actually relate in part to the protease–polypeptide inhibitor complexes described in Section III,C.) In 0.01 M phosphate (pH 7) protease C is bound to TEAE-cellulose and DEAE-Sephadex at 0.1 M NaCl and is eluted at about 0.3 M NaCl (Hata *et al.*, 1967a,b, 1972; Hayashi *et al.*, 1973a; Saheki and Holzer, 1974a; Kuhn *et al.*, 1974). Protease C is bound to hydroxyapatite in 1 mM phosphate (pH 7) and is eluted at slightly higher phosphate concentrations (Katsunuma *et al.*, 1972; Saheki and Holzer, 1974a).

2. OTHER CARBOXYPEPTIDASES

The peptidase α described by Felix and Brouillet (1966) seems to be a second yeast carboxypeptidase. In contrast to protease C, peptidase α is inhibited by chelating agents such as EDTA and 1,10-phenanthroline (activity can be restored by the addition of Zn^{2+} or Co^{2+}), but is not inhibited by DFP. Peptidase α has a molecular weight of roughly 90,000, is inhibited by PCMB, and is active in the neutral pH range on several peptide substrates that have free carboxyl groups. In contrast to protease C, it is much more active against Cbz-gly-leu than against Cbz-phe-leu. It is unfortunately not clear whether peptidase α is actually active against protein (as opposed to oligopeptide) substrates. There is also mention of a metal-dependent, DFP-resistant exopeptidase in Maddox and Hough (1970).

3. AMINOPEPTIDASES

Several aminopeptidase activities have been observed in cell extracts of yeast (Johnson and Berger, 1942; Hagenmaier, 1967; Matile and Wiemken,

1967; Matile *et al.*, 1971). Unfortunately, detailed information about the catalytic properties (including substrate specificity and methods for inhibition), stability, and behavior during separation procedures of these enzymes is not yet available. In particular, it is not yet clear which, if any, of the enzymes attack protein substrates. One enzyme (the vacuole-localized species; see Section III,C) seems to be of high molecular weight, is active between pH 5 and 9 (with an optimum at pH 7.6), and is apparently neither inhibited by EDTA nor activated by Co^{2+} (Matile *et al.*, 1971). At least one other aminopeptidase activity seems to be inhibited by chelating agents and activated by divalent cations (Johnson and Berger, 1942; Matile and Wiemken, 1967).

C. Intracellular Localization and Polypeptide Inhibitors

1. PROTEASES A, B, AND C

It has become clear that in a crude extract at pH 7 each of these three major proteases exists largely in an inactive form, which can be activated by incubation at lower pH or (at least for proteases B and C) under mildly denaturing conditions (Lenney, 1956, 1973; Lenney and Dalbec, 1967, 1969; Hata *et al.*, 1967b, 1972; Hayashi *et al.*, 1967, 1968a,b, 1969, 1972; Juni and Heym, 1968; Hayashi and Hata, 1972b). [Some conflicting evidence has indicated that protease A is mostly already active in crude extracts at neutral pH (Hayashi *et al.*, 1967, 1968a; compare Lenney, 1956, 1973; Hata *et al.*, 1967b; Lenney and Dalbec, 1967; Juni and Heym, 1968). Perhaps this is the case in some strains, or perhaps the observation is explained by an activation of protease A during the assay procedure (Juni and Heym, 1968).] The explanation of this situation seems to be as follows. In the living cell each of these proteases is largely or entirely sequestered in the vacuoles, or in similar membrane-bounded vesicles (Matile and Wiemken, 1967; Juni and Heym, 1970, also personal communication; Cartledge and Lloyd, 1972; Holley and Kidby, 1973; Lenney, 1973; Cabib *et al.*, 1973, 1974; Lenney *et al.*, 1974; Hasilik *et al.*, 1974; Matern *et al.*, 1974a,b). The extravacuolar cytoplasm contains polypeptide inhibitors for each protease (Manney, 1968; Juni and Heym, 1968; Lenney and Dalbec, 1969; Cabib and Farkas, 1971; Ferguson *et al.*, 1973; Betz and Holzer, 1973; Lenney, 1973; Betz *et al.*, 1974; Saheki *et al.*, 1974; Lenney *et al.*, 1974; Hasilik *et al.*, 1974; Matern *et al.*, 1974a,b; Maier *et al.*, 1974; Ulane and Cabib, 1974; Lenney, 1975), and the complexes that form on cell (and vacuole) disruption are rather stable at pH 7 (Hayashi *et al.*, 1967, 1968a,b, 1969, 1972; Lenney and Dalbec, 1969; Lenney, 1973; Betz and Holzer, 1973; Betz *et al.*, 1974; Maier *et al.*, 1974; Saheki *et al.*, 1974; Lenney, 1975). Alteration of the conditions allows dissociation of the inhibitors and/or their digestion by the various proteases.

Evidence has been presented that protease A is involved in the inactivation of protease-B inhibitor (Juni and Heym, 1968; Maier *et al.*, 1974; Betz *et al.*, 1974; Lenney, 1975), and of protease-C inhibitor (Hayashi *et al.*, 1967, 1968b; Hayashi and Hata, 1972b; Matern *et al.*, 1974c); that protease B is involved in the inactivation of its own inhibitor (Maier *et al.*, 1974; Lenney, 1975), of protease-C inhibitor (Matern *et al.*, 1974c), and of protease-A inhibitor (Saheki *et al.*, 1974); and that protease C is involved in the inactivation of protease-B inhibitor (Lenney and Dalbec, 1969; Lenney, 1975).

The inhibitors of proteases A and B are stable to boiling and to precipitation with trichloroacetic acid (Manney, 1968; Lenney and Dalbec, 1969; Cabib and Farkas, 1971; Ferguson *et al.*, 1973), and have been purified to homogeneity (Betz and Holzer, 1973; Lenney, 1973; Betz *et al.*, 1974; Maier *et al.*, 1974; Saheki *et al.*, 1974; Lenney, 1975). The protease-C inhibitor, in contrast, is very labile (Lenney, 1973; Maier *et al.*, 1974; Matern *et al.*, 1974c; Lenney, 1975). The complexes of the proteases with their inhibitors seem to be similar to the proteases themselves in their behavior during fractionation procedures, although protease C was definitely separable from its inhibited complex by chromatography on DEAE-cellulose (Hayashi *et al.*, 1968a, 1969; Lenney and Dalbec, 1969; Lenney, 1975).

2. THE OTHER PROTEASES

The high-molecular-weight protease (Section III,A,3) was said to be localized in the extravacuolar cytoplasm (Juni and Heym, 1970; also personal communication). Most of the aminopeptidase activity of the cell seems to be localized in the periplasmic space (Matile, 1969); of the four aminopeptidase species identified by Matile *et al.* (1971), one was definitely vacuolar, one was definitely periplasmic, and the others were either cytoplasmic or periplasmic.

IV.   Antiartefact Methodology

A.   Preventing Attack on Native Proteins

There are two facets of the problem of preventing proteolytic artefacts during the isolation, purification, and study of a native protein. First, what can one do from the outset to minimize the chances of proteolytic attack? Second, how can one decide if the measures employed have been sufficient? In Sections A,1 through A,7, we describe several useful antiproteolysis procedures. In Section A,8 we give an optimal general strategy for avoiding artefacts, and then return to the second facet of the problem. It cannot be emphasized too strongly that the optimal general strategy cannot be applied in all cases with a blind faith in its efficacy. Some proteins of interest may be so sensitive to attack that the general strategy is not sufficient; some proteins may themselves be sensitive to PMSF, or to some other feature of the general strategy; in some cases the attacking protease may be one that is not inactivated by the conditions described. Thus the efficacy of the general strategy, or of any alternative approach, must be evaluated for each protein to be studied on a case-by-case basis. In almost all cases, however, some combination of the methods described should be successful in preventing proteolytic artefacts.

1. Growth, Harvesting, and Disruption of Cells

There is good evidence that stationary-phase cells (Halvorson, 1958; Fukuhara, 1967; Cabib and Farkas, 1971; Matile *et al.*, 1971; Katsunuma *et al.*, 1972), sporulating cells (Chen and Miller, 1968; Hopper *et al.*, 1974; A. Singh and H. Halvorson, personal communication, 1974; H. Betz, personal communication, 1974), and cells in the throes of respiratory adaptation (Matile *et al.*, 1971) have higher rates of protein turnover *in vivo*, and higher proteolytic activities in cell extracts, than do cells from the exponential phase of growth. Moreover, there is good evidence that cells grown in minimal medium contain substantially more proteolytic activity than do cells grown in rich medium (Manney, 1968; Cabib and Farkas, 1971; Katsunuma *et al.*, 1972; Tsai *et al.*, 1973). Thus it is highly desirable to begin protein studies with cells harvested from exponential-phase cultures in rich medium. This rule cannot be applied, of course, when the protein to be studied is specific to another growth phase, or is drastically reduced in amount in rich medium–grown cells; in studies of such proteins a special caution seems to be in order. Indeed, studies of ribosomal proteins (A. Hopper, personal communication, 1972) and of RNA polymerase (P. T. Magee, personal communication, 1974; J. Sebastian, personal communication, 1974; L. D. Schultz, personal communication, 1974) have already provided two examples in which proteolysis is more troublesome during the study of a protein derived from sporulating cells than during the study of the same protein from exponentially growing cells.

What the investigator sees as the beginning of cell harvesting the cells may see as the beginning of stationary phase, and they may accordingly begin to increase their protease levels and/or their *in vivo* protein degradation rates. Thus it is highly desirable to conduct cell harvesting as rapidly, and (except in the special case of cold-labile proteins) as close to 0°C, as possible.

Yeast cells have tough cell walls and are somewhat difficult to disrupt. This situation led to the classic extraction of yeast enzymes by prolonged periods of autolysis, a process in which the hydrolytic activity of the yeast's own proteases plays a prominent part. There may be circumstances under which the use of autolysis is justified, but in general it should be avoided. If autolysis *must* be used, it should be conducted at pH 7, where the protease–polypeptide inhibitor complexes are relatively stable. It should be noted that Lazarus *et al.* (1966) have reported that addition of DFP to autolyzing extracts very seriously impedes the release of hexokinase; thus too much control of protease activity prevents the autolysis from working normally.

The importance of avoiding autolysis has also been stressed by Lazarus *et al.* (1966), Clark and Jakoby (1970), Fink (1970), and Ramel *et al.* (1971), and these investigators have described other effective methods for disrupting yeast cells and releasing their proteins. We have found grinding the cells with acid-washed glass beads, in any of several apparatuses, to be extremely effective; disruption seems more thorough when beads of smaller diameter are used. This process can be carried out at low temperatures and in the presence of protease inhibitors.

2. THE USE OF POWERFUL SEPARATION METHODS

The use of powerful separation methods, particularly at early stages in purification procedures, is highly recommended. This in general significantly reduces the exposure of the protein of interest to proteolytic attack. Some information on the behavior of proteases A, B, and C during commonly used separation procedures is given in Section III; this information may facilitate the design of effective purification protocols. It should be kept in mind, however, that these three proteases are probably not the only ones present in a crude extract (Section III), that the proteases may exist in modified forms with altered chromatographic properties (see the comments in Section III,A,2 on the gel filtration behavior of protease B), that protease–polypeptide inhibitor complexes may be distributed in different fractions from the proteases themselves, that the proteases may adsorb to other proteins and be difficult to eliminate for that reason (Diezel *et al.*, 1972), and that some proteolytic artifacts may be produced by very small amounts of contaminating protease (Section II). Whatever the reasons, the weight of empirical evidence is heavily against the idea that powerful separation procedures alone are adequate to eliminate proteolytic artefacts. Too many highly purified proteins, of too many different types, have been found to contain residual traces of protease.

3. PMSF AND DFP

PMSF and DFP are potent and highly specific irreversible inhibitors of many proteases, including at least two of those from yeast (Section III). Thus it is not surprising that these reagents have proved highly useful in preventing proteolytic artefacts in yeast studies (Section II). In using these reagents the following points should be kept in mind:

1. PMSF does not go readily into aqueous solution. Thus the solid should first be dissolved at relatively high concentration in an organic solvent (ethanol, dimethyl sulfoxide, propanol, and acetonitrile have been used). By adding portions of this concentrate to the aqueous buffer, aqueous solutions approximately 1 mM (or slightly higher) in PMSF can be obtained (Gold, 1967). We have always made up a 40 mM solution in 95% ethanol and then added it to protein solutions (with good stirring, to avoid high local ethanol concentrations) at 1 part in 20.

2. PMSF is hydrolyzed with a half-life of about 100 minutes at pH 7; hydrolysis is even faster at higher pH values (Gold, 1967). Thus, solutions should be made up shortly before use.

3. Although PMSF and DFP react with high specificity with the unusual seryl residues at the active sites of certain proteases (Balls and Jansen, 1952; Fahrney and Gold, 1963; Gold and Fahrney, 1964; Oosterbaan and Cohen, 1964; Gold, 1967;Chaiken and Smith, 1969; Hayashi et al., 1972, 1973b), this specificity is not complete (Balls and Jansen, 1952; Oosterbaan and Cohen, 1964; Whitaker and Perez-Villaseñor, 1968; Chaiken and Smith, 1969). In particular, many esterases also have reactive seryl residues and thus are also sensitive to the reagents. DFP seems more versatile in its reactions than does PMSF (Fahrney and Gold, 1963; Gold, 1967). There is also some danger that impurities in the reagent preparations will lead to nonspecific reactions (Doi et al., 1967; Chaiken and Smith, 1969). Thus one should keep in mind the possibility that the protein of interest may itself react with the reagents. On at least one occasion, the activity of the protein of interest should be measured carefully immediately before, and immediately after, a first exposure to the reagent. A parallel purification using other protease inhibitors (Sections IV,A,4 and A,5) would be a very useful control. It is also possible to check the protein of interest for incorporation of label from radioactive reagents; radioactive DFP, at least, is commercially available (Cohen et al., 1967; Chaiken and Smith, 1969; Hayashi et al., 1973b). [Gold and Fahrney (1964) have described the preparation and use of PMSF-[14]C.]

4. There have been several reports that DFP is more effective than PMSF in inhibiting proteolytic activity in yeast extracts (Clark and Jakoby, 1970; Diezel et al., 1973). This may well be true, although it has not yet been documented with isolated proteases; it is also possible that PMSF was not used under optimal conditions (see points 1 and 2) in these studies. In other studies PMSF has seemed highly effective (Schulze and Colowick, 1969; Pringle, 1970b; Lederer and Jacq, 1971).

5. PMSF has a low toxicity (Fahrney and Gold, 1963; Gold, 1967), but

DFP is a highly toxic reagent. Many people have worked with DFP and survived, but it is in fact dangerous enough that its routine or large-scale use presents a significant safety problem. Thus, in spite of point 4, we are reluctant to recommend its universal use, especially since we have had no personal experience with it. The best strategy may be to use PMSF for routine purposes, such as large-scale purifications, and to reserve DFP for special small-scale controls and for particular cases in which PMSF seems to be not fully effective. If the use of DFP is indicated, information about procedures, and about the necessary safety precautions, can be found in Cohen *et al.* (1967), Potts (1967), Clark and Jakoby (1970), and Lazarus *et al.* (1966).

6. It is empirically clear (Section II) that a single treatment of a crude or partly purified extract with PMSF or DFP is not sufficient to eliminate artefacts due to PMSF- and DFP-sensitive proteases. Presumably, the explanation is that the protease–polypeptide inhibitor complexes (Section III) are resistant to the reagents and may then subsequently release active protease (Hayashi *et al.*, 1972). It is also conceivable that reactivation of PMSF- and DFP-inhibited protease may sometimes also play a role (Gold and Fahrney, 1964; Gold, 1967; Cohen *et al.*, 1967).

4. THE YEAST'S OWN POLYPEPTIDE PROTEASE INHIBITORS

An approach to the protease artefact problem that has never been utilized, but that has considerable promise, is the use of the yeast's own polypeptide protease-inhibitors (see Section III) to minimize proteolysis. The very effective inhibitors of proteases A and B can be obtained in large amounts, free of proteolytic activity, simply by boiling extracts (even autolysates) of commercial yeast. Trichloroacetic acid precipitation gives some further purification. Protease-C inhibitor is unfortunately more difficult to separate from proteases, but progress is being made (Matern *et al.*, 1974c; Lenney, 1975). Such impure inhibitor preparations can then simply be added to crude and partly purified preparations of the protein of interest. This approach has two main advantages:

1. It offers an effective way to minimize the activity of the PMSF- and DFP-resistant protease A.

2. It should be free of "reagent artefacts," and thus should be particularly useful in studying proteins that may react with PMSF and DFP.

The approach also has several significant disadvantages:

1. The inhibition is reversible, and there is a continual tendency to release active protease. This is no problem so long as the inhibitors are present in excess, but it is difficult to keep the inhibitors in excess at all times (e.g., during separation procedures).

2. The inhibition is only effective within a certain pH range (roughly pH 6.5–8.5).

3. The method involves the repeated addition of protein impurities to the protein preparation one may be trying to purify. If these impurities can easily be separated from the protein of interest (probably a common situation, since the impurities are proteins of somewhat unusual properties), this disadvantage is not too serious.

5.   OTHER INHIBITORS

Pepstatin and chymostatin are potent and selective inhibitors of proteases A and B, respectively (Section III). Pepstatin should be very useful, as the inhibition of protease A has been a problem. Chymostatin may be useful in the occasional cases in which the protein of interest is PMSF-sensitive. These inhibitions do have the disadvantage of being reversible. It is anticipated that these substances will become commercially available; in the meantime, it may be possible to obtain samples from H. Umezawa.

PCMB is a potent inhibitor of proteases B and C. Unfortunately, it reacts with so many proteins that its usefulness in preventing artefacts is extremely limited.

The following references have information on several protease inhibitors that apparently have never been tried on yeast proteases, but which are highly effective in other systems: Erlanger and Cohen (1963), Geratz (1969), Shaw and Ruscica (1971), and Prouty and Goldberg (1972).

6.   CONTROL OF pH AND OF DIVALENT CATIONS

The activity of each protease is pH-dependent; the precise pattern of this dependence varies from substrate to substrate (Section III). Likewise, the stability (including resistance to proteolysis) of any particular protein of interest is also a function of pH. Thus proper choice of pH in individual cases can contribute to minimizing proteolytic attack. This approach has been regarded mainly as a weapon against protease A, whose activity is low above pH 6 and for which effective inhibitors have not been available. It may even be possible to achieve a permanent inactivation of protease A by means of short-term incubations at elevated pH and temperature (Section III,A,1); this approach, however, has obvious associated dangers. If pepstatin proves effective and readily obtainable, or if other good inhibitors of protease A become available, pH control may prove more useful as an antiaminopeptidase weapon, since as yet no inhibitors for these enzymes are known. Unfortunately, there is also no detailed information about the pH dependence of activity for these enzymes.

Two further points about pH control deserve mention. The intracellular pH of yeast cells may be rather low; thus, if it is intended to keep the pH above 7, to minimize protease-A activity, a strong buffer must be used at the time of crude extract preparation. Also, it is important that manipulations of pH be coordinated with other methods being used. For example, if the polypeptide inhibitors are being used, all steps must be conducted in the neutral pH range, where these inhibitors are effective.

Since yeast very likely contains one or more divalent cation-dependent exopeptidases (Section III), the inclusion in all buffers of strong chelating agents [EDTA or 1,10-phenanthroline; the latter may be somewhat more effective than EDTA (Vallee, 1960)] is frequently a useful antiproteolysis procedure. (The use of chelators also helps to avoid heavy-metal poisoning and sulfhydryl oxidation.) Of course, many proteins to be studied are themselves dependent on divalent cations for activity or stability; in particular, some enzymes are sensitized to proteolytic attack by the removal of bound cations (Stein and Fischer, 1958). In such cases it may be necessary to avoid the use of chelating agents. EDTA and 1,10-phenanthroline are generally effective in inhibiting sensitive enzymes when present at about 5 mM. Like PMSF, 1,10-phenanthroline should first be dissolved in an organic solvent such as ethanol.

7. Other Methods

It is possible to prepare cell extracts in which the vacuoles remain mostly intact and can be separated from other cell components by centrifugation (Matile and Wiemken, 1967; Indge, 1968; Holley and Kidby, 1973; Wiemken, this volume, Chapter 7). Since most of the cellular proteolytic activity is localized in the vacuoles (Section III,C,1), such extracts contain much less proteolytic activity than do whole-cell homogenates, and may thus be useful as a starting point for the purification of extravacuolar proteins. This approach has several significant disadvantages: it is laborious (especially if large amounts of material must be processed), expensive [although the expense can be reduced by using the same batch of snail enzyme on successive batches of cells (E. Cabib, personal communication, 1974)], and incompletely effective (since some vacuoles always break and release their contents). Also, the preparation of protoplasts involves incubating the cells for some time under conditions in which *in vivo* protein degradation rates may increase. However, the method may have some utility, especially in combination with the use of protease inhibitors, in studies of some especially sensitive proteins, or in situations in which small-scale experiments are sufficient.

It has been reported that high concentrations of glycerol, sodium phosphate, and $(NH_4)_2SO_4$ are all effective in reducing the proteolytic activity of yeast extracts (Juni and Heym, 1968; Diezel et al., 1973). This approach to the artefact problem may be quite useful in conjunction with the use of more specific protease inhibitors, or in situations in which the more specific inhibitors cannot be applied.

There is a possibility that affinity chromatography, perhaps with the polypeptide inhibitors incorporated in a Sepharose column (E. Cabib, personal communication, 1974), may be effective in removing protease from cell extracts. We are pessimistic, however, since the large fraction of the protease already complexed with inhibitors presumably will not bind to the column.

It may soon be possible to begin protein studies with extracts derived from mutant yeast strains that specifically lack one or more of the proteolytic enzymes; this approach to the artefact problem seems to have considerable promise. (One successful application of this approach, in a study of *B. subtilis* phosphoglucomutase, has been reported. See Maino and Young, 1974.) Such mutants are currently being sought in several laboratories, at least two of which have obtained promising results (E. W. Jones, personal communication, 1974; D. Wolf and G. R. Fink, personal communication, 1974).

8. A GENERAL STRATEGY; RECOGNITION OF PERSISTENT ARTEFACTS

Although every protein to be studied represents a special case, the basic problem at the beginning of any study is always the same: How can the opportunity for proteolytic attack be minimized without simultaneously damaging the protein of interest? The advantages of working with exponential-phase cells, and of conducting harvesting and cell breakage quickly and at a low temperature, are clear. Similarly, it is clearly desirable to purify quickly (substitution of Sephadex G-25 filtration for dialysis is helpful), at low temperatures, and with the most powerful separation methods available.

In addition, it is clear that an effort must be made to minimize the activity of each of the yeast proteases. The most useful tool here is PMSF (or DFP), which is a potent irreversible inhibitor of at least two prominent yeast proteases, and which is specific enough in its action that it can be applied in most cases without affecting the protein of interest. PMSF should in general be added to the cell slurry immediately before beginning breakage, and fresh PMSF should then be repeatedly added (being careful not to generate excessive concentrations of organic solvent in the protein solutions) during the course of purification and study. It may be desirable to combine PMSF treatment with one of the other measures discussed above, but it should be noted that the polypeptide inhibitors of proteases B and C protect these enzymes from the more permanent inactivation provided by PMSF. When the protein of interest is sensitive to PMSF, it may be possible to minimize the activities of proteases B and C by using the polypeptide inhibitors, Cu^{++} ions, chymostatin, or PCMB.

Protease-A action can probably best be minimized by a combination of pH control with pepstatin treatment. The yeast polypeptide inhibitor provides an alternative. The activity of metal-dependent exopeptidases can in many cases be minimized by the use of chelating agents; unfortunately, not enough is known about these enzymes to recommend an alternative for cases in which the protein of interest is itself adversely affected by the chelating agents. The aminopeptidases remain a special problem, since at this time no method of inhibition is known. We can suggest nothing better than to check especially carefully for small changes at the amino terminal end of the protein of interest.

No matter what precautions are taken, there remains a chance that proteolytic attack on the protein of interest may occur. [These might be persistent effects due to known proteases, or effects due to other proteases which have not yet been characterized (and which can thus not be covered by current antiproteolysis schemes).] Thus it is crucial always to be alert to indications that the antiproteolysis measures employed have not been fully successful. Particularly pertinent danger signals are:

1. Demonstrable residual protease activity. The classic methods for protease assay (e.g., Kunitz, 1947) are not sensitive enough to provide a very rigorous control, but several more sensitive assays have been described in recent years (Lin et al., 1969; De Lumen and Tappel, 1970; Roffman et al., 1970; Roth et al., 1971). Diezel et al. (1972, 1973) made good use of the assay of Roth et al. A purposeful look for proteolysis in SDS (see Fig. 1, gels f to j) is probably the most sensitive way to monitor residual activity of the SDS-resistant protease(s).

2. Certain stability patterns. If the apparent stability of the protein of interest varies greatly with the state of purification, or decreases as the protein concentration is increased, or if inactivation has a biphasic course, proteolytic involvement is a likely possibility.

3. Any change in properties (such as in the ratio of two catalytic activities) during purification or storage, or any variation in properties (or in the relative amounts of two isoenzymes) from preparation to preparation. (Such effects may of course also have other bases.)

4. The appearance of low-molecular-weight peptide material during dialysis or gel filtration.

A very useful approach to the detection of proteolytic (and other) artefacts is to purify the protein of interest by two different methods (including the use of two different approaches to minimizing proteolysis), and to compare the resulting products. An extreme form of this approach is probably the most powerful method currently available for monitoring for proteolytic artefacts. One first purifies the protein of interest by conventional methods, making every effort to limit proteolysis. One then makes antibodies against this purified protein. Even if the purified protein has been slightly modified during purification, these antibodies should recognize the genuine native protein. One then uses the antibodies to precipitate the protein of interest from a freshly prepared (in the presence of protease inhibitors) crude extract, dissociates the precipitate with SDS, and separates the immunoglobulin polypeptides from the polypeptides of the protein of interest by SDS–polyacrylamide gel electrophoresis. If the polypeptides of the protein of interest have the same molecular weights, and the same amino terminal and carboxy terminal amino acids (see Weiner *et al.*, 1972, for the relevant microtechniques), as the protein purified by conventional methods, one has a very strong argument that the genuine native protein has been isolated. Useful references on this technique are Horwitz and Scharff (1969), Schubert and Cohn (1970), Platt *et al.* (1970), Dannies and Tashjian (1973), Greenleaf *et al.* (1973), Mason *et al.* (1973), Fox and Pero (1974), Tjian and Losick (1974), Linn *et al.* (1973), MacGregor *et al.* (1974), and Klier and Lecadet (1974). The last three of these references provide illustrations of the power of the method for detecting persistent proteolytic artefacts.

B.   Preventing Attack at the Time of Denaturation

Even a protein that has been successfully protected from proteolysis during isolation and purification can be seriously degraded when it is denatured in the course of studies of its structure (Section II,B,4). It should, however, always be possible to avoid artefacts of this type by attending to the following principles:

1. When possible, treat with protease inhibitors immediately before denaturation.

2. Denature as abruptly as possible. The crux of the problem is that the protease(s) may stay active temporarily as other proteins become denatured. If the initial denaturing conditions are sufficiently harsh, it should be possible to denature the protease itself so quickly that it has no time to degrade the other proteins present. We have described elsewhere (Weber *et al.*, 1972) the details of several denaturation procedures found to be effective in preventing proteolysis. It should be noted that it is possible to overdo the harshness of the denaturing treatment, giving rise to the occurrence of nonenzymatic peptide bond hydrolysis. For example, extended treatment of hexokinase samples at 100°C in Tris–HCl buffer (pH 7.6) containing 1% SDS and 1% 2-mercaptoethanol gave a progressive generation of many different low-molecular-weight fragments (Pringle, 1970b).

3. Denature by several methods and compare the results. If proteolysis is a factor, its extent will almost certainly be a function of the particular denaturing conditions. Thus, if several different denaturation procedures, differing in harshness and in the nature of the denaturing conditions, give rise to exactly the same denatured polypeptide(s), one may be reasonably certain that proteolysis is not a factor. See Weber *et al.* (1972) for further discussion of these problems in the specific context of SDS–polyacrylamide gel electrophoresis.

V.   Concluding Remarks

It seems appropriate to conclude with the following three disclaimers:

1. It is not impossible to deal effectively with proteolysis problems in studies of yeasts. Powerful antiartefact measures are already available, and further improvements in technique are almost certainly on the way.

2. Not all conclusions reached to date in studies of yeast proteins are invalid. The danger in a particular study obviously depends on the susceptibility of the protein of interest to attack, on the efficacy of the purification methods used, and on the type of conclusion reached. (For example, there are undoubtedly cases in which a little exopeptidase action does not affect estimations of the molecular weight or subunit structure, but profoundly affects the catalytic properties and isoenzyme pattern observed.) Also, there exist already several cases in which the protease problem has been recognized and carefully attended to. However, it is certain (see Section II) that proteolysis problems are widespread, and can be difficult to eliminate even when recognized. It seems clear that any study of a yeast protein should begin with the assumption that proteolysis will be a problem.

3. One cannot avoid worrying about proteolysis problems by avoiding working on yeasts. Such problems appear to be more severe in studies of yeast than in many other systems, but this is probably in part due simply to better recognition of the problem. Proteolytic artefacts have also been encountered in studies of systems as diverse as (this is *not* a comprehensive list) erythrocyte membranes (Fairbanks *et al.*, 1971), bovine milk proteins (Magee *et al.*, 1973), calf thymus histones (Stellwagen *et al.*, 1968), human immunoglobulins (Griffiths and Gleich, 1972), oat and rye phytochromes (Briggs and Rice, 1972), *Pseudomonas aeruginosa* DNase (Bryan *et al.*, 1972), and *E. coli* DNA polymerase (Setlow *et al.*, 1972), RNA polymerase (Hermier *et al.*, 1973), protein synthesis transfer factors (Hollis and Furano,

1968), and methionyl- and leucyl-tRNA synthetases (Cassio and Waller, 1971; Fayat and Waller, 1974; Rouget and Chapeville, 1971). The extended difficulties in attempting to determine the true structures of mammalian aldolase (Penhoet *et al.*, 1967, 1969; Lacko *et al.*, 1970) and of the RNA polymerase from sporulating *B. subtilis* (Linn *et al.*, 1973; Klier and Lecadet, 1974) are particularly instructive (and sobering) examples. (The references given are in both cases only keys to rather extensive series of publications.)

ACKNOWLEDGMENTS

I thank Guido Guidotti, who made it possible for me to begin this work, and Armin Fiechter, who made it possible for me to finish it. I am also indebted to R. Weil, K. Weber, W. R. Briggs, E. Juni, H. Holzer, and B. S. Mitchell, for suggestions and encouragement at various stages, as well as to the many people who have shared with me their unpublished results.

REFERENCES

Afting, E.-G., Katsunuma, T., Holzer, H., Katunuma, N., and Kominami, E. (1972). *Biochem. Biophys. Res. Commun.* **47**, 103.

Ahearn, D. G., Meyers, S. P., and Nichols, R. A. (1968). *Appl. Microbiol.* **16**, 1370.

Aibara, S., Hayashi, R., and Hata, T. (1971). *Agr. Biol. Chem.* **35**, 658.

Anson, M. L., and Mirsky, A. E. (1933). *J. Gen. Physiol.* **17**, 151.

Aoyagi, T., Kinimoto, S., Morishima, H., Takeuchi, T., and Umezawa, H. (1971). *J. Antibiot. Ser. A.* **24**, 687.

Awad, W. M., Ochoa, M. S., and Toomey, T. P. (1972). *Proc. Nat. Acad. Sci. U.S.* **69**, 2561.

Balls, A. K., and Jansen, E. F. (1952). *Advan. Enzymol.* **13**, 321.

Betz, H., and Holzer, H. (1973). *Proc. Int. Spec. Symp. Yeasts, 3rd, 1973*, Part II, p. 67.

Betz, H., Hinze, H., and Holzer, H. (1974). *J. Biol. Chem.* **249**, 4515.

Briggs, W. R., and Rice, H. V. (1972). *Annu. Rev. Plant Physiol.* **23**, 293.

Bryan, L. E., Razzel, W. E., and Campbell, J. N. (1972). *J. Biol. Chem.* **247**, 1236.

Bühner, M., and Sund, H. (1969). *Eur. J. Biochem.* **11**, 73.

Cabib, E., and Farkas, V. (1971). *Proc. Nat. Acad. Sci. U.S.* **68**, 2052.

Cabib, E., and Ulane, R. (1973). *Biochem. Biophys. Res. Commun.* **50**, 186.

Cabib, E., Ulane, R., and Bowers, B. (1973). *J. Biol Chem.* **248**, 1451.

Cabib, E., Ulane, R., and Bowers, B. (1974). *Curr. Top. Cell. Regul.* **8**, 1.

Cartledge, T. G., and Lloyd, D. (1972). *Biochem. J.* **126**, 755.

Cassio, D., and Waller, J.-P. (1971). *Eur. J. Biochem.* **20**, 283.

Cassman, M., and Schachman, H. K. (1971). *Biochemistry* **10**, 1015.

Chaiken, I. M., and Smith, E. L. (1969). *J. Biol. Chem.* **244**, 4247.

Chapman, C., and Bartley, W. (1968). *Biochem. J.* **107**, 455.

Chen, A. W., and Miller, J. J. (1968). *Can. J. Microbiol.* **14**, 957.

Clark, J. F., and Jakoby, W. B. (1970). *J. Biol. Chem.* **245**, 6065.

Cohen, J. A., Oosterbaan, R. A., and Berends, F. (1967). *In* "Methods in Enzymology" (C. H. W. Hirs, ed.), Vol. 11, p. 686. Academic Press, New York.

Collins, K. D. (1974). *J. Biol. Chem.* **249**, 136.

Colowick, S. P. (1973). *In* "The Enzymes" (P. D. Boyer, ed.), 3rd ed., Vol. 9, Part B, p. 1. Academic Press, New York.

Dannies, P. S., and Tashjian, A. H. (1973). *J. Biol. Chem.* **248**, 6174.

Darrow, R. A., and Rodstrom, R. (1970). *J. Biol. Chem.* **245**, 2036.

De Lumen, B. O., and Tappel, A. L. (1970). *Anal. Biochem.* **36**, 22.

Derechin, M., Ramel, A. H., Lazarus, N. R., and Barnard, E. A. (1966). *Biochemistry* **5**, 4017.

Diezel, W., Nissler, K., Heilmann, W., Kopperschläger, G., and Hofmann, E. (1972). *FEBS (Fed. Eur. Biochem. Soc.) Lett.* **27**, 195.

Diezel, W., Böhme, H.-J., Nissler, K., Freyer, R., Heilmann, W., Kopperschläger, G., and Hofmann, E. (1973). *Eur. J. Biochem.* **38**, 479.

Doi, E., Hayashi, R., and Hata, T. (1967). *Agr. Biol. Chem.* **31**, 160.

Duntze, W., Neumann, D., and Holzer, H. (1968). *Eur. J. Biochem.* **3**, 326.

Easterby, J. S., and Rosemeyer, M. A. (1969). *FEBS (Fed. Eur. Biochem. Soc.) Lett.* **4**, 84.

Easterby, J. S., and Rosemeyer, M. A. (1972). *Eur. J. Biochem.* **28**, 241.

Erlanger, B. F., and Cohen, W. (1963). *J. Amer. Chem. Soc.* **85**, 348.

Fahrney, D. E., and Gold, A. M. (1963). *J. Amer. Chem. Soc.* **85**, 997.

Fairbanks, G., Steck, T. L., and Wallach, D. F. H. (1971). *Biochemistry* **10**, 2606.

Fayat, G., and Waller, J.-P. (1974). *Eur. J. Biochem.* **44**, 335.

Félix, F., and Brouillet, N. (1966). *Biochim. Biophys. Acta* **122**, 127.

Ferguson, A. R., Katsunuma, T., Betz, H., and Holzer, H. (1973). *Eur. J. Biochem.* **32**, 444.

Ferguson, J. J., Boll, M., and Holzer, H. (1967). *Eur. J. Biochem.* **1**, 21.

Fessenden-Raden, J. M., and Hack, A. M. (1972). *Biochemistry* **11**, 4609.

Fink, G. R. (1970). *In* "Methods in Enzymology" (H. Tabor and C. W. Tabor, eds.), Vol. 17A, p. 59. Academic Press, New York.

Fosset, M., Muir, L. W., Nielsen, L. D., and Fischer, E. H. (1971). *Biochemistry* **10**, 4105.

Fox, T. D., and Pero, J. (1974). *Proc. Nat. Acad. Sci. U.S.* **71**, 2761.

Fukuhara, H. (1967). *Biochim. Biophys. Acta* **134**, 143.

Gazith, J., Schulze, I. T., Gooding, R. H., Womack, F. C., and Colowick, S. P. (1968). *Ann. N. Y. Acad. Sci.* **151**, 307.

Geratz, J. D. (1969). *Experientia* **25**, 1254.

Givol, D., Craven, G. R., Steers, E., and Anfinsen, C. B. (1966). *Biochim. Biophys. Acta* **113**, 120.

Gold, A. M. (1967). *In* "Methods in Enzymology" (C. H. W. Hirs, ed.), Vol. 11, p. 706. Academic Press, New York.

Gold, A. M., and Fahrney, D. (1964). *Biochemistry* **3**, 783.

Greenleaf, A. L., Linn, T. G., and Losick, R. (1973). *Proc. Nat. Acad. Sci. U.S.* **70**, 490.

Griffiths, R. W., and Gleich, G. J. (1972). *J. Biol. Chem.* **247**, 4543.

Groot, G. S. P., Rouslin, W., and Schatz, G. (1972). *J. Biol. Chem.* **247**, 1735.

Gross, E. (1967). *In* "Methods in Enzymology" (C. H. W. Hirs, ed.), Vol. 11, p. 254. Academic Press, New York.

Guidotti, G. (1960). *Biochim. Biophys. Acta* **42**, 177.

Hagenmaier, H. (1967). Ph.D. Thesis, Cornell University, Ithaca, New York (cited by Gross, 1967).

Halsey, Y. D., and Neurath, H. (1955). *J. Biol. Chem.* **217**, 247.

Halvorson, H. (1958). *Biochim. Biophys. Acta* **27**, 255, 267.

Harris, C. E., Kobes, R. D., Teller, D. C., and Rutter, W. J. (1969). *Biochemistry* **8**, 2442.

Hasilik, A. (1974a). *Proc. Int. Symp. Yeasts, 4th, 1974.* Part 1, p. 15.

Hasilik, A. (1974b). *Arch. Mikrobiol.* **101**, 295.

Hasilik, A., and Holzer, H. (1973). *Biochem. Biophys. Res. Commun.* **53**, 552.

Hasilik, A., Müller, H., and Holzer, H. (1974). *Eur. J. Biochem.* **48**, 111.

Hata, T., Hayashi, R., and Doi, E. (1967a). *Agr. Biol. Chem.* **31**, 150.

Hata, T., Hayashi, R., and Doi, E. (1967b). *Agr. Biol. Chem.* **31**, 357.

Hata, T., Hayashi, R., Doi, E., Minami, Y., and Aibara, S. (1972). *In* "Fermentation Technology Today" (G. Terui, ed.), Proc. 4th Int. Ferment. Symp., p. 279. Soc. Ferment. Technol., Japan.

Hayakawa, T., Perkins, J. P., and Krebs, E. G. (1973). *Biochemistry* **12**, 574.

Hayashi, R., and Hata, T. (1972a). *Biochim. Biophys. Acta* **263**, 673.

Hayashi, R., and Hata, T. (1972b). *Agr. Biol. Chem.* **36**, 630.

Hayashi, R., Oka, Y., Doi, E., and Hata, T. (1967). *Agr. Biol. Chem.* **31**, 1102.

Hayashi, R., Oka, Y., Doi, E., and Hata, T. (1968a). *Agr. Biol. Chem.* **32**, 359.

Hayashi, R., Oka, Y., Doi, E., and Hata, T. (1968b). *Agr. Biol. Chem.* **32**, 367.

Hayashi, R., Oka, Y., and Hata, T. (1969). *Agr. Biol. Chem.* **33**, 196.

Hayashi, R., Aibara, S., and Hata, T. (1970). *Biochim. Biophys. Acta* **212**, 359.

Hayashi, R., Minami, Y., and Hata, T. (1972). *Agr. Biol. Chem.* **36**, 621.

Hayashi, R., Moore, S., and Stein, W. H. (1973a). *J. Biol. Chem.* **248**, 2296.

Hayashi, R., Moore, S., and Stein, W. H. (1973b). *J. Biol. Chem.* **248**, 8366.

Hermier, B., Pacaud, M., and Dubert, J.-M. (1973). *Eur. J. Biochem.* **38**, 307.

Hermodson, M. A., Kuhn, R. W., Walsh, K. A., Neurath, H., Eriksen, N., and Benditt, R. (1972). *Biochemistry* **11**, 2934.

Holley, R. A., and Kidby, D. K. (1973). *Can. J. Microbiol.* **19**, 113.

Hollis, V. W., and Furano, A. V. (1968). *J. Biol. Chem.* **243**, 4926.

Holzer, H. (1974). *Advan. Enzyme Regul.* **12**, 1.

Holzer, H., Katsunuma, T., Schött, E. G., Ferguson, A. R., Hasilik, A., and Betz, H. (1973). *Advan. Enzyme Regul.* **11**, 53.

Holzer, H., Betz, H., and Ebner, E. (1974). *Curr. Top. Cell. Regul.* (in press).

Hopper, A. K., Magee, P. T., Welch, S. K., Friedman, M., and Hall, B. D. (1974). *J. Bacteriol.* **119**, 619.

Horwitz, M. S., and Scharff, M. D. (1969). In "Fundamental Techniques in Virology" (K. Habel and N. P. Salzman, eds.), Vol. 1, pp. 253*ff*. and 297*ff*. Academic Press, New York.

Huang, K.-P., and Cabib, E. (1974). *J. Biol. Chem.* **249**, 3851, 3858.

Indge, K. J. (1968). *J. Gen. Microbiol.* **51**, 441.

Jacq, C., and Lederer, F. (1974). *Eur. J. Biochem.* **41**, 311.

Johnson, M. J., and Berger, J. (1942). *Advan. Enzymol.* **2**, 69.

Juni, E., and Heym, G. A. (1968). *Arch. Biochem. Biophys.* **127**, 79, 89.

Juni, E., and Heym, G. A. (1970). *Bacteriol. Proc.* p. 123.

Kaji, A. (1965). *Arch. Biochem. Biophys.* **112**, 54.

Kaji, A., Trayser, K. A., and Colowick, S. P. (1961). *Ann. N. Y. Acad. Sci.* **94**, 798.

Katsunuma, T., Schött, E., Elsässer, S., and Holzer, H. (1972). *Eur. J. Biochem.* **27**, 520.

Kempe, T. D., Gee, D. M., Hathaway, G. M., and Noltmann, E. A. (1974). *J. Biol. Chem.* **249**, 4625.

Kenkare, U. W., and Colowick, S. P. (1965). *J. Biol. Chem.* **240**, 4570.

Kenkare, U. W., Schulze, I. T., Gazith, J., and Colowick, S. P. (1964). *Abstr. Int. Congr. Biochem. 6th, 1964*, Vol. 32, p. 477, VI-S10.

Klier, A., and Lecadet, M.-M. (1974). *Eur. J. Biochem.* **47**, 111.

Kopperschläger, G., and Hofmann, E. (1969). *Eur. J. Biochem.* **9**, 419.

Kuhn, R. W., Walsh, K. A., and Neurath, H. (1974). *Biochemistry* **13**, 3871.

Kunitz, M. (1947). *J. Gen. Physiol.* **30**, 291.

Lacko, A. G., Brox, L. W., Gracy, R. W., and Horecker, B. L. (1970). *J. Biol. Chem.* **245**, 2140.

Lazarus, N. R., Ramel, A. H., Rustum, Y. M., and Barnard, E. A. (1966). *Biochemistry* **5**, 4003.

Lazarus, N. R., Derechin, M., and Barnard, E. A. (1968). *Biochemistry* **7**, 2390.

Lederer, F., and Jacq, C. (1971). *Eur. J. Biochem.* **20**, 475.

Lee, E. Y. C., Carter, J. H., Nielsen, L. D., and Fischer, E. H. (1970). *Biochemistry* **9**, 2347.

Lenney, J. F. (1956). *J. Biol. Chem.* **221**, 919.
Lenney, J. F. (1973). *Fed. Proc., Fed. Amer. Soc. Exp. Biol.* **32**, 659.
Lenney, J. F. (1975). *J. Bacteriol.*, in press.
Lenney, J. F., and Dalbec, J. M. (1967). *Arch. Biochem. Biophys.* **120**, 42.
Lenney, J. F., and Dalbec, J. M. (1969). *Arch. Biochem. Biophys.* **129**, 407.
Lenney, J. F., Matile, P., Wiemken, A., Schellenberg, M., and Meyer, J. (1974). *Biochem. Biophys. Res. Commun.* **60**, 1378.
Liebe, S., Kopperschläger, G., Diezel, W., Nissler, K., Wolff, J., and Hofmann, E. (1970). *FEBS (Fed. Eur. Biochem. Soc.) Lett.* **8**, 20.
Lin, Y., Means, G. E., and Feeney, R. E. (1969). *J. Biol. Chem.* **244**, 789.
Lindell, T. J., and Stellwagen, E. (1968). *J. Biol. Chem.* **243**, 907.
Linn, T. G., Greenleaf, A. L., Shorenstein, R. G., and Losick, R. (1973). *Proc. Nat. Acad. Sci. U.S.* **70**, 1865.
MacGregor, C. H., Schnaitman, C. A., Normansell, D. E., and Hodgins, M. G. (1974). *J. Biol. Chem.* **249**, 5321.
Maddox, I. S., and Hough, J. S. (1970). *Biochem. J.* **117**, 843.
Magee, S. C., Mawal, R., and Ebner, K. E. (1973). *J. Biol. Chem.* **248**, 7565.
Maier, K., Saheki, T., Matern, H., Betz, H., and Holzer, H. (1974). *Proc. Int. Symp. Yeasts, 4th, 1974*, Part 1, p. 27.
Maino, V. C., and Young, F. E. (1974). *J. Biol. Chem.* **249**, 5169.
Manney, T. R. (1968). *J. Bacteriol.* **96**, 403.
Mason, T. L., Poynton, R. O., Wharton, D. C., and Schatz, G. (1973). *J. Biol. Chem.* **248**, 1346.
Matern, H., Betz, H., and Holzer, H. (1974a). *Biochem. Biophys. Res. Commun.* **60**, 1051.
Matern, H., Hasilik, A., Betz, H., and Holzer, H. (1974b). *Proc. Int. Symp. Yeasts, 4th, 1974*, Part 1, p. 29.
Matern, H., Hoffmann, M., and Holzer, H. (1974c). *Proc. Nat. Acad. Sci. U.S.*, in press.
Matile, P. (1969). *In* "Yeasts: The Proceedings of the 2nd Symposium on Yeasts, Bratislava, 1966" (A. Kocková-Kratochvílová, ed.), p. 503. Slovak Acad. Sci., Bratislava.
Matile, P., and Wiemken, A. (1967). *Arch. Mikrobiol.* **56**, 148.
Matile, P., Wiemken, A., and Guyer, W. (1971). *Planta* **96**, 43.
Molano, J., and Gancedo, C. (1974). *Eur. J. Biochem.* **44**, 213.
Neeff, J., Mecke, D., and Hasilik, A. (1974). In preparation.
Nelson, C. A. (1971). *J. Biol. Chem.* **246**, 3895.
Noltmann, E. A., Gubler, C. J., and Kuby, S. A. (1961). *J. Biol. Chem.* **236**, 1225.
Northrup, J. H., Kunitz, M., and Herriot, R. M. (1948). "Crystalline Enzymes," 2nd ed. Columbia Univ. Press, New York.
Oosterbaan, R. A., and Cohen, J. A. (1964). *In* "Structure and Activity of Enzymes" (T. W. Goodwin, J. I. Harris, and B. S. Hartley, eds.), p. 87. Academic Press, New York.
Penhoet, E., Kochman, M., Valentine, R., and Rutter, W. J. (1967). *Biochemistry* **6**, 2940.
Penhoet, E., Kochman, M., and Rutter, W. J. (1969). *Biochemistry* **8**, 4391 and 4396.
Perlman, G. E. (1956). *Arch. Biochem. Biophys.* **65**, 210.
Piñon, R. (1970). *Biochemistry* **9**, 2839.
Platt, T., Miller, J. H., and Weber, K. (1970). *Nature (London)* **228**, 1154.
Potts, J. T. (1967). *In* "Methods in Enzymology" (C. H. W. Hirs, ed.), Vol. 11, p. 648. Academic Press, New York.
Pringle, J. R. (1970a). *Biochem. Biophys. Res. Commun.* **39**, 46.
Pringle, J. R. (1970b). Ph.D. thesis, Department of Biology, Harvard University, Cambridge, Massachusetts.
Prouty, W. F., and Goldberg, A. L. (1972). *J. Biol. Chem.* **247**, 3341.

Ramel, A. H. (1964). Habilitation Thesis, University of Basel, Basel, Switzerland (quoted by Lazarus *et al.*, 1966).

Ramel, A. H., Stellwagen, E., and Schachman, H. K. (1961). *Fed. Proc., Fed. Amer. Soc. Exp. Biol.* **20**, 387.

Ramel, A. H., Rustum, Y. M., Jones, J. G., and Barnard, E. A. (1971). *Biochemistry* **10**, 3499.

Roffman, S., Sanocka, U., and Troll, W. (1970). *Anal. Biochem.* **36**, 11.

Roth, J. S., Losty, T., and Wierbicki, E. (1971). *Anal. Biochem.* **42**, 214.

Rouget, P., and Chapeville, F. (1971). *Eur. J. Biochem.* **23**, 459.

Rustum, Y. M., Massaro, E. J., and Barnard, E. A. (1970). *Fed. Proc., Fed. Amer. Soc. Exp. Biol.* **29**, 334 (abstr.).

Rustum, Y. M., Massaro, E. J., and Barnard, E. A. (1971). *Biochemistry* **10**, 3509.

Saheki, T., and Holzer, H. (1974a). *Eur. J. Biochem.* **42**, 621.

Saheki, T., and Holzer, H. (1974b). *Biochim. Biophys. Acta*, in press.

Saheki, T., Matsuda, Y., and Holzer, H. (1974). *Eur. J. Biochem.* **47**, 325.

Sasaki, R., Sugimoto, E., and Chiba, H. (1966). *Arch. Biochem. Biophys.* **115**, 53.

Schmidt, J. J., and Colowick, S. P. (1970). *Fed. Proc., Fed. Amer. Soc. Exp. Biol.* **29**, 334 (abstr.).

Schmidt, J. J., and Colowick, S. P. (1973a). *Arch. Biochem. Biophys.* **158**, 458.

Schmidt, J. J., and Colowick, S. P. (1973b). *Arch. Biochem. Biophys.* **158**, 471.

Schött, E. H., and Holzer, H. (1974). *Eur. J. Biochem.* **42**, 61.

Schubert, D., and Cohn, M. (1970). *J. Mol. Biol.* **53**, 305.

Schulze, I. T., and Colowick, S. P. (1969). *J. Biol. Chem.* **244**, 2306.

Schulze, I. T., Gazith, J., and Gooding, R. H. (1966). *In* "Methods in Enzymology" (W. A. Wood, ed.), Vol. 9, p. 376. Academic Press, New York.

Setlow, P., Brutlag, D., and Kornberg, A. (1972). *J. Biol. Chem.* **247**, 224.

Shaffer, B., Edelstein, S., and Fink, G. R. (1972). *Brookhaven Symp. Biol.* **23**, 250.

Shaw, E., and Ruscica, J. (1971). *Arch. Biochem. Biophys.* **145**, 484.

Stein, E. A., and Fischer, E. H. (1958). *J. Biol. Chem.* **232**, 867.

Stellwagen, R. H., Reid, B. R., and Cole, R. D. (1968). *Biochim. Biophys. Acta* **155**, 581.

Tjian, R., and Losick, R. (1974). *Proc. Nat. Acad. Sci. U.S.* **71**, 2872.

Trayser, K. A., and Colowick, S. P. (1961). *Arch. Biochem. Biophys.* **94**, 177.

Tsai, H., Tsai, J. H. J., and Yu, P. H. (1973). *Eur. J. Biochem.* **40**, 225.

Ulane, R. E., and Cabib, E. (1974). *J. Biol. Chem.* **249**, 3418.

Ulm, E. H., Böhme, R., and Kohlhaw, G. (1972). *J. Bacteriol.* **110**, 1118.

Umezawa, H., Aoyagi, T., Morishima, H., Matsuzaki, M., Hamada, M., and Takeuchi, T. (1970a). *J. Antibiot., Ser. A.* **23**, 259.

Umezawa, H., Aoyagi, T., Morishima, H., Kunimoto, S., Matsuzuki, M., Hamada, M., and Takeuchi, T. (1970b). *J. Antibiot., Ser. A.* **23**, 425.

Vallee, B. L. (1960). *In* "The Enzymes" (P. D. Boyer, H. Lardy, and K. Myrbäck, eds.), 2nd ed., Vol. 3, Part B, p. 225. Academic Press, New York.

van Rijn, J., and van Wijk, R. (1972). *J. Bacteriol.* **110**, 477.

Viswanatha, T., and Liener, I. E. (1955). *J. Biol. Chem.* **215**, 777.

Wais, U., Gillmann, U., and Ullrich, J. (1973). *Hoppe-Seyler's Z. Physiol. Chem.* **354**, 1378.

Weber, K., Pringle, J. R., and Osborn, M. (1972). *In* "Methods in Enzymology" (C. H. W. Hirs and S. N. Timasheff, eds.), Vol. 26, Part C, p. 3. Academic Press, New York.

Weiner, A. M., Platt, T., and Weber, K. (1972). *J. Biol. Chem.* **247**, 3242.

Whitaker, J. R., and Perez-Villaseñor, J. (1968). *Arch. Biochem. Biophys.* **124**, 70.

Wilgus, H., Pringle, J. R., and Stellwagen, E. (1971). *Biochem. Biophys. Res. Commun.* **44**, 89.

Witt, I., Kronau, R., and Holzer, H. (1966). *Biochim. Biophys. Acta* **118**, 522.

Womack, F. C., Welch, M. K., Nielsen, J., and Colowick, S. P. (1973). *Arch. Biochem. Biophys.* **158**, 451.

Chapter 10

Induction of Haploid Glycoprotein Mating Factors in Diploid Yeasts

MARJORIE CRANDALL[1] AND JOAN H. CAULTON[2]

Department of Microbiology,
Indiana University,
Bloomington, Indiana

[1] *Present address:* T. H. Morgan School of Biological Sciences, University of Kentucky, Lexington, Kentucky 40509.
[2] *Present address:* Department of Zoology, Indiana University, Bloomington, Indiana 47401.

I.   Biology of the Yeast *Hansenula wingei*

A.   Life Cycle

Figure 1 illustrates the steps in the alternation of generations of *H. wingei*, a heterothallic yeast. The two opposite mating types are haploid strains 5 and 21 (Wickerham, 1956). Both haploids are constitutively agglutinative and, while cell suspensions of each are stable separately, they agglutinate immediately when mixed together. Following agglutination, the cells conjugate pairwise, producing a 5 × 21 diploid hybrid which is nonagglutinative and a nonmater (Crandall and Brock, 1968a). The diploid can be either grown vegetatively or sporulated if transferred to special media. The sporulation process results in a tetrad of meiotic products: two spores of mating type 5 and two spores of mating type 21, indicating that these mating types are allelic. Agglutination type and mating type segregate

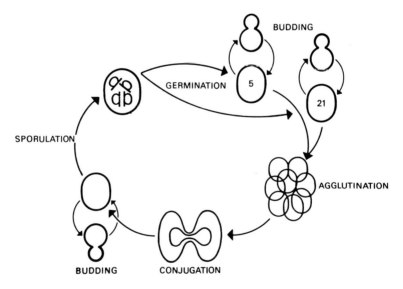

FIG. 1.   Life cycle of *H. wingei*. Conditions and media for growth, agglutination, conjugation, and sporulation are given in Section II.

together (Herman *et al.*, 1966). Therefore the mating-type locus and the agglutination-type locus are either identical or closely linked. The haploids can be grown vegetatively, as can the diploid, so that effects of gene dosage and dominance can be studied. All the morphogenetic steps in the life cycle can be controlled by environmental conditions (see Section II).

In this chapter methods are presented that alter one step of the normal life cycle—the mating response of the diploid. We have discovered conditions that induce the diploid to become sexually agglutinative and therefore able to participate in the first step in conjugation—cellular recognition.

B. Agglutination

The ability of the haploids to recognize the complementary mating type is attributable to the presence of complementary glycoproteins on the respective cell surfaces. The glycoprotein from strain 5 is called 5-factor (5f), and the glycoprotein from strain 21 is called 21-factor (21f). The activities and properties of 5f and 21f are summarized in Table I. Purified 5f and 21f form a neutralized complex which is soluble, because 21f is univalent and cannot form cross-links as in antibody reactions. The 5f–21f complex can be assayed by recovery of 5f agglutination activity after treatment of

TABLE I

COMPARISON OF THE PROPERTIES OF HAPLOID GLYCOPROTEIN MATING FACTORS[a]

Property	5f	21f
Activity	Agglutinates cells of strain 21 and forms a soluble complex with 21f *in vitro*	Inhibits the agglutination activity of 5f by forming a neutralized 5f–21f complex
Number of combining sites	Six or more	One
Isolation	Cytoplasmic extracts or subtilisin digestion of whole cells	Cytoplasmic extracts or trypsin digestion of whole cells
Molecular size	Heterogeneous—the sedimentation coefficient varies with the preparation. Values reported: $S_{20,W} = 3.5, 6.5, 9.0, 15.4, 16.7, 31$	Homogeneous, $S_{20,w} = 2.9$
Molecular weight	Varies from 15,000 to 10^8 daltons	About 40,000 daltons
Chemical composition	Mannan–protein	Mannan–protein
Carbohydrate content	50–96%	25–35%
Inactivated by	Reducing agents	Heat, alkali, and other protein denaturants

[a] These data are summarized from results and literature presented in Crandall *et al.* (1974), Taylor and Orton (1971), and Yen and Ballou (1974).

the complex with alkali, which destroys 21f activity (Crandall, *et al.*, 1974). The neutralized complex, when purified and assayed in this manner, was found to be large and heterogeneous, reflecting the molecular-weight heterogeneity of the 5f. This same assay was used to look for 5f and 21f in diploid extracts, and neither was found (see Section III,B).

Certain yeastlike fungi have recently been found to produce large numbers of external hairs or filaments similar to the fimbriae or pili found in gram-negative bacteria. These filaments are about 0.007 μm in diameter and vary in length from about 0.5 μm in the ascomycete *Saccharomyces cerevisiae* to over 10 μm in some species of the basidiomycete genus *Ustilago*. They are easily observed under the electron microscope after the cells have been thoroughly washed and shadowed with tungsten. The *Ustilago* filaments are composed of protein, but they may have carbohydrate molecules attached as part of a secondary structure. The filaments of *Ustilago violacea* have been demonstrated to be involved in conjugation, and those of *S. cerevisiae* to be important for flocculation (A. W. Day, N. H. Poon and G. G. Stewart, 1975). Similar filaments have been observed in strains 5 and 21 of the ascomycete *H. wingei* (Fig. 2a,b). In view of the results with *Saccharomyces* and *Ustilago*, it is possible that the filaments of *Hansenula* cells may be involved in sexual agglutination or conjugation (N. H. Poon and A. W. Day, Department of Plant Sciences, University of Western Ontario, London, Ontario, Canada, personal communication). Filaments on the cell surface of *H. wingei* have also been observed by K. Aufderheide (Department of Zoology, Indiana University, Bloomington, Indiana; personal communication). Cells were prepared for electron microscopy using procedures similar to those of Conti and Brock (1965), but with minimal staining to increase visualization of the fibrous coat. The thickness of this coat on sexually agglutinative strains of *H. wingei* is approximately 0.2 μm. No filaments were observed on nonagglutinative strains of *S. cerevisiae*. When mating types 5 and 21 of *H. wingei* were mixed and allowed to agglutinate, only the outer fibrous coats were seen to be in contact, not the other cell wall layers. Furthermore, only the distal portions of the fibers appeared to be in contact, producing a region of greater electron density between the cell walls, perhaps as the result of an overlap of fibers. It is suggested by K. Aufderheide that the agglutination factors from strains 5 and 21 may be associated with these fibers and may possibly be located on their distal portions.

C.   Conjugation

Mating in *H. wingei* occurs best with stationary-phase cells resuspended in glucose-containing buffer lacking a nitrogen source (Brock, 1961; Cran-

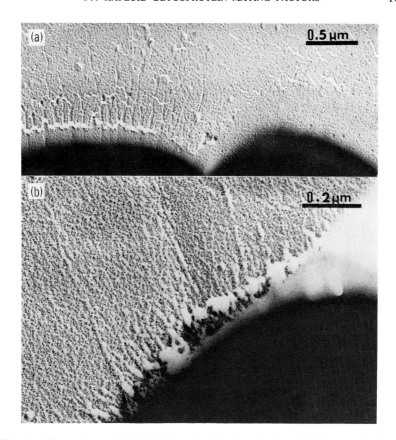

FIG. 2. Filaments associated with the cell walls of strains 5 and 21. Tungsten-shadowed cells of *H. wingei* strain 5–9A (a) and strain 21–9D (b) show evidence of short surface hairs. Cells were washed four times with particle-free distilled water, once with ether, and twice more with water before air-drying on grids and shadowing at an angle of 18°–22°. Similar filaments can be observed but less clearly, on cells washed only with water. (Courtesy of N. H. Poon and A. W. Day, Department of Plant Sciences, University of Western Ontario, London, Ontario, Canada.)

dall and Brock, 1968a). Typically, 80% conjugation occurs in a nonsynchronized population. Mating could presumably be synchronized and increased to 95% or better if competent cells were selected on sucrose gradients, as has been reported for *S. cerevisiae* (Sena *et al.*, 1973; Sena *et al.*, 1975). Although experiments have been performed in *H. wingei* to detect diffusible sex hormones as have been reported in *S. cerevisiae* (Duntze *et al.*, 1973), no evidence for such hormones has been obtained yet. During aeration of a mixed suspension of strains 5 and 21, agglutination occurs immediately and increases to a maximum after 1 hour, at which time the first zygotes are observed. The diploid zygotes remain clumped, indicating that the cell wall mating factors are not destroyed during conjugation. It is not until the newly formed zygote has undergone several generations of growth in fresh medium that the diploid culture becomes nonagglutinative.

D. Sporulation

Meiosis is induced in *Hansenula* by growth on malt extract (Wickerham, 1951) and *not* by growth on potassium acetate, which induces sporulation in *Saccharomyces* (Fowell, 1969). Strain 5 × 21 hybrids and diploid Y2340 sporulate very poorly (< 1%), whereas VIA × VIB hybrids and diploid YB-4662 sporulate well (∼50%). Crosses between good sporulators and poor sporulators (5 × VIA or 21 × VIB) yield diploids that sporulate weakly (∼1–10% with many three-spored asci), but well enough to perform a tetrad analysis. From such crosses recombinant ascosporic isolates could conceivably be selected that combine both good agglutinability with good sporulation ability for future genetic studies of the mating-type locus. Tetrad dissection in this yeast is hampered by the fact that the spores stick together tightly even after they are released from the ascus. The ascospores have a brim of cell wall material and very often stick to sister ascospores brim to brim. Thus extensive reduction and enzymatic digestion are required before dissection is performed (Crandall and Richter, 1973). The spores germinate well, and cultures are tested for mating type simply by determining whether they agglutinate with strain 5 or 21.

II. Methods

A. Media

Complex media are autoclaved for 15 minutes at 121°C, but synthetic media are autoclaved only 10 minutes. For slants and petri plates, 2% Bacto agar is added unless stated otherwise. The following were purchased from Difco Laboratories, Detroit, Michigan: yeast extract (YE), Bacto agar, yeast nitrogen base without amino acids (YNB), malt extract (ME), and Bacto peptone. The compositions of media and their use are given in Table II.

B. Standard Growth Conditions

Since *Hansenula* is an obligate aerobe (Wickerham, 1970), liquid cultures are grown in Erlenmeyer flasks filled to one-fifth capacity and aerated on

TABLE II

MEDIA COMPOSITION AND USE[a]

Use	Abbreviation	Composition
Growth, complex	YG	0.7% yeast extract (YE) plus 3% glucose (G) plus 0.5% KH₂PO₄
	YM	0.3% YE plus 0.3% malt extract (ME) plus 0.5% Bacto peptone plus 1% glucose
Growth, minimal	3% G-YNB	3% glucose plus 0.67% yeast nitrogen base (YNB)
	0.5% G-YNB	0.5% glucose plus 0.67% YNB
	0.5% G-SM 1 × vitamins	0.5% glucose plus synthetic medium of the same same composition as YNB
	0.5% G-SM 0.15 × vitamins	Same as above, but with all nine vitamins reduced to 0.15 × the concentration present in YNB
Conjugation	PMG	10 mM KH₂PO₄ (pH 5.5) plus 0.1% MgSO₄ · 7H₂O plus 0.5% glucose
	RGM	Restricted growth medium; 0.02% YE plus 0.02% Bacto peptone plus 0.1% glucose
Sporulation	ME	2.5% ME plus 3% agar
Induction	PG	10 mM KH₂PO₄ (pH 4.9 unadjusted) plus 0.2% glucose; in some experiments 1 mM MgSO₄ was added.

[a]Compositions taken from Crandall and Brock (1968a) and Crandall and Caulton (1973).

a rotary shaker at 200 rpm at 30°C unless stated otherwise. Stocks are grown on YG slants (see Table II) overnight and then refrigerated.

C. Yeast Strains

The origins of *H. wingei* strains are given in Table III.

D. Conjugation

Stationary-phase cultures of strains 5 and 21 (3–5 × 10⁸ cells/ml) are washed twice in water or PMG (see Table II) and resuspended in PMG at about 10⁸ cells/ml. Equal numbers of cells (as determined by either cell counts or optical density measurements) are mixed in an Erlenmeyer flask (about 2.5 ml of each suspension per 50-ml flask) and aerated at 200 rpm at 30°C for 5 hours. To break up the clumps, a representative aliquot (0.1 ml) from the suspension is added to 1 ml of 8 M urea and autoclaved. To determine the percentage conjugation, a differential count is performed in which all the single cells and zygotes in each field are counted. The percentage

TABLE III

ORIGINS OF *H. wingei* STRAINS[a]

Strain number	Ploidy	Description
Y2340	Diploid	Isolated from frass of bark beetles living in conifer trees in the western United States; poor sporulation; gift from L. J. Wickerham to T. D. Brock
5–9A and 21–9D	Haploid	Ascosporic isolates of opposite mating type from one tetrad of Y2340; good agglutination; gift from A. I. Herman
VIA and VIB	Haploid	Ascosporic isolates of opposite mating type from one tetrad of YB-4662 (a good sporulating diploid isolated in the eastern United States); VIA and VIB agglutination is weaker than 5 and 21; gift from A. I. Herman
5 cyh lys	Haploid	Spontaneous mutation of strain 5–9A to cyclo- heximide resistance (100 μg/ml) plus a nitrous acid– induced mutation to lysine auxotrophy (requires 200 μg/ml)
21 ade his	Haploid	Leaky adenine auxotrophic mutation derived from strain 21-9D (gift from A. I. Herman) plus an ultra- violet-induced mutation to histidine auxotrophy
44	Diploid	Prototrophic hybrid of strains 5 cyh lys and 21 ade his

[a]Information summarized from Crandall (1973).

conjugation is equal to two times the total number of conjugants times 100, divided by the total number of single cells, plus two times the total number of conjugants (Brock, 1961). If viable counts are to be made, repeated water washes and sonication will eventually remove the counterions required for agglutination, and a suspension of single unmated cells and zygotes may be obtained (Crandall and Richter, 1973). Addition of disodium ethylene- diaminetetraacetic acid (disodium EDTA) to the washes is not useful, since even it acts as a counterion and promotes agglutination. Declumped zygotes can then be purified on sucrose gradients to obtain a synchronous diploid population of zero generation, since no growth can occur in the nitrogen- free conjugation buffer. Such a population would be useful in the analysis of mitochondrial recombination.

E. Sporulation

For sporulation, the two haploids are pregrown separately in YM broth (see Table II) overnight, and then mixed together on RGM agar (Herman, 1971; see Table II) where they conjugate. The hybrid is allowed to grow for 2 days, and then a loopful of cells is streak-purified on ME agar (see Table II) and incubated for 7–9 days at 25°C (sporulation is decreased at 30°C). If the diploid is already available, it is sporulated in the same manner. The best sporulation is observed with crowded colonies on ME agar. Ripe asci appear about the same time as pseudohyphae develop. The asci rupture spontaneously, the tetrads remain intact and may be separated by micromanipulation. Enzymatic digestion at this point facilitates tetrad dissection. Techniques for a preliminary genetic analysis in *H. wingei* were described in detail in Crandall and Richter (1973). However, these techniques are presently being revised in a large scale effort to map the chromosomes in *H. wingei* in the laboratory of M. Crandall.

F. Preparation of Haploid Tester Cells

Heating haploid cells causes them to become more agglutinative, because a nonspecific inhibitor is removed from the cell surface by this treatment (Crandall and Brock, 1968a). To prepare tester cells, cultures of strains 5 and 21 are grown to stationary phase in YG, harvested, washed once in distilled water, and resuspended in water at a 2 × cell concentration (one-half the volume of culture). Suspensions are steamed in 200-ml aliquots in 250-ml polypropylene centrifuge bottles for 30 minutes and then cooled quickly by swirling in an ice-water bath. Cells are harvested immediately and washed twice more in water. The cells are resuspended at 40 × (by adding 10 ml of water to the pellet from 200 ml of 2 × cells) and stored in the refrigerator after the addition of a few drops of chloroform.

G. Agglutination Assay of Induced Diploid Cells

Under special conditions the diploids can be induced to synthesize either 5f or 21f. Diploid cells are assayed for agglutination with each haploid tester at time intervals during induction by harvesting 1 ml (5 ml from exponential cultures), resuspending the pellet in 0.1 ml of 0.88% saline on a wrist-action shaker, and scoring for self-agglutination by visual observation. To interpret results of diploid agglutination with the haploid testers correctly, it is important to note first whether the diploids are clumpy. When self-agglutination is present, it is usually weak, and very small clumps are seen to climb up the walls of the test tube. After scoring for self-agglutination, one-half the suspension is pipetted into a second test tube. To one set of tubes is added 0.1 ml of heated strain-5 tester (diluted 1:10 in saline from the 40 × stock), and to the other set of tubes is added 0.1 ml of heated strain-21 tester (diluted similarly). The tubes are shaken on the wrist-action shaker for 5 minutes and scored for agglutination. Scoring is from no agglutination (−) to complete agglutination (+ + + +). A score of ± indicates weak agglutination. A score of + indicates positive agglutination with small clumps. Scores of + +, + + +, and + + + + represent increasingly strong agglutination, as judged visually by an increasing size of clumps and a decreasing number of nonagglutinated cells remaining in suspension.

An attempt was made to relate visual scoring to a physical measurement. A standard curve was constructed using mixtures giving varying strengths of agglutination, starting with equal cell numbers of the two types being mixed (Crandall and Caulton, 1973). However, this curve was never used routinely, because of the difficulty in adjusting the cell density of each culture to be equal to the tester for each assay.

III. Control of Haploid Mating Functions in the Diploid

A. The Normal, Nonagglutinative Diploid

In the 5 × 21 hybrid, the genes for sporulation are functional and the genes for conjugation are nonfunctional. Apparently, the control mechanism that turns sporulation genes on and conjugation genes off is heterozygosity at the mating-type locus. Since this locus controls so many functions, it must be a complex locus containing some structural and some regulatory elements. The question was asked: Do the regulatory genes code for repressors of the synthesis of the mating-type factors, or are the complementary factors both synthesized but present as the neutralized 5f–21f complex in the diploid (Crandall and Brock, 1968b)? Evidence that neither 5f nor 21f is present in the diploid is given in the following discussion.

B. Absence of Glycoprotein Mating Factors in the Uninduced Diploid

1. Studies of Diploid Cells

When whole diploid cells were treated with reagents that specifically inactivated one or the other haploid agglutination, no recovery of the complementary agglutination was found, i.e., the nonagglutinative diploid could not be converted to agglutinability with either tester after treatment with trypsin which destroys the agglutinability of strain 21, or β-mercaptoethanol which destroys the agglutinability of strain 5 (Crandall and Brock, 1968b).

2. STUDIES OF DIPLOID CYTOPLASMIC EXTRACTS

Since both 5f and 21f are present in the cytoplasm of the respective haploid strains, cell-free extracts of the diploid were prepared and assayed for both 5f (by agglutination of heated 21 cells) and 21f (by inhibition of added 5f activity; see Table I). Neither 5f nor 21f was detected. Because it was conceivable that the neutralized 5f–21f complex might be present, alkali was added to the diploid extract to dissociate the complex, inactivate the 21f, and allow recovery of 5f activity. No 5f activity was recovered, indicating that neither factor is synthesized in the normal, nonagglutinative diploid (Crandall and Brock, 1968b).

Although the diploid does not normally synthesize either mating factor, under certain conditions it can be induced to synthesize either 5f (see Section IV) or 21f (see Section V).

IV.   Induction of 5f in the Diploid

A.   Under Growing Conditions

1.   YEAST EXTRACT AS 5f INDUCER

Induction of 5f synthesis occurs when the diploid is grown to late stationary phase in YG broth (see Table II). This is called the diploid-to-5 or D → 5 transition, because the diploid cells now behave as strain 5 in that they agglutinate with strain 21 and 5f is present in cytoplasmic extracts. The D → 5 transition is a physiological change, *not* a genetic change. There were no ascospores present in the D → 5 culture, and all the subclones derived from this culture were nonagglutinative (Crandall and Brock, 1968b).

2.   VANADATE PLUS MOLYBDATE AS 5f INDUCERS

Since the D → 5 transition does not occur in minimal medium, but does occur in minimal medium containing YE, it was suggested (E. D. Weinberg, personal communication) that metal ions present in YE might be responsible for this induction. After comparing the concentrations of various metal ions in the minimal medium (3% G-YNB; see Table II) to the estimated concentration of metal ions contributed by YE, it was determined that the metal ions present at higher concentrations in 3% G-YNB + 0.63% YE are copper, iron, vanadium, and zinc. A mixture of 0.4 mM of each of these added to 3% G-YNB induced the D → 5 transition. When vanadium was left out, no induction occurred. When vanadium was tested alone in 3% G-YNB, the D → 5 transition occurred on the second day of growth, just as with the

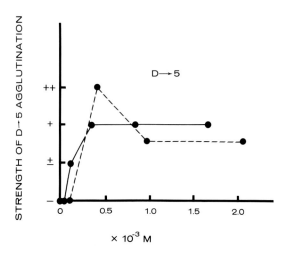

Fig. 3. Optimal concentration of vanadium-molybdenum for the D → 5 transition. To 25 ml of 3% G-YNB in a 125-ml Erlenmeyer flask was added 0.25 ml of a stationary-phase preculture of diploid 44 plus various amounts of NaVO₃ (solid line). To a duplicate series of flasks was also added various amounts of Na₂MoO₄ (broken line). Agglutination with strain 5 and 21 testers was assayed after 48 hours of aeration at 200 rpm at 30°C. [Reproduced from Crandall and Caulton (1973) by permission of Academic Press, Inc.]

YG culture (Fig. 3). Vanadyl sulfate is just as active in induction as sodium metavanadate. In some experiments, maximal induction occurred with both 0.4 mM NaVO₃ and 0.4 mM Na₂MoO₄ but routinely only vanadium is used in induction experiments.

A recent study of the effect of aeration on D → 5 induction by vanadium has shown that increasing the degree of aeration (from 50 ml of culture/ 125-ml flask to 5 ml/flask) serves to decrease the degree of agglutination induction. This result is contrary to expectation for an obligate aerobe but is in agreement with the hypothesis (see Section VI,C) that growth-limiting or stress conditions perturb the normal regulatory mechanisms of the diploid cell resulting in induction of the mating factors.

3. 5f Induction by Growth in Calcium- and Magnesium-Free Medium

Since metal ions such as vanadium had been shown to play a role in 5f synthesis in the diploid, further experiments were performed with synthetic medium prepared without added metal ions. It was found that 5f was induced when the diploid was grown in 0.5% G-SM (see Table II) lacking both calcium and magnesium. This experiment was performed by growing the diploid to stationary phase in 0.5% G-SM and then inoculating 1:100

into calcium- and magnesium-free 0.5% G-SM. The 5f appeared in the culture when growth slowed as a result of calcium and magnesium exhaustion. Maximal D → 5 agglutination occurred as the culture entered stationary phase. Adenine, guanine, cytosine, thymine, and uracil inhibited the D → 5 transition when added together to the calcium- and magnesium-free medium. When the diploid inoculum was washed first and then grown in media with varying concentrations of either calcium or magnesium, it was found that small amounts of both calcium and magnesium (equivalent to 0.01 × the concentration in 0.5% G-SM) are required for D → 5 induction.

Thus low concentrations of calcium and magnesium (0.009 mM CaCl$_2$ and 0.042 mM MgSO$_4$, respectively) induced 5f synthesis, high concentrations (0.9 mM CaCl$_2$ and 4.2 mM MgSO$_4$) inhibit 5f synthesis, but this inhibition may be overcome by adding either 0.4 mM NaVO$_3$ or 0.4 mM VOSO$_4$ or 0.4 mM (V + Na$_2$MoO$_4$). The D → 5 inducer may be the vanadyl ion (VO^{2+}) which has the same charge as Mg^{2+} and Ca^{2+}.

B.   Under Nongrowing Conditions

1.   INFLUENCE OF VITAMINS ON 5f SYNTHESIS

a. *Discovery of pH and Vitamin Effects.*   In a series of experiments designed to detect sex hormones in *H. wingei*, the nonagglutinative diploid was used as the inducible system, since the haploids are constitutively agglutinative. Stationary-phase diploid cells grown in 0.5% G-YNB were harvested and resuspended in an equal volume of culture filtrate from cultures of strains 5, 21, and 5 × 21 of *H. wingei* and strains *a* and *α* of *S. cerevisiae*. The pH was raised from 2.5 to 5.5 with K$_3$PO$_4$, and the suspensions were aerated. All the culture filtrates caused the 5 × 21 hybrid to become agglutinative with strain 21. This D → 5-inducing activity, while being specific in inducing only the D → 5 and not the D → 21 transition, could not be called a sex hormone, since it was not sex-specific with respect to strain of origin.

When the concentration of D → 5 inducer in diploid cultures was measured as a function of growth stage (Fig. 4), it was seen that there was a peak of D → 5-inducing activity in exponential phase, but even at zero time there was some induction, indicating that the inducer might be a component of the synthetic medium—such as a vitamin. Growth in this medium must then modify it, causing first an increase in the inducer activity in exponential phase, and then a decrease in stationary phase. This modification might be the consumption of some vitamins that inhibit the D → 5 transition. To test this idea stationary-phase diploid cells were resuspended in PG buffer (see Table II) containing 1 mM MgSO$_4$, and different vitamins were added

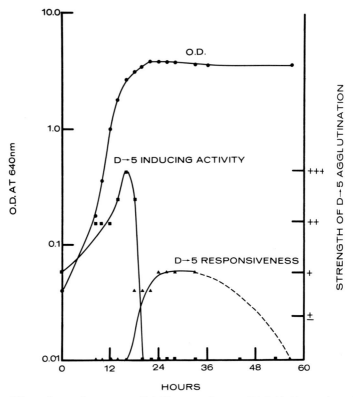

FIG. 4. Effect of growth stage on diploid responsiveness. Diploid 44 was inoculated 1:25, grown to stationary phase, and inoculated 1:100 into a 3-year-old batch of 0.5% G-YNB which was later found to be vitamin-limiting. The culture was aerated, and aliquots were taken at intervals to measure optical density, responsiveness of cells, and inducing activity of the medium. The optical density at 640 nm was measured in a Bausch and Lomb Spectronic 20. Cells were tested for responsiveness by harvesting and resuspending at the same cell density in culture filtrate from an exponential diploid culture. The suspension was adjusted to pH 5 with K_3PO_4, aerated for 4 hours, and assayed for agglutination with strain 5 and 21 testers. The culture filtrate was sampled at different times and assayed for D → 5-inducing activity by resuspending responsive cells from a 24-hour diploid culture at the same cell density in the culture filtrate, adjusting the pH to 5, and assaying as above.

at different concentrations and in different combinations. The suspensions were adjusted to pH 5, and the flasks were aerated for 4 hours. The control for this experiment resulted in weak ($\pm \rightarrow +$) D → 5 agglutination without added vitamins. Thus simply adjusting the pH of resuspended diploid cells is sufficient to trigger 5f synthesis. However, certain vitamins enhanced this synthesis, whereas others inhibited it. Pyridoxine, one of the three required vitamins (the others are biotin and thiamine), appeared to be the most active in enhancing 5f synthesis, and therefore was used in all later experiments. Since the D → 5 agglutination induced by resuspending cells in pyridoxine-containing buffer is weaker (only + +) than that induced by resuspending cells in culture filtrates (+ + +), as yet unidentified cellular products in cultures may also act as inducers. It should be noted that the same peak of D → 5-inducing activity observed in Fig. 4 is also present in cultures grown in the presence of vanadium-molybdenum. At the time the cells in a vitamin-limited culture become responsive (24–36 hours), the cells in a vanadium–molybdenum culture start to become agglutinative with strain 21 (see Fig. 4 in Crandall and Caulton, 1973).

A second vitamin effect was also discovered in such a serendipitous fashion. When a fresh lot of YNB was used, the diploid was no longer responsive to induction after 24 hours of growth. It was thought that perhaps the fresher batch had a higher level of vitamin activity. To test this idea synthetic medium of the same composition as 0.5% G-YNB was prepared with different vitamin concentrations, and in fact it was found that vitamin limitation during growth results in diploid responsiveness to induction (see the following discussion).

b. *Effect of Vitamin Limitation on Diploid Responsiveness during Growth.* Stationary-phase diploid cells grown in synthetic medium with less than the concentration of vitamins present in YNB ($< 1 \times$ vitamin concentration) became responsive to D → 5 induction in buffer when the vitamin concentration became growth-limiting, i.e., when the terminal optical density of a 24-hour culture was lower (Fig. 5). Thus normally the diploid remains nonagglutinative at all growth stages, because vitamins are in excess in standard media and therefore the cells never become responsive. Even if the cells were responsive, the pH of spent medium is too low for induction by the vitamins present (see the following discussion).

c. *Optimal pH for D → 5 Induction (Fig. 6.).* There is a broad pH optimum for D → 5 induction in pyridoxine-containing buffer; pH values between 5.5 and 6.0 are the most effective.

d. *Pyridoxine Concentration for Maximal Induction.* Maximal induction occurs at very low pyridoxine concentrations (0.04 μg/ml). (The concentration of pyridoxine in growth medium is 0.4 μg/ml.)

e. *Effect of an Energy Source on Induction.* Addition of 0.1% glucose to the phosphate buffer increased the degree of induction to a maximum.

f. *Standard Conditions for D → 5 Induction in Buffer Suspensions.* A preculture of diploid 44 is inoculated 1:25 into 0.5% G-YNB ($1 \times$ vitamin concentration), grown for 24 hours, and diluted 1:100 into 0.5% G-SM containing vitamins at 0.15 $\times$ concentration; the vitamin-limited culture is grown for 24 hours. (Note: When new solutions of all the vitamins were prepared and the experiment in Fig. 5 was repeated, maximal responsive-

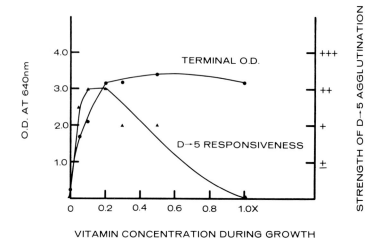

VITAMIN CONCENTRATION DURING GROWTH

FIG. 5. Effect of vitamin limitation during growth on diploid responsiveness. A preculture of diploid 44 was inoculated 1:25 into 0.5% G-YNB, grown to stationary phase, and then inoculated 1:100 into 0.5% G-SM containing different vitamin concentrations. After 24 hours of growth, cells were harvested and resuspended in PG buffer at pH 5.5 containing 4 μg pyridoxine/ml plus 1 mM MgSO$_4$. This suspension was aerated for 4 hours and then assayed for agglutination with strain 5 and 21 testers.

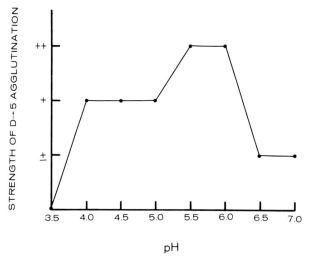

pH

FIG. 6. Optimal pH for D $\rightarrow$ 5 induction. A preculture of diploid 44 was inoculated 1:25 into 0.5% G-YNB, grown to stationary phase, and inoculated 1:100 into 0.5% G-SM containing vitamins at 0.15× concentration. Cells were grown for 24 hours, harvested, and resuspended in water at 2× cell concentration. Then 2.5 ml of this suspension was pipetted into 50-ml Erlenmeyer flasks to which had been added 2.5 ml of 2× PG buffer containing 8 μg pyridoxine/ml plus 2 mM MgSO$_4$ and previously adjusted to different pH values with either HCl or K$_3$PO$_4$. The unadjusted pH was 4.5. After 2 hours of aeration, the cells were assayed for agglutination with strain 5 and 21 testers.

ness occurred between 0.2 and 0.3 ×.) Cells from this vitamin-limited culture are harvested and resuspended in the same volume of PG buffer containing 1 mM MgSO$_4$ plus 4 μg pyridoxine per milliliter. The pH is increased to 5.5–6.0 by the addition of 100 mM K$_3$PO$_4$, and the suspensions are aerated at 30°C. Aliquots are withdrawn every hour or two to be assayed for agglutination with strain 5 and 21 testers. The peak of 5f induction in the diploid occurs usually between 2 and 4 hours. At this point the cells appear normal. There is no cell wall deformation after D → 5 induction, as reported for vanadium–molybdenum cultures (Crandall and Caulton, 1973).

   g. *Vitamin Specificity for Responsiveness and Induction.* If one of the three required vitamins (biotin, thiamine, and pyridoxine) is left out of the medium, growth is diminished and the cells are not responsive. If any or all of the other six vitamins present in YNB (folic acid, *p*-aminobenzoic acid, inositol, niacin, riboflavin, and pantothenate) are added at 0.25 × to medium containing biotin, thiamine, and pyridoxine at 0.25×, there is no enhancement of responsiveness. Therefore the same vitamins required for growth are also required for the production of diploid cells capable of synthesizing 5f. Similarly, pyridoxine is the most active in the induction of responsive cells resuspended in buffer.

V.   Induction of 21f in the Diploid

A.   Under Growing Conditions

1.   YE PLUS EDTA AS 21f INDUCERS

When it was found that the metal ions in YE induced 5f synthesis, it was predicted that the addition of a chelating agent would prevent the D → 5 transition, which it did. However, under these same growth conditions, the D → 21 transition occurred (Fig. 7). This experiment shows that, while metal ions are important for D → 5 induction, they inhibit D → 21 synthesis and, furthermore, that there is a substance in YE that induces the D → 21 transition, since EDTA added to minimal medium has no inductive effect (Crandall and Caulton, 1973). This D → 21 inducer has eluded all attempts to purify it from YE. It is stable to autoclaving, inactivated by alkali, and probably a cation because it is adsorbed to Dowex-50W but not to Dowex-1.

2.   TRACE ELEMENT LIMITATION AS 21f INDUCER

A variation of the above conditions allows D → 21 induction to occur in defined medium. Precultures of the diploid are grown in YG broth to stationary phase, harvested, washed, resuspended at the same cell con-

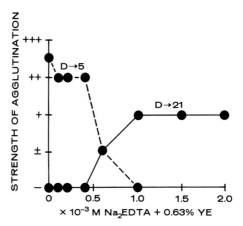

Fig. 7.   Induction of 21f in the diploid by adding EDTA to growth medium containing yeast extract. To 25 ml of 3% G-YNB was added 0.25 ml of a stationary-phase preculture of diploid 44, plus 2.5 ml of 7% YE, plus different amounts of disodium EDTA. Agglutination with strain 5 tester (D → 21) occurred after 24 hours (solid line). Agglutination with strain 21 tester (D → 5) occurred after 48 hours (broken line). [Reproduced from Crandall and Caulton (1973) by permission of Academic Press, Inc.]

centration in 0.5% G-SM lacking added trace elements, and diluted 1:100 into this same trace element–limited medium (TELM). D → 21 agglutination occurs maximally as the cells enter stationary phase. The addition of amino acids, excess vitamins, or bases to TELM inhibited D → 21 induction.

B.   Under Nongrowing Conditions

The D → 21 transition has occurred sporadically in buffer suspensions, but it is transitory and more experiments must be performed before the conditions can be reported.

VI.   Significance of Induction of Sex-Specific Cell Surface Glycoproteins in the Diploid

A.   Role of Mating Factors as the First Step in Conjugation

The mating factors on the cell surfaces of strains 5 and 21 promote conjugation by bringing the cells together into intimate contact. Solubilized 5f and 21f act out their same functional roles *in vitro* by forming a neutralized complex. The complex is soluble and stable in solution and may be purified using conventional methods in protein chemistry (Crandall *et al.*, 1974). Furthermore, 5f may be recovered from the complex by denaturing the 21f. Thus the two factors do not inactivate each other, nor do they appear to have any enzymatic activity (Crandall and Brock, 1968b). The next series of steps in conjugation (cell wall degradation and synthesis of the conjugation tube) may be induced as a consequence of the cell wall deformation observed during agglutination (Conti and Brock, 1965), or may be induced by, as yet unidentified, sex hormones. It would be interesting to determine whether the 5 × 21 hybrid has surface filaments like those observed on strains 5 and 21 either before or after induction of the mating factors in the diploid.

B. Formation and Genetic Analysis of Triploids

Since methods were found that induced one or the other agglutination type in the diploid, the question was asked: Is this one step sufficient to induce the next series of steps in conjugation? The answer is that the induced diploid does *not* mate with enhanced frequency compared to the frequency of mating of the uninduced control. Thus agglutination is necessary but not sufficient to trigger the conjugation process. Conjugation of induced and uninduced diploids was studied by mating an auxotrophic diploid with an auxotrophic haploid and then determining the prototrophic frequency in the mixture (Table IV). The same series of experiments presented in Table IV was repeated, except that matings were done on RGM agar instead of in PMG buffer. The results were the same in that induction of agglutination did not increase the mating frequency in the diploid, but the frequency of triploid formation in both the induced and uninduced cultures on RGM agar was higher (about 5×10^{-3}) and the frequency of revertants in the controls was lower (about 10^{-8} to 10^{-7}) than in the results presented in Table IV. Six prototrophs were isolated from the mating mixtures and all were found to be triploid as determined by the higher DNA content per cell. The total cellular DNA content in terms of micrograms of DNA per 10^8 cells plus or minus the standard deviation was: 2.75 ± 0.55 and 2.30 ± 0.71 for haploids 5 and 21; 4.46 ± 0.45 and 4.23 ± 0.13 for different diploids; 6.35 ± 0.67, 5.80 ± 0.00, 6.15 ± 0.89, 5.60 ± 0.30, 5.30 ± 1.04 and 7.96 ± 0.56 for different triploids. The haploid value for *H. wingei* reported here is the same as that reported for *S. cerevisiae* (2.4 μg DNA per 10^8 cells; Parry and Cox, 1972). Yet the chromosome number is 4 in *H. wingei* (Crandall and Robinow, 1973) and 17 in *S. cerevisiae* (Mortimer and Hawthorne, 1973).

The triploids were nonagglutinative and nonmaters, as expected, but they differed among themselves in their relative degree of 5f and 21f syn-

TABLE IV

No Effect of D → 5 Induction on Diploid Mating

Strain	Media in which cells were grown or induced	Prototrophic frequencies[c]	
		Before induction	After induction
Controls for determination of prototrophic reversion frequencies:			
21 ade his	3% G-YNB plus 0.63% YE	0.5, 1.9 × 10⁻⁶	—
Diploid 83	3% G-YNB plus lysine[a]	< 0.01, 0.14, 0.45, 1.0 × 10⁻⁷	—
	3% G-YNB plus lysine plus 0.4 mM (V + Mo)	1.1 × 10⁻⁶ 6.0 × 10⁻⁶ 2.3 × 10⁻⁶	17.0 × 10⁻⁶ 0.1 × 10⁻⁶ 1.2 × 10⁻⁶
	3% G-YNB plus 0.63% YE	18.0 × 10⁻⁶ 0.4 × 10⁻⁶	37.0 × 10⁻⁶ 0.1 × 10⁻⁶
Matings for determination of frequency of triploid formation:			
21 ade his plus diploid 83	3% G-YNB plus lysine	0.2, 0.2, 1.5, 3.3 × 10⁻⁴	—
	3% G-YNB plus lysine plus 0.4 mM (V + Mo)[b]	0.9 × 10⁻⁴ 0.3 × 10⁻⁴ 6.6 × 10⁻⁴	1.4 × 10⁻⁴ 6.0 × 10⁻⁴ 10.0 × 10⁻⁴
	3% G-YNB plus 0.63% YE[b]	4.0 × 10⁻⁴ 1.2 × 10⁻⁴	2.8 × 10⁻⁴ 2.6 × 10⁻⁴

[a]Diploid 83 is a lysine auxotroph [5 cyh lys etb36 × 21 cyh lys etb36; see Crandall and Richter (1973) for strain descriptions] and therefore was grown in media containing 200 μg lysine/ml.

[b]Diploid 83 was induced for 5f by growing it in media containing either 0.4 mM (V + Mo) or 0.63% YE for 1, 2, or 3 days and then mated overnight in PMG buffer (see Section II) with 21 ade his.

[c]The frequency of prototrophs is calculated by dividing the number of colonies on 3% G-YNB agar by the number of colonies on YG agar and multiplying this quotient by the dilution factor. Prototrophic frequencies across the same line were determined from the same culture before and after the D → 5 transition. Duplicate or triplicate experiments are presented on the lines following.

thesis on induction, indicating differences in genetic constitution. Preliminary tetrad analyses of three different triploid isolates gave poor survival of the ascospores and segregation of both diploid and haploid spores (as judged by agglutination and sporulation abilities).

C. Control of Synthesis of Mating Factors

Since the haploids synthesize the mating factors constitutively, there appears to be no control at the haploid level. At the diploid level there is a control mechanism, because the synthesis of the mating factors is inducible. At present, it is not known whether the factors synthesized during D → 5 and D → 21 transitions are of the same molecular weight, carbohydrate composition, and so on, as those same activities synthesized in the respective haploids. But assuming they are, the synthesis of these mating factors in the diploid is considered repressed in the genetic model proposed by Crandall and Brock (1968b). This model predicts that there are two regulatory genes, for example, *5R* carried by strain 21 and *21R* carried by strain 5, which would function only in apposition in the hybrid. Genetic evidence for the existence of regulatory genes affecting the expression of sexual agglutination has already been reported (Herman and Griffin, 1967).

The physiological results just presented demonstrate that synthesis of both 5f and 21f in the diploid is under the control of metal ions. However, the data presented in Table V show that the conditions that cause a differential synthesis of one or the other mating factor in the diploid have no differential

TABLE V

No Differential Effect on Haploid Agglutination by Conditions Which Induce Agglutination in the Diploid

Medium in which strains were grown (0.5% G-YNB) plus:	24-Hour assay of:			46-Hour assay of:		
	Strain 5 with 21 tester[a]	Strain 21 with 5 tester	Hybrid 5 × 21 with both testers	Strain 5 with 21 tester	Strain 21 with 5 tester	Hybrid 5 × 21 with both testers
No addition	+ + +	+ + +	−	+ + + +	+ + +	−
+ 0.63% YE	+ + +	+ + +	−	+ + +	+ + +	D → 5 (+ + to + + + with 21 tester)
+ 0.63% YE plus 1 mM EDTA	+ + +	+ + +	D → 21 (+ to + + with 5 tester)	+ + +	+ + +	−
+ 0.4 mM (V + Mo)	+ + + +	+ + +	D → 5 (± with 21 tester)	+ + + +	+ + +	D → 5 (+ + to + + + with 21 tester)
0.5% G-SM lacking Ca and Mg instead of YNB	+ + + +	+ +	D → 5 (+ + with 21 tester)	+ + + +	+ + +	D → 5 (+ + with 21 tester)

[a] Strains 5–9A, 21–9D and diploid 44 were inoculated 1:100 into the media listed from a stationary-phase preculture grown in 0.5% G-YNB medium. Agglutination with strain 5 and 21 testers was assayed at 24 and 46 hours. The controls (5 plus 5 tester and 21 plus 21 tester) were all negative for agglutination, and therefore not included in the table.

effect on haploid agglutination. For example, if metal ions were involved directly in the synthesis of the mating factors in the haploids as well, then it might be predicted that growth in YG + EDTA would destroy agglutinability specifically in strain 5, which it did not, and growth in vanadium–molybdenum would destroy agglutinability specifically in strain 21, which it did not. Thus the components of these media may act at a site in the diploid that is not present or functional in either haploid. Since the D → 21 transition occurs only in exponential phase, and the D → 5 transition occurs only in stationary phase or under nongrowing conditions, there may also be a cell cycle control mechanism affecting the synthesis of these mating factors in the diploid. But whatever the control mechanism is, it is certainly clear that nutrient limitation or stress conditions are disturbing the normal regulatory events in the cell. In summary, 5f is induced in the diploid by limiting calcium plus magnesium, oxygen or vitamins or by adding excess metal ions whereas 21f is induced in the diploid by limiting trace metals. As a consequence of these physiological disturbances, cell surface glycoproteins normally repressed are now synthesized.

D.   Comparison of Induction of Glycoprotein Mating Factors in Diploid Yeast to Malignant Transformation

The induction of cell surface glycoproteins reported here is relevant to the induction of surface glycoproteins that occurs during malignant transformation of animal cells with oncogenic viruses. Both inductions occur in eukaryotic diploid organisms and involve the appearance of a new glycoprotein that changes the surface properties and aggregation functions of the cell. However, these two systems differ in the following important aspects. (1) The yeast system is unicellular; (2) induction is mediated by a change in physiological conditions; (3) the carbohydrate portion of the induced glycoprotein is primarily mannose; (4) the glycoprotein is incorporated into the cell wall; and (5) after induction, the cell can adhere to an unlike but complementary cell type. However, (1) the mammalian system is multicellular; (2) induction is mediated by a genetic change caused by incorporation of the viral genome into the host genome (the new surface antigen is presumably a viral protein); (3) the carbohydrate portion of the induced glycoprotein is primarily sialic acid (Warren et al., 1972); (4) the protein is incorporated into the cell membrane; and (5) after induction, cell division is no longer inhibited by contact with a neighboring cell and, furthermore, the malignant cell is less "sticky" and can now slide over other cells and spread throughout the body (metastasis). Although these two systems are not biochemically similar, they are analogous, and it seems worthwhile to compare studies of gene expression and surface glycoproteins in both of these systems to learn more about genetic regulation in eukaryotic cells.

ACKNOWLEDGMENTS

The work reported in this article was supported by USPHS Grant GM18708-02 and by a grant from the Research Corporation to M. C. Gordon Hornbaker performed the experiments on trace element and calcium and magnesium limitation. This article is dedicated to Mrs. Helen Arthur who, as head of the microbiology stockroom at Indiana University, Bloomington, Indiana, has greatly facilitated the research efforts of M. C. over the last 10 years.

REFERENCES

Brock, T. D. (1961). *J. Gen. Microbiol.* **26**, 487.

Conti, S. F., and Brock, T. D. (1965). *J. Bacteriol.* **90**, 524.

Crandall, M. (1973). *J. Gen. Microbiol.* **75**, 363.

Crandall, M., and Brock, T. D. (1968a). *Bacteriol. Rev.* **32**, 139.

Crandall, M., and Brock, T. D. (1968b). *Nature (London)* **219**, 533.

Crandall, M., and Caulton, J. H. (1973). *Exp. Cell Res.* **82**, 159.

Crandall, M., and Richter, R. H. (1973). *Mol. Gen. Genet.* **125**, 279.

Crandall, M., and Robinow, C. F. (1973). *Can. J. Genet. Cytol.* **15**, 874.

Crandall, M., Lawrence, L. M., and Saunders, R. M. (1974). *Proc. Nat. Acad. Sci. U.S.* **71**, 26.

Day, A. W., Poon, N. H., and Stewart, G. G. (1975). *Can. J. Microbiol.* **21**, 558.

Duntze, W., Stotzler, D., Bucking-Throm, E., and Kalbitzer, S. (1973). *Eur. J. Biochem.* **35**, 357.

Fowell, R. R. (1969). *In* "The Yeasts" (A. H. Rose and J. S. Harrison, eds.), Vol. 1, pp. 303–385. Academic Press, New York.

Herman, A. (1971). *Antoniè van Leeuwenhoek, J. Microbiol. Serol.* **37**, 275.

Herman, A., and Griffin, P. (1967). *Genetics* **56**, 564 (abstr.).

Herman, A., Wickerham, L. J., and Griffin, P. (1966). *Genetics* **55**, 339 (abstr.)

Mortimer, R. K., and Hawthorne, D. C. (1973). *Genetics* **74**, 33.

Parry, E. M., and Cox, B. S. (1972). *J. Gen. Microbiol.* **70**, 129.

Sena, E. P., Radin, D. N., and Fogel, S. (1973). *Proc. Nat. Acad. Sci. U.S.* **70**, 1373.

Sena, E. P., Radin, D. N., Welch, J., and Fogel, S. (1975). *In* "Methods in Cell Biology" (D. M. Prescott, ed.), Vol. XI, pp. 71–88. Academic Press, New York.

Taylor, N. W., and Orton, W. L. (1971). *Biochemistry* **10**, 2043.

Warren, L., Fuhrer, J. P., and Buck, C. A. (1972). *Proc. Nat. Acad. Sci. U.S.* **69**, 1838.

Wickerham, L. J. (1951). *U.S., Dep. Agr., Tech. Bull.* **1029**.

Wickerham, L. J. (1956). *C. R. Trav. Lab. Carlsberg, Ser. Physiol.* **26**, 423.

Wickerham, L. J. (1970). *In* "The Yeasts: A Taxonomic Study" (J. Lodder, ed.), pp. 226–315. North-Holland Publ., Amsterdam.

Yen, P. H., and Ballou, C. B. (1974). *Biochemistry* **13**, 2428.

Chapter 11

Mutagenesis in Yeast

B. J. KILBEY

Institute of Animal Genetics,
Edinburgh, Scotland

I. Introduction

Yeasts display many properties that make them ideal for the study of mutagenesis. They grow quickly as single, uninucleate cells to provide the large genetically homogeneous populations necessary for the study of rare genetic changes; the means are available for the genetic analysis of induced mutants, so their nature as inherited alterations can be easily validated. Some species can be maintained in either the diploid or the haploid state at will, and there is a broad background of biochemical and genetic knowledge—two important requirements for the correct interpretation of some aspects of mutagenesis. Recently, biophysical techniques have been successfully applied to yeasts for the preparation of synchronous cultures. This opens up the way for precise studies of mutation induction patterns throughout the cell cycle. However, for some investigators the major attraction of yeasts is that they bring together all these advantages in a cell with eukaryotic organization. This may mean that the results obtained with them are more relevant to higher organisms than those obtained with prokaryotes.

In this chapter we describe first some of the methods used to detect and select mutations in yeasts. We then go on to describe some of the studies made of various aspects of mutagenesis using yeasts, "clothing the skeleton" with a brief consideration of the results obtained. For space reasons we have unfortunately had to restrict the subject matter to nuclear mutations.

II. Forward Mutations

A.   Total Isolation

The mutant clone is distinguished primarily by its altered phenotype. Such phenotypic differences are sometimes immediately visible and can be used to screen out mutant individuals (see Section II,B). More often, however, the mutant clone is morphologically indistinguishable from nonmutant clones, and other phenotypic criteria must be used in its recognition. In the earliest experiments made to isolate new mutants, treated cells were plated and allowed to form clones on complete medium before being tested individually for growth on minimal medium. Nongrowers were tested further to see which substance or combination of substances they would grow on. This tedious process can be speeded up in yeasts by the use of bacterial replication techniques (Cavalli-Sforza and Lederberg, 1953). Mutants isolated in this way are not a random sample of those induced; partial requirers can be overlooked because of their limited ability to grow on minimal medium, and special care must be taken to include them. This difficulty is also a feature of the enrichment procedures to be described below.

B. Visible Screening Methods

Occasionally, the loss of a particular biosynthetic function is associated with a visible change in the mutant clone. The best known of these are mutations that block the biosynthesis of adenine in *Saccharomyces cerevisiae* (*ade1* and *ade2* loci) and *Schizosaccharomyces pombe* (*ade6* and *ade7* loci). In both fungi these loci control the conversion of 5-aminoimidazole ribotide to 5-amino-4-imidazole-*N*-succinyl carboxamide ribotide. Interruption of these functions leads to the accumulation of a reddish-purple pigment. (Woods and Bevan, 1966; Leupold, 1956, 1957). Mutants at these loci can thus be detected as red colonies on medium containing a low level of adenine. Even partial requirers can be easily detected.

A modification of the method suggested by Roman (1956) has proved extremely valuable. By utilizing a red adenine mutant as the starting material, it is possible to detect white variants. Some of these may be reversions of the original mutant, and some may be suppressors. Both are readily distinguished by their ability to grow without adenine. A large proportion of them, however, are mutations blocked at one of the several steps in the adenine biosynthetic pathway before the mutational blocks producing red pigmentation. Since they prevent pathway intermediates from passing through the chain of reactions, none can accumulate to form the red pigment. In *S. pombe* five loci are involved, *ade1*, *ade3*, *ade4*, *ade5*, and *ade9*, and the second mutant can subsequently be located at one or another of these loci by using complementation and recombination tests with known mutants. Because five loci are involved, the total mutation rates are often quite high.

Horne and Wilkie (1966) found that auxotrophs growing on complete medium containing the dye Magdala red stained deep pink, and they attempted to use this method to facilitate the detection of induced auxotrophs. Unfortunately, respiratory-deficient mutants also stain with the dye and, because they are often frequent, constitute a source of error. The use of glycerol as sole carbon source in the scoring medium eliminates petites, but it also prevents development of the color (Queiroz, 1970). The method does, however, allow leaky mutants to be isolated, since they too take on color with the dye.

C. Enrichment Techniques

Various means have been tried in attempts to enrich the proportion of auxotrophic mutants in treated populations. All of them depend on the creation of conditions in which growing prototrophic cells are discriminated against while the nongrowing, newly induced auxotrophs are able to survive.

Thus, by using certain drugs, e.g., endomycin and amphotericin B, Moat et al. (1959) selectively destroyed growing prototrophs and enriched the proportion of nongrowing auxotrophs in the final population. Megnet (1965) used 2-deoxyglucose, and Snow (1966) used nystatin to the same end. Henry (1973) used conditions that allowed only the prototrophs to grow and to take up tritiated uridine or tritiated amino acids from the medium. After 2–5 hours of labeling, the cells were washed and stored at 4°C for about 12 days, during which time survival fell from 100 to less than 10%. The nongrowing auxotrophs failed to take up the tritiated compounds and were consequently not killed to the same extent by their decay. A 10-fold enrichment was recorded.

Another procedure is to use strains which are themselves predisposed to death when starved of some required supplement. Inositolless death is an example of this. (Ridgeway and Douglas, 1958; Megnet, 1964). In the presence of a second auxotrophic mutation, death is prevented. Obviously, this property can be used to enrich the proportions of new auxotrophs appearing in a population of inos⁻ cells treated with a mutagen.

All these methods work well enough when reconstruction experiments using preexisting auxotrophic mutants are tried. However, in practice they are not nearly as successful. This is probably because the mutagenic treatment itself often leads to a delay in the onset of cell division, reducing the differences between prototrophs and nongrowing auxotrophs in this regard. Where it has been examined, it is also clear that the efficiency of the technique can be markedly affected by slight differences in the experimental conditions, e.g., cell density and pH.

Forward-mutation experiments have also been conducted using the selective potential provided by drug resistance and analog resistance mutations. Canavanine, an analog of arginine, has been used widely in S. cerevisiae (Magni et al., 1964). Wild-type cells are able to tolerate $4–10 \mu g/ml$, while resistant mutants can grow on media containing $30 \mu g/ml$. Since canavanine resistance is a recessive mutation, the system can only be used in haploid cells. Mutations have also been obtained conferring resistance to cyclohexamide (Wilkie and Lee, 1965) and to nystatin (e.g., Patel and Johnston, 1968), although these compounds have not been used in quantitative studies of mutagenesis.

III.   The Detection of Reverse Mutations

A.   The Back-Mutation Technique

This type of test was developed with Neurospora independently by Giles (1951) and Kølmark and Westergaard (1949). It involves the selection of rare mutational events that abolish a particular auxotrophy. Since 10^6 to 10^7 cells can be plated on each petri dish, the resolution of the method is high and it has been widely used not only in *Neurospora* but also with many other microorganisms including yeasts. Useful though it is, certain precautions must be observed in its use. It is important, for instance, to ensure that increases in the proportion of revertants among the population surviving the treatment are really indicative of induction and not simply of selection operating to favor the survival of revertant rather than nonrevertant individuals. This usually entails ensuring that the treatment increases the *absolute* number of reversions scored. However, because some mutagens only act at highly toxic levels, this may be difficult, and it is then necessary to conduct reconstruction experiments to eliminate the possibility of selection (Auerbach, 1962). Another problem connected with the reversion test arises from the possible overcrowding of the mutation plates by the background cells. In extreme instances this can prevent the growth of completed mutant cells by exhausting the nutrients in the medium and reducing the number of colonies that will attain visible dimensions. By simply lowering the number of viable cells on a plate, a toxic agent can cause more mutants to appear by relieving this overcrowding. A simple way to test for this is simply to plate the suspension at several different dilutions. If the revertants are diluted out by the same factor, the crowding or Grigg effect (Grigg, 1952) cannot be operating.

In order to express the counted mutants in some meaningful way, an estimate must be made of the number of viable cells plated to produce the observed number of mutants. To do this the treated suspension is diluted by a known factor and a known volume plated on medium which should support the growth of all viable cells in the suspension. From the number of colonies appearing on these plates and the known dilution factor, the number of viable cells on the mutation plates can be estimated. In practice this can sometimes lead to large sources of error. If the auxotroph being studied for its ability to revert is, for example, *inos⁻*, the cells plated for reversion on medium lacking inositol will immediately start to die. The viability estimate therefore only applies at the time the cells are plated for mutation selection. If there is some delay in mutant expression, an underestimate of the proportion of mutants induced may be obtained, because some of them will die before they can be expressed phenotypically. Such a source of error clearly varies from one type of auxotroph to another, and may distort estimates of the reversional response of different auxotrophs to different degrees. One way of overcoming this difficulty is to ensure that all potential mutants are expressed before applying selective conditions. This can be achieved by allowing a few cell divisions on nonselective medium before transferring the cells to selective media.

B.   Suppressor Mutations

A phenotypic reversion may result from a mutational change at the precise site of the original mutation, or from a change elsewhere in the genome. The technique for differentiating between these possibilities simply consists in outcrossing "revertant" clones to a wild-type strain. If the progeny are all nonmutant, the reversion must have resulted from a change at or very near the original mutant site. The appearance of auxotrophic progeny from such a cross means that a suppressor mutation has occurred and that the original mutant remains unaltered.

Some auxotrophs appear to revert mainly via suppressor mutations elsewhere in the genome. Methionine auxotrophs, for example, are often found to behave in this way. As Heslot (1962) has pointed out, the methionine reversion system is thus essentially a selective forward-mutation system involving an unknown number of suppressor loci. The suppressors involved may often be of unknown function, or they may be of the nonsense type which appear in yeasts, as in bacteria, to be mutations at tRNA loci (Gilmore et al., 1968).

IV.   Lethal Mutations

The earliest mutation studies to be put on a quantitative basis involved the use of sex-linked recessive lethal mutations in *Drosophila*. Since then, sophisticated methods of a quantitative nature have been developed for fungi, notably *Neurospora* (de Serres and Osterbind, 1962). In some yeasts the existence of a stable diploid phase has made qualitative studies of dominant and recessive lethals possible. Quantitative studies are difficult, however, because of the high frequency of mitotic recombination. As a result, recessive lethals become homozygous and are eliminated from the population. However, demonstration of both types of lethality has been achieved. Diploid cells of *S. cerevisiae* were exposed to ionizing radiation and made to sporulate either at once or after a few generations had elapsed. Dissection of the four meiotic products showed that in several cases two out of the four ascospores failed to divide (Mortimer and Tobias, 1953). In the control sample all the asci dissected showed four viable ascospores. The frequency of recessive lethal mutations fell after several divisions. In general, diploid cells have been shown to be more resistant to radiation than haploid cells, and this has been attributed to some extent to the induction of recessive lethals. However, estimates of the recessive lethal frequency required to account for the difference between haploid and diploid sensitivities are always in excess of those observed, even allowing as far as one can for the possibility of mitotic crossing-over and the premeiotic elimination of lethals (Magni, 1956).

Dominant lethality has been demonstrated in the following way. Irradiated haploid cells of *S. cerevisiae* may be mated with untreated cells of the opposite mating type. Dominant lethal mutations, presumably chromosomal breakage events, induced in the treated parent are expressed as death of the diploid zygote formed (Mortimer, 1955). Variations in this method have been used to demonstrate that the frequency of induced dominant lethals is proportional to the ploidy of the irradiated parent (Owen and Mortimer, 1956). Unfortunately, cytological verification of chromosome breakage has not been achieved in yeasts to date. The rest of this chapter is concerned with alterations in the information content of particular genes, i.e., point mutations.

V. Molecular Mutagenesis

One of the major objectives of mutation research has always been an understanding of the alterations in the DNA molecule that underlie heritable phenotypic changes. The study of patterns of reversion exhibited by mutants when subjected to supposedly specific mutagens has played an important part in such studies. Reversion tests have been made very sophisticated, as we shall see, but in essence they are strictly biological. Subsequently, biochemical methods of analysis have become available and have been used very effectively to supplement and provide verification for the simpler reversion analyses. Although to separate these approaches is artificial, it is convenient to consider them separately.

A. The Biological Approach

Early experiments involved the study of forward and reverse mutation at the *rII* loci in bacteriophage T4. As an outcome of the work first of Benzer and Freese (1958), Freese (1959), and later of Brenner *et al.* (1961), two classes of mutational events were defined: base pair substitution and frameshift or sign mutation. The two groups are induced by and revert with mutually exclusive classes of chemical mutagen.

The same methods were quickly applied to other microorganisms, and they have been used with yeasts. However, it has always been doubtful whether the same degree of precision attaches to the tests in these organisms as it does in bacteriophage, and other supplementary criteria have been used in attempts to strengthen the conclusions reached from reversion studies. These concern the phenotypes of the mutants under study, their leakiness, pH sensitivity, temperature sensitivity, and osmotic remediability. Mutants displaying one or another of these characteristics are assumed to possess a mutant enzyme altered, in all probability, by a single amino acid substitution which has been made in turn by a missense mutation in the DNA. Some missense mutations cause a complete loss of function of course, but these can sometimes be evaluated on the basis of their complementation characteristics with known tester mutants. Complementation criteria are probably more satisfactory when they can be applied, since the proportion of missense mutants that are leaky, pH-sensitive, temperature-sensitive, etc., may be small. A nonpolarized complementation pattern is usually taken to indicate the occurrence of a missense mutation which, although abolishing functionality of the enzyme molecule, nevertheless gives a product that can participate in the complementation interaction. Complete failure to complement can clearly result from missense- and nonsense-producing base pair substitution events, as well as from frame-shift mutations. Polarized complementation behavior is also difficult to interpret, since it can result from frame shifts as well as from base pair substitution changes giving nonsense mutations. It is sometimes possible to use nonsense suppressors to distinguish the different causes of polarized complementation and nonpolarized complementation but, as Leupold and his associates point out (Munz and Leupold, 1970), weak suppression may be masked by the presence of excess numbers of faulty multimers in the cell.

It is interesting to note that, whereas these criteria identify base pair substitution events in a positive way, frame-shift mutations are identified on a negative basis (lack of complementation, nonleakness, nonsuppressibility, etc.). It is therefore likely that frequently base pair substitutions are likely to be underestimated, while frame shifts are likely to be overestimated.

In spite of these difficulties, several studies have been reported in which attempts have been made to classify the molecular events underlying the mutant phenotype. Loprieno *et al.* (1969b) used the criteria of pH sensitivity, temperature sensitivity, and leakiness coupled with responsiveness to supersuppressors and an ability to complement in a classification of mutants induced by hydroxylamine (HA), ethylmethane sulfonate (EMS), methylmethane sulfonate (MMS), and nitrous acid (NA) at the *ade6* and −7 loci in *S. pombe*. HA, EMS, and NA produced 95–99% mutants which according to these criteria could be called base pair substitutions. MMS produced rather fewer, 85%. Possibly the remaining 15% represent deletion events arising from depurination associated with MMS-mediated alkylation. Nashed

(1968) also showed that all of a group of 208 mutants induced at the *ade2* locus in *S. cerevisiae* by NA and nitrosoimadazolidone (NIL) were able to complement, suggesting, here too, a high proportion of base pair substitution events had been induced. The reports of Leupold and his co-workers are most illuminating (Munz and Leupold, 1970). They worked with mutants induced by the presumptive frame-shift mutagen acridine half-mustard (ICR-170) at the *ade6* and −7 loci of *S. pombe*. On the basis of their nonrevertability with nitrosoguanidine (NTG), low ICR-170 revertability, failure to complement or to show only polarized complementation ability, nonosmotic remediability, and nontemperature sensitivity, only 19% of the *ade6* mutants proved to be unambiguous frame shifts. In a later study using mutants induced by ICR-170, NTG, and EMS at the *ade1A* and −*B* loci in *S. pombe*, only 9% of the ICR-170-induced mutants at the *ade1A* locus and 12% at the *ade1B* locus were shown to meet all the requirements for classification as frame shifts (Segal *et al.*, 1973). Because ICR-170 is an acridine half-mustard, it is possible that its alkylating potential may be responsible for the apparent ICR-170-induced base pair substitution mutants. Reversion tests may also be inefficient indicators of base pair substitution mutations; 26% of a sample of EMS-induced mutants studied by Segal and co-workers were not classifiable as base pair substitutions on the basis of their specific revertibility. In another study, this time with spontaneously derived mutants at the *ade1* locus of *S. pombe*, Friis *et al.* (1971) showed that only about half of a sample of mutants that behaved as base pair substitutions according to other criteria were able to revert with NTG, a known base pair substitution mutagen.

Friis and his co-workers were in fact interested in comparing the types of spontaneous mutations occurring at meiosis and mitosis. Loprieno and his collaborators (Loprieno *et al.*, 1969a) had shown earlier that the mutations produced spontaneously at mitosis at the *ade6* locus in *S. pombe* could be classified almost entirely as base pair substitution events. They used the same types of phenotypic means of classifying the mutants as described earlier, but did not perform any reversion studies. Magni, in contrast, showed that the frequency of mutation was several times as great at meiosis as compared with mitosis in *S. cerevisiae* and that the mutations appearing more frequently, i.e., showing the meiotic effect, appeared to be frame-shift mutations (Magni, 1964). Because the meiotic effect required the presence of both homologous chromosome segments, and because the occurrence of a mutation was often correlated with recombination (Magni, 1963), he came to the conclusion that the spontaneous mutants occurring at meiosis were frame shifts resulting from inaccurate recombination. Friis *et al.* found that in *S. pombe* also frame-shift mutations were produced only at meiosis, however, a high proportion of the mutants recovered were base pair substitutions also. The difference between mutation frequencies at meiosis and mitosis proved to be rather less than in *S. cerevisiae*, probably because higher mitotic rates were encountered in *S. pombe*.

B.   The Use of Chain-Termination Codons

Apart from a tendency to overestimate frame-shift mutants, the methods of analysis so far described allow a reasonably unambiguous classification of point mutations into their two basic types. It was also found possible with bacteriophage T4 to determine the direction of particular base pair substitutions by using suitable diagnostic mutagens. One such is HA, which has a high specificity for cytosine. Similar attempts have been made in fungi but, because the specificity of HA is much less pronounced in almost any *in vivo* situation than in bacteriophage treated *in vitro*, their reliability is doubtful. Other means have therefore been used to try to determine the direction of base pair substitution changes. In particular, chain-termination codons and their suppressors have been important here.

There are in principle two ways of utilizing nonsense or nonsense suppressors to detect specific molecular events without recourse to detailed biochemical analyses. An alteration of amber to ochre nonsense requires a $G \rightarrow A$ transition (UAG $\rightarrow$ UAA). Thus, in the presence of an ochre-specific suppressor, a change of this type should appear as a revertant. Provided the means are available for excluding other phenotypically similar changes, such as reversions at the nonsense codon, this method can be a valuable way of detecting mutagens which preferentially change G to A.

An alternative way is to score changes occurring presumably in the anticodon of the suppressor tRNA itself such that it no longer recognizes UAG but recognizes UAA. This way of doing the experiment is less ambiguous than the first, since a range of suppressible ochre or amber mutants can be incorporated into one strain and the supersuppressor can be shown first to suppress one block of these mutants while, after mutation, it should suppress the other block.

In practice neither method appears to have been used extensively in yeasts with the object of identifying GC $\rightarrow$ AT-producing mutagens. Hawthorne (1969), using *S. cerevisiae*, started with the assumption that EMS and HA specifically induced changes from GC to AT when he identified the mutant allele *ade5-7-100* as an amber mutant, showing that it could be mutated to give an alternative allele which was recognizable by an ochre-specific suppressor. Gilmore *et al.* (1968) provided independent evidence that eight different class I suppressors (of the same type that suppressed the mutated form of *ade5-7-100*) all substituted tyrosine for glutamic acid in iso-1-cytochrome-c, suggesting very strongly that they suppress ochre codons.

In only one case has the method been used in yeasts to verify the action of a mutagen. Sora *et al.* (1973) recently devised conditions under which 2-aminopurine (2AP) acted as a mutagen in yeasts. From studies with bacteriophages; 2AP has always been suspected of inducing transitions from AT to CG. In yeasts these workers first showed that no nonsense mutations were induced in a sample of 24 randomly isolated auxotrophs induced after treatment with this agent. Since nonsense codons of the UAA and UAG types can arise from sense codons by transitions from GC to AT and transversions, but never by AT → GC transitions, this suggests that 2AP cannot induce transversions and GC → AT transitions. However, better evidence than this was provided by the finding that, whereas ochre nonsense could be mutated by 2AP to amber nonsense (UAA → UAG), the reverse was not observed. Ultraviolet light, the positive control, produced transitions in both directions.

C. The Biochemical Approach

The most satisfactory method of analyzing the molecular basis for mutational changes at present is the study of the primary structure of protein molecules. From a knowledge of the code and information concerning the amino acid substitutions in the mutated protein, it is possible with little ambiguity to infer the sequence of bases at the DNA level in wild-type and mutant forms. The most extensive work along these lines has been that of Sherman and his collaborators using the iso-1-cytochrome-c enzyme mutants of *S. cerevisiae* (Sherman and Stewart, 1971).

The amino terminal end of iso-1-cytochrome-c appears to be virtually redundant, however, mutants that prevent initiation of protein synthesis or cause premature termination of translation lead to a loss of function. Such mutants can be detected in several ways. They are, for example, unable to use lactate as a sole carbon source in contrast to petites. Colorimetric methods may also be used. The protein is rather small, consisting of approximately 108 amino acid residues, and can be easily sequenced. Reversions of these mutants can also be obtained. These act either by restoring the original initiation signal or by creating new ones elsewhere on the one hand, or by reverting the nonsense codons on the other. After sequencing the revertant enzyme molecule, precise indications can be obtained concerning the particular substitution that has occurred. The sequence of amino acids has been determined for between 50 and 60 revertants of the ochre mutant *cyc1-9* and the amber mutant *cyc1-179*. All possible substitutions have been observed in these revertants except lysine. It should be noted that any agent acting on AT should be able to revert both codons, but agents acting specifically on GC can revert only the amber codon of *cyc1-179*.

Amino acid substitutions also provide a way of discriminating between amber and ochre codons since, only in the former case, can tryptophan be inserted as the result of a single base pair substitution to give UGG.

Chain initiation requires the codon for formylmethionine, AUG. AUG can mutate to give nine possible triplets which code for five different amino acids. In fact, no protein is produced at all in these mutants, but reversions are found which may be of three basic types.

1. Chains that are longer than the wild-type protein and which result from initiation at a new initiation codon located before the original AUG (now mutated) codon. In these the amino acid inserted instead of formyl-methionine can be ascertained, and the base change in the old initiation codon can be elucidated.

2. Alternatively, true reversions to AUG can be detected.

3. The third type of change involves the creation of another new initiation codon from the AAG codon of the fourth residue, lysine. This becomes AUG, presumably by a A → U transversion, and a protein is produced that is four amino acid units shorter than the normal wild-type polypeptide.

Recently, the set of mutants used in these studies has been extended— notably by the inclusion of three frame-shift mutants *cyc1-31*, *-183*, and *-239* (Prakash *et al.*, 1975).

D.   Mutagen Specificity with the Iso-1-Cytochrome-c System

By this means it has been possible to investigate with certainty the muta-genic specificities of several mutagens for the first time in an eukaryotic organism. In the first of these (using only base pair substitution *cyc1* mutants, the majority of which occurred at the initiation codon), two classes of mutagen were defined; those mutagens that act selectively to produce GC → AT transitions (EMS, DES, NTG, NIL, NA, uridine-5-^{3}H and β-propriolactone) and those that lack any marked specificity [MMS, dimethyl sulfoxide (DMS), half-nitrogen mustard (HN$_2$), x-rays, and ultraviolet light] (Prakash and Sherman, 1973).

A very important consideration in the interpretation of such data is the certainty with which any one revertant can be said to have arisen by induc-tion rather than to have occurred spontaneously. In these experiments treatments yielding large numbers of revertants were used. Another potential difficulty arises from the possibility of suppressors. However, these can be excluded on the basis of color, size, and cytochrome c content, as well as on genetic criteria.

Recently Sherman and his collaborators have published data on the potent carcinogen 4-nitroquinoline-1-oxide (Sherman and Stewart, 1975). By using the battery of mutants referred to above and determining the amino acid replacement in the revertants induced, it was possible to infer that this compound, like EMS, induces GC → AT changes but, in addition, and unlike EMS, it also converts GC to TA. Unfortunately, the degeneracy of the genetic code prevented these investigators from detecting GC → CG changes.

E. Complexities in the Patterns of Molecular Specificity

The advantage of this approach to the study of specificity is that it is possible to verify the specificity of particular mutagenic agents. For example, Sherman and his collaborators have demonstrated that glutamine replacements occur almost exclusively in ultraviolet-induced revertants of the ochre mutant *cyc1-9* (UAA → CAA) (Sherman and Stewart, 1975), while in revertants of other ochre mutants glutamine is only one of the observed infections. The specificity of NA is also dependent on the position of the ochre codon in the gene. In one, *cyc1-72*, lysine replacements occurred almost exclusively, while in others no such predominance was found. In these cases the results reveal that mutation of AT base pairs by ultraviolet light and NA is affected not only by the location of the codon within the gene, but also by the position of the base pair within the codon. That is, the specificity of these agents is affected by the adjacent nucleotide sequences. For other examples of this type, the reader is referred to Sherman's papers.

Another factor playing a role in the pattern of insertion of amino acids in ultraviolet-induced revertants of *cyc1-9* is repair activity. As already stated, glutamine is inserted exclusively when ultraviolet is used to revert this codon. In the mutant *uvs9*, a presumed excision-defective strain, this predominance is upheld, but in the x-ray-sensitive mutant *rad6* the restriction is relaxed and leucine, serine, and glutamic acid are also found, although it should perhaps be pointed out that reversion frequency is greatly reduced in this case (Lawrence *et al.*, 1970).

F. Conclusions

The results so far lead us to conclude the following. Attempts to categorize particular mutational events by reversion alone are in general reliable insofar as they help to decide whether a particular mutant results from a base pair substitution or a frame-shift change. By using nonsense codons in the mutational system, the activities of certain mutagens can be established with more certainty, but even here the marked position effects that exist may distort the response so much that negative findings have little real meaning. Studies using amino acid substitutions in iso-1-cytochrome c have so far shown clearly that, at least in eukaryotes, it is naive to believe either that a particular mutagen always acts on a particular base pair and has a preferred direction of action, or that tester strains can be used to ascertain the detailed mutagenic activity of a new mutagen. Position of the target base and the repair activity of the cell are two factors that have so far been found to modify mutagenic responses to particular mutagens.

VI.   The Mutation Process

Mutagenesis is often thought of exclusively in terms of events at the level of DNA. These are of course crucially important for a mutation to occur, but it is as well to remind ourselves that we never observe these events; our level of observation is at the phenotypic level, and our conclusions concerning events at the DNA level are inferences from these observations. Even the elegant studies just described for iso-1-cytochrome-c tell us only about the events that reach phenotypic expression. They say little, if anything, concerning what proportion of revertants is lost because they fail to survive repair and become fixed, or fail, once fixed, to become expressed. These latter aspects of mutagenesis are becoming increasingly important, and it would be imprudent to ignore the studies and the techniques used for their investigation.

A.   Repair and Mutagenesis

The most dramatic advances in our appreciation of the events that convert mutational prelesions to altered DNA base sequences has come from the study of bacterial repair. In yeasts, rightly or wrongly, the results with bacteria have inevitably colored our thinking. The main types of approach have been to study the effects of repair in cells held in nongrowth conditions for various times after treatment, to use putative repair inhibitors, e.g., caffeine, and to study repair-deficient mutants and their effects on mutagenesis. On occasions these different approaches have been combined.

1.   LIQUID HOLDING TECHNIQUES

In this type of experiment the treated cells are stored in a nonnutritive milieu for several hours. Mutant yields are assayed at various times during the storage. From the results obtained with this method in bacteria, it is to be expected that, during liquid holding, repair activity will have a higher probability of reaching completion than in an actively dividing cell culture.

Parry and Parry (1972) showed that in a homozygous adenine-requiring diploid of S. cerevisiae liquid holding brought about an increase in survival and a decline in the proportion of ad^+ revertants induced by given doses of ultraviolet light. After storage for some days the population was exposed to further ultraviolet treatments, and it was found that, in keeping with an earlier report by Patrick and Haynes (1968), the culture had become more resistant to ultraviolet light. They also reported that no further mutations could be induced by this second dose. Parry and Parry interpreted this finding strictly in terms of known bacterial repair systems. Two repair pathways are visualized in yeast, one that is essentially error-proof and produces no mutations, and one that is error-prone and therefore has the ability to produce mutants. The observed results can be interpreted as a change in the proportion of lesions repaired by the error-proof and error-prone processes such that, as liquid holding proceeds, fewer lesions are repaired by the error-prone system. The failure of the second ultraviolet dose to elicit reversions after storage suggests to them that the error-prone activity may be lost during storage. It is interesting that, having lost a repair system, the cells become more *resistant* to the second ultraviolet dose, but this can be explained on the basis of an interaction between the two systems. Resnick (1969) also showed that in haploid yeasts carrying the mutant *uvs9-3*, liquid holding reduces the frequency of reversions induced by ultraviolet light. In this strain excision is probably at a low level, and he ascribes the decline in mutations either to the leakiness of the *uvs9-3* mutation or to the activity of some other unspecified repair system. It is interesting to note that, in contrast to the conditions in most liquid holding experiments, biochemical evidence for excision in yeasts can only be obtained in growing cells, unlike the situation in *Escherichia coli.*

2. Repair Inhibition

Caffeine has been used in bacteria on several occasions to create repairless conditions in otherwise repair-competent cells. *In vitro* studies have established that bacterial excision repair can be prevented at low concentrations of caffeine (quoted in Sideropoulos *et al.*, 1968), while Witkin and Farquharson (1969) produced evidence of an antimutagenic effect of caffeine at higher concentrations in *E. coli.*

Clarke (1968), using *S. pombe*, showed that at concentrations of 0.5 mg/ml caffeine acted as an antimutagen at low doses of ultraviolet light, while at higher doses, i.e., 5%, survival comutagenic tendencies were apparent. Other studies with the same organism have revealed similar effects. Loprieno and Schüpbach (1971) showed that, for forward mutations induced by ultraviolet light and NG at the *ade6* and *-7* loci, caffeine reduces the frequency of mutation per survivor drastically. At the same time, they demonstrated a reduction in intragenic recombination. Loprieno and Guglielminetti (1969) studied the effect of caffeine also in an ultraviolet-sensitive *uvs1* mutant of *S. pombe* with the interesting result that, for survival after ultraviolet light and NG, caffeine had no effect, but with x-rays it did reduce viability. In contrast, mutation frequency at the *ade6* and *-7* loci in both wild-type and *uvs1* strains was reduced by the presence of caffeine in the plating medium after ultraviolet treatment. These results suggest that caffeine differentiates between mechanisms of lethality and mutation by ultraviolet light, as does *uvs1*. The preliminary conclusion reached by Loprieno is that the error-prone repair system for ultraviolet light at any rate is caffeine-sensitive but, in view of the many points at which caffeine can interact with cellular metabolism, this can only be treated as a tentative conclusion.

1. REPAIRLESS MUTANTS

Again, under the impetus of bacterial studies, successful attempts have been made to isolate yeast mutants that are sensitive to radiation. Not all of these are deficient in known repair mechanisms, but some of them have the properties expected of such mutants. The effects of these mutants on induced and spontaneous mutation frequencies have been studied in a few cases. In keeping with the bacterial results, although possibly differing in small details, it was shown that certain of these mutants lacked the excision repair function and at equal doses of ultraviolet light many more mutations were found in the sensitive strain compared with the wild-type material (Resnick, 1969). Lemontt (1971a) took this investigation further when he isolated mutants able to confer the loss of ultraviolet mutability on the strains carrying them. Mutations at three loci, *rev1*, *rev2*, and *rev3*, were found, and later a fourth locus was demonstrated, by Lemontt, *umr⁻* (Lemontt, 1973). In the presence of *rev1–1* and *rev 3–1* reversion, mutation is reduced to a very low level. This is true whether missense mutants, nonsense mutants, or forward mutations are considered. When combined with *uvs9–2*, a presumptive excision-deficient mutant, the mutants induced by ultraviolet light fall drastically, supporting the contention that the lesions left unrepaired by the absence or reduction of excision are repaired via a mutation-generating system the steps in which are controlled by *rev⁺* alleles. Combinations of different ultraviolet alleles with *rev3* produce strains of similar ultraviolet sensitivity, as do triple recombinants. Lemontt suggests (1971b,c) that two pathways converge, one controlled by *rev1*, and the other by *rev2*. The products of their action are dealt with by a common pathway, controlled by *rev3*.

B. Effects Other Than Repair

The experiments in the preceding section have an obvious relationship to repair, although it is rare that much is known of the repair system concerned.

The remaining examples of work in which the factors affecting mutation yield have been studied cannot be divorced from repair effects with certainty, but are separated from them here because of no obvious connection. They form a heterogeneous collection of studies which for the sake of convenience may be classified as (1) those in which pre- and postexposure conditions are manipulated for all or part of the experimental period, and (2) those of the effects of genetic background on mutation behavior.

1. EXPERIMENTAL CONDITIONS BEFORE AND AFTER TREATMENT

Clarke showed that an apparent case of mutagen specificity in *S. pombe* could be attributed to an effect of methionine in the plating medium (Clarke, 1962). Methionine revertants were obtained at several hundred times the frequency of adenine revertants when the strain *adn1–199; met4–D19* was treated with ultraviolet. A similar although less marked effect was found with NA. In the absence of *met⁻*, an *adn1* mutant responded well to ultraviolet light unless methionine was included gratuitously in the plating medium. The depressive effect of methionine in the plating medium does not appear to be restricted to any specific period during postirradiation incubation, and the biochemical details of this interesting result have never been analyzed, although Clarke believes S-adenosylmethionine to be implicated (Clarke, 1965).

Zimmermann and his collaborators (1966a,b) used similar techniques with *S. cerevisiae*. Temperature of incubation proved to be an important factor in determining the yield of adenine reversions from treatments with NA and alkylating nitrosamides. Incubation at elevated temperatures resulted in a reduction in mutational yield. Residual growth was not involved. In contrast to the methionine effect, it was possible to show that the temperature-sensitive phase was short and confined to the period of a few hours between treatment and cell division, and possibly occurred at the point at which new DNA is synthesized.

The same workers (Zimmermann *et al.*, 1966c) also showed that the number of residual divisions possible after treatment can modify the yield of NA-induced adenine reversions in this material. Cells starved of adenine and then treated with 25.4 mM NaNO$_2$ or 20 mM 1-nitroso-imidazolidione-2 were plated either on minimal medium or on media supplemented with 10 mg adenine per liter for various times before being transferred to minimal media. Only 30 minutes' exposure to adenine was sufficient to produce a maximum or near-maximum mutational response. The reverse experiment, transfer from minimal to medium with adenine, shows a slow decline in the effect of adenine as the period between treatment and transfer to adenine is extended.

The role of the plating medium has been demonstrated in some other instances also. Supersuppressors have been the subject of one such study. In a strain bearing the five ochre mutations *try5–48*, *lys1–1*, *his5–2*, *arg4–17*, and *ade2–1*, the recovery of suppressor mutations suppressing all five simultaneously varied with the plating medium used to select them (Gilmore, 1966). On complete minus tryptophan the yield was highest of all, while on complete less adenine low yields or even no suppressors were found at all. In order to investigate this variable recovery, Queiroz (1973) compared recovery of supersuppressors of the five ochre mutants on all possible omission media, i.e., if complete medium in this context is minimal supplemented with lysine, arginine, tryptophan, histidine and adenine. The prepared complete minus four supplements, complete minus three supplements, complete minus two supplements, complete minus one supplement, and minimal medium alone, 31 media in all. As a result, two different problems were exposed: (1) the cause of the high response on complete minus tryptophan, and (2) the inhibition of suppressor expression when histidine but not adenine was in the medium and the reversal of this effect by adenine. Queiroz believes that the second of these arises from an interaction between the biosynthetic pathways of adenine and histidine such that, in the absence of adenine but in the presence of histidine, the histidine pathway is inhibited by feedback and cannot feed adenine intermediates into the adenine pathway. When this occurs, the synthesis of new tRNA is thought to be retarded, and the expression of nonsense suppressors cut back. No satisfactory explanation of the first problem has yet been found and established.

2. GENETIC BACKGROUND EFFECTS

Although Clarke's study mentioned above appeared to be a genetic background effect of the *met*⁻ mutation on the reversion of the *adn*⁻ allele, this is in reality a plating medium effect, since it could be produced in the absence of the *met*⁻ mutation by adding gratuitous methionine. A case that appears to be a bona fide effect of the genetic background was described by Silhankova (1969, 1972). Rough colony mutants appear to result from alterations at three loci in *S. cerevisiae*, and suppressors of the rough phenotype are known. Silhankova noticed that no rough mutants were obtained in a tryptophan auxotroph and later showed that mutants at two other tryptophan loci were similarly unable to produce rough mutants. This is not primarily an effect of tryptophan in the medium although, when the tryptophan level was raised to 100–200 mg/ml, several times the normal level, the roughness returned. Revertants of *try*⁻ in a genotypically rough strain were all rough, as were segregants from crosses in which the rough mutant and the *try*⁺ mutant segregated. Another mutant, *lys*⁻*1–1*, also suppressed rough, but several other auxotrophs tested, namely, *his5–2* and *ade2–1*, were ineffective.

The explanation of this effect is not known to date, but it is interesting that the rough phenotype is associated with extensive elongation and branching of the individual cells. Indole-3-acetic acid (IAA) has been implicated in the control of cell elongations in yeast, and IAA can be derived from tryptophan. IAA additions did not restore the rough phenotype in rough mutants bearing *try⁻*.

VII. Special Problems

There are particular areas of mutation research in which yeasts have proved to be far superior to many of the other eukaryotic organisms traditionally used. We consider two of them briefly.

A. Mutation during the Cell Cycle

An extremely useful attribute of many yeast cells is the close relationship between cell age or position reached in the cell cycle and cell size. By making use of techniques of centrifugation, cells of different sizes, and, therefore of different ages, can be segregated from each other. Centrifugal techniques are particularly useful, since they undoubtedly subject the cells to far less metabolic stress than other methods of attaining synchrony and thus, from the point of view of mutagenesis, are likely to provide a clearer picture of the relationship between cellular events and susceptibility to mutagenic treatments than can other methods. If cells of a growing population are exposed to a mutagen and then separated on a zonal rotor, fractions from such a distribution can easily be assayed for mutation and the susceptibilities of different age classes can be correlated with other cellular events such as the DNA synthetic period. Dawes and Carter (1974) used this technique in an attempt to determine whether NTG acts at the time of DNA replication in *S. cerevisiae*. Evidence for this has been reported for bacteria (Cerda-Olmedo *et al.*, 1968; Guerola *et al.*, 1971) and also for the alga *Chlamydamonas reinhardtii* (Avner *et al.*, 1973). Dawes and Carter also sought evidence on the time of replication of mitochondrial DNA in the cell cycle from temporal changes in the sensitivity to NTG-induced mitochondrial mutagenesis. Their data show that, for erythromycin and oligomycin resistance, known to have both nuclear and cytoplasmic inheritance characteristics, mutations occur at higher frequencies at a definite period during the cell cycle corresponding to the period of nuclear DNA synthesis.

On analysis most of these mutants show Mendelian inheritance characteristics, and there are indications that mutants occupying slightly different
peaks of response represent mutations at distinct loci. Later in the cell cycle
a second peak is found, the mutants induced here being almost exclusively of
a cytoplasmic nature. These results appear to support the suggestion that
NTG acts at the DNA replication fork, and that mitochondrial DNA replicates at a late stage in the cell cycle. Cytoplasmic petites can be induced
throughout the cycle, which probably reflects their distinctness from other
cytoplasmic variants.

B.   Mutational Mosaicism

In many organisms mutant clones arise as mosaics, i.e., some of the des-
cendants of the treated cell remain unmutated while the remainder manifest
the mutation. Mosaics are to be expected, since the DNA molecule is a duplex structure and most mutagens attack only one base in a pair. Completely
mutant clones are thus not readily explained, and yeasts have been extremely useful in analyzing this problem. Their advantages are:

1. Sectored (mosaic) colonies can be easily observed along with unsec-
tored ones by using the red pigmentation produced by certain adenine
mutants.

2. The ease with which yeast cells can be manipulated under the microscope makes it possible to ensure that mosaics arise from single cells. It is
also possible to construct cell pedigrees in which lethal sectoring and even
the mode of origin of nonlethal mutations can be followed.

3. Methods are available that permit $S.$ $pombe$ to be grown to stationary
phase and to produce a population of G_1 (pre-DNA synthetic) cells. This
reduces the possibility of a mosaic arising from a cell possessing two copies
of its genome.

By using these methods it has been possible to show that the most likely
explanation for pure mutant clones lies in the error-correction system
of the cell (Abbondandolo and Bonatti, 1970). The data agree at present with
an interpretation which holds that damage administered to one of the two
strands can be transferred to the other strand by the mediation of repair
enzymes.

C.   Replicating Instabilities

Another source of sectored colonies is replicating instability. When a
sectored colony is resuspended and replated, the majority of the colonies
are either mutant or nonmutant. Occasionally, however, further mosaics
are formed. The replating of such mosaics gives further mosaics (Nasim,

1967). Experimental artefacts, in particular, adherence of mutant and non-mutant cells, can be excluded in a variety of ways, and it appears that an unstable state is induced which can replicate as such, giving more than one such instability. Evidence has been produced suggesting that such unstable cells exist in their own right and are not associated with either the mutant or the nonmutant phenotype (James *et al.*, 1972). Those induced by EMS in *S. pombe* appear to mutate to a stable form at 10^4 times the spontaneous mutation rate (Nasim and James, 1971). In the mosaic colony such cells are located centrally (Nasim and James, 1972). From the genetic aspect instabilities are interesting, since they appear to be completely specific: The same unstable line always produces stable mutants which map at the same locus and have the same phenotypic properties (Loprieno *et al.*, 1968). Different lines are similarly specific for other locations.

There are still many questions concerning replicating instabilities. What is the molecular basis for the instability? Do all mutagens produce them? Are they all highly unstable? Or are there degrees of instability? To what extent are mosaics the product of instabilities? Replicating instabilites certainly represent an area of great potential genetic interest, and yeasts may continue to be a useful tool for their analysis.

REFERENCES

Abbondandolo, A., and Bonatti, S. (1970). *Mutat. Res.* **9**, 59–69.
Auerbach, C. (1962). *In* Mutation—An Introduction to Research on Mutagenesis, Part 1, pp. 102–104. Oliver & Boyd, Edinburgh.
Avner, P. R., Coen, D., Dujon, B., and Sloninski, P. P. (1973). *Mol. Gen. Genet.* **125**, 9–52.
Benzer, S., and Freese, E. (1958). *Proc. Nat. Acad. Sci. U.S.* **44**, 112–119.
Brenner, S., Barnett, L., Crick, F.H.C., and Orgel, A. (1961). *J. Mol. Biol.* **3**, 121–124.
Cavalli-Sforza, L. L., and Lederberg, J. (1953). *In* "Inhibitori di creseta e chemioterapia," Ist. Super. Sanita, Rome.
Cerda-Olmedo, E., Hanawalt, P. C., and Guerola, N. (1968). *J. Mol. Biol.* **33**, 705–719.
Clarke, C. H. (1962). *Z. Vererbungslehre* **93**, 435–440.
Clarke, C. H. (1965). *J. Gen. Microbiol.* **39**, 21–31.
Clarke, C. H. (1968). *Mutat. Res.* **5**, 33–40.
Dawes, I. W., and Carter, B. L. A. (1974). In press.
de Serres, F. J., and Osterbind, R. S. (1962). *Genetics* **47**, 793–796.
Freese, E. (1959). *Proc. Nat. Acad. Sci. U.S.* **45**, 622–633.
Friis, J., Flury, F., and Leupold, U. (1971). *Mutat. Res.* **11**, 373–390.
Giles, N. H., Jr. (1951). *Cold Spring Harbor Symp. Quant. Biol.* **16**, 283–313.
Gilmore, R. A. (1966). Ph.D. Thesis, University of California, Berkeley
Gilmore, R. A., Stewart, J. W., and Sherman, F. W. (1968). *Biochim. Biophys. Acta* **161**, 270–272.
Grigg, G. W. (1952). *Nature (London)* **169**, 98–100.
Guerola, N., Ingraham, J. L., and Cerda-Olmedo, E. (1971). *Nature (London)* **230**, 122–125.
Hawthorne, D. C. (1969). *J. Mol. Biol.* **43**, 71–75.
Henry, S. A. (1973). *J. Bacteriol.* **116**, 1293–1303.
Heslot, H. (1962). *Abh. Dent. Akad. Wiss. Berlin, Kl. Med.* pp. 193–228.

Horne, P., and Wilkie, D. (1966). *J. Bacteriol.* **91**, 1388.

James, A. P., Nasim, A., and McCullough, R. S. (1972). *Mutat. Res.* **15**, 125–133.

Kølmark, G., and Westergaard, M. (1949). *Hereditas* **35**, 490–506.

Lawrence, C. W., Stewart, J. W., Sherman, F., and Thomas, F. L. X. (1970). *Genetics* **64**, Suppl., 536–37.

Lemontt, J. F. (1971a). *Genetics* **68**, 21–33.

Lemontt, J. F. (1971b). *Mutat. Res.* **13**, 311–317.

Lemontt, J. F. (1971c). *Mutat. Res.* **13**, 319–326.

Lemontt, J. F. (1973). *Genetics* **73**, Suppl. 153–159.

Leupold, U. (1956). *Arch. Julius Klaus-Stift. Vererbungs-forsch., Sozialanthropol. Rassenyg.* **30**, 506–516.

Leupold, U. (1957). *Schweiz. Z. Allg. Pathol.* **20**, 535–544.

Loprieno, N., and Guglielminetti, R. (1969). *Antonie van Leeuwenhok; J. Microbiol. Serol.* **35**, Suppl., c15.

Loprieno, N., and Schüpbach, M. (1971). *Mol. Gen. Genet.* **110**, 348–354.

Loprieno, N., Abbondandolo, A., Bonatti, S., and Guglielminetti, R. (1968). *Genet. Res.* **12**, 45–54.

Loprieno, N., Bonatti, S., Abbondandolo, A., and Guglielminetti, R. (1969a). *Mol. Gen. Genet.* **104**, 40–50.

Loprieno, N., Guglielminetti, R., Bonatti, S., and Abbondandolo, A. (1969b). *Mutat. Res.* **8**, 31–36.

Magni, G. E. (1956). *C. R. Trav. Lab. Carlsberg, Ser. Physiol.* **26**, 279–282.

Magni, G. E. (1963). *Proc. Nat. Acad. Sci. U.S.* **50**, 975–980.

Magni, G. E. (1964). *J. Cell. Comp. Physiol.* **64**, Suppl. 1., 165–172.

Magni, G. E., von Borstel, R. C., and Sora, S. (1964). *Mutat. Res.* **1**, 227–230.

Megnet, R. (1964). *Experientia* **20**, 320–321.

Megnet, R. (1965). *Mutat. Res.* **2**, 328–331.

Moat, A. G., Peters, N., Jr., and Srb, A. M. (1959). *J. Bacteriol.* **77**, 673–677.

Mortimer, R. K. (1955). *Radiat. Res.* **2**, 361–368.

Mortimer, R. K., and Tobias, C. A. (1953). *Science* **118**, 517–518.

Munz, P., and Leupold, U. (1970). *Mutat. Res.* **9**, 199–212.

Nashed, N. (1968). *Mol. Gen. Genet.* **102**, 348–352.

Nasim, A. (1967). *Mutat. Res.* **4**, 753–763.

Nasim, A., and James, A. P. (1971). *Genetics* **69**, 513–516.

Nasim, A., and James, A. P. (1972). *Can. J. Genet. Cytol.* **14**, 979–982.

Owen, M. E., and Mortimer, R. K. (1956). *Nature* (*London*) **177**, 625.

Parry, J. M., and Parry, E. M. (1972). *Genet. Res.* **19**, 1–16.

Patel, P. V., and Johnston, J. R. (1968). *Appl. Microbiol.* **16**, 164–165.

Patrick, M. H., and Haynes, R. H. (1968). *J. Bacteriol.* **95**, 1350–1354.

Prakash, L., and Sherman, F. (1973). *J. Mol. Biol.* **79**, 65–82.

Prakash, L., Stewart, J. W., and Sherman, F. (1975). *J. Mol. Biol.* (in press).

Queiroz, C. (1970). Ph.D. Thesis, Edinburgh University.

Queiroz, C. (1973). *Biochem. Genet.* **8**, 85–100.

Resnick, M. A. (1969). *Mutat. Res.* **7**, 315–332.

Ridgeway, G. J., and Douglas, H. C. (1958). *J. Bacteriol.* **76**, 163–166.

Roman, H. (1956). *C. R. Trav. Lab. Carlsberg, Ser. Physiol.* **26**, 299–314.

Segal, E., Munz, P., and Leupold, U. (1973). *Mutat. Res.* **18**, 15–24.

Sherman, F., and Stewart, J. W. (1971). *Annu. Rev. Genet.* **5**, 257–296.

Sherman, F., and Stewart, J. W. (1975). In press.

Sideropoulos, A. S., Johnson, R. C., and Shankel, D. M. (1968). *J. Bacteriol.* **95**, 1486–1488.

Silhankova, L. (1969). *Antonie van Leeuwenhoek*; *J. Microbiol. Serol.* **35**, Suppl., C11.

Silhankova, L. (1972). *Folia Microbiol.* (*Prague*) **17**, 161–169.

Snow, R. (1966). *Nature* (*London*) **211**, 206–207.

Sora, S., Panzeri, L., and Magni, G. E. (1973). *Mutat. Res.* **20**, 207–213.

Wilkie, D., and Lee, B. K. (1965). *Genet. Res.* **6**, 130–138.

Witkin, E. M., and Farquharson, E. L. (1969). *Mutat. Cell. Process*, *Ciba Found. Symp.* pp. 36–49.

Woods, R. A., and Bevan, E. A. (1966). *Heredity* **21**, 121–130.

Zimmermann, F. K., Schwaier, R., and von Laer, U. (1966a). *Mutat. Res.* **3**, 90–92.

Zimmermann, F. K., Schwaier, R., and von Laer, U. (1966b). *Z. Vererbungslchre* **98**, 152–166.

Zimmermann, F. K., Schwaier, R., and von Laer, U. (1966c). *Mutat. Res.* **3**, 171–173.

Chapter 12

Induction, Selection, and Experimental Uses of Temperature-Sensitive and Other Conditional Mutants of Yeast

JOHN R. PRINGLE

Institute of Microbiology,
Swiss Federal Institute of Technology
Zürich, Switzerland

I. Introduction

A. The Power of Genetic Analysis

In some areas of biological investigation (e.g., studies of the pathways of small-molecule metabolism), the contribution of genetic analysis has clearly been great; in other areas (e.g., studies of development), one still hears more of the potential than of the triumphs of genetic methods. Yet few biologists seriously doubt that this potential will be realized. The intrinsic power of genetic analysis seems to be compounded from the following properties.

1. *Range.*   By the introduction of appropriate mutations, one can, in principle, interfere with the synthesis or function of *any* cellular protein or RNA species. Since virtually everything that happens in a cell is a consequence of protein and/or RNA function, virtually any cellular process can be disrupted by an appropriate mutation.

2. *Specificity.*   A single mutation directly affects, in general, the synthesis or function of a *single* protein or RNA species. Thus the primary disruption introduced by a mutation is in general limited to one, or a few, cellular processes.

3. *The possibility of isolating interesting mutants in the absence of any prior knowledge about the functions that will be affected.*   Although such knowledge may facilitate the isolation of a mutant, isolation schemes *need* demand nothing more than that the mutation prevent growth, or prevent cell division, or prevent protein synthesis, etc.

Properties (1) and (2) lead to the unequaled power of genetic analysis to define the precise biological roles played by particular molecules or functions; one can ask about the exact biochemical, physiological, and developmental consequences to a cell of a specific lack of function x. Properties (1) and (3) lead to the unequaled power of genetic analysis to identify new functions. The properties of a mutant defective in a previously unknown function frequently suggest the direction that molecular analysis of that function should take, but it is an important aspect of the power of genetic analysis that not only the isolation, but also several of the uses, of a mutant can take place in blissful ignorance of its specific molecular defect. (See also Section V.)

Formerly, the range of genetic analysis was severely limited by the fact that mutants defective in a variety of essential functions could not be kept alive long enough to study their properties. The solution to this problem was provided by Horowitz (1948, 1950; Horowitz and Leupold, 1951), who suggested that the so-called temperature mutants, which displayed a mutant phenotype at some temperatures but not at others, provided a way to extend the range of genetic analysis to genes controlling essential functions. Subsequent work on a variety of systems, including the bacteriophages λ and T4 (Campbell, 1961; Hershey, 1971; Herskowitz, 1973; Epstein *et al.*, 1963; Edgar and Lielausis, 1964; Wood *et al.*, 1968; Eiserling and Dickson, 1972), bacteria (Neidhardt, 1964; Kohiyama *et al.*, 1966; Beyersmann *et al.*, 1974; Wickner *et al.*, 1974; Tait and Smith, 1974; Konrad and Lehman, 1974;

Hiraga and Saitoh, 1974), yeast (as discussed below), flies (Suzuki, 1970; Tasaka and Suzuki, 1973), and mammalian somatic cells in culture (Thompson and Baker, 1973; Thompson et al., 1973), has amply demonstrated the validity, and value, of Horowitz's suggestion. In the course of this work, the original concept of temperature-conditional mutants has been generalized to include other types of conditional mutants also. The bacteriophage work cited above provides particularly impressive examples of the use of various types of conditional mutants to analyze processes which are at least prototypically developmental and organismal.

B. Miscellaneous Background and Terminology

In this article we have tried to provide useful suggestions and references for those who may be contemplating the use of conditional mutants of yeasts. (Although much of the discussion is in fact more generally applicable, this article is based on experience with *Saccharomyces cerevisiae*, and refers specifically only to that organism.) It seems wise to deal at the outset with some possible sources of terminological or conceptual confusion.

By a *conditional mutant* is meant an organism that resembles the wild type under some environmental conditions (termed *permissive*), but displays a characteristic mutant phenotype under other environmental conditions (termed *nonpermissive* or *restrictive*). The mutation responsible for this phenotype is a *conditional mutation*. If a conditional mutant is incapable of prolonged vegetative growth under the restrictive conditions, the mutation it harbors is a *conditional lethal* mutation, and the gene and function that are defective are referred to as *indispensable*, or *essential*. (Although convenient and consistent with general usage, this definition is in fact somewhat naïve biologically; other functions, such as those necessary for survival under starvation conditions, may also be in a real sense essential to the organism.) Various conditional lethal mutants can lose *viability* (the ability to resume vegetative growth on return to the permissive conditions) at very different rates during a period of exposure to the restrictive conditions.

The most widely studied conditional mutants are those in which the permissive and restrictive conditions are, respectively, relatively low and relatively high temperatures within the normal growth range for the organism. Such mutants are termed *temperature-sensitive* (abbreviated *ts*). Conditional mutants for which the permissive condition is a relatively high temperature and the restrictive condition a relatively low temperature are termed *cold-sensitive*. Neither *ts* nor cold-sensitive mutations have any necessary effect on the organism's ability to cope with temperatures outside its normal growth range. In other potentially important types of conditional mutants of yeasts (see Section VI), the permissive and restrictive conditions are determined by the pH or osmolarity of the medium. For simplicity, and also because *ts* mutations are the only conditional mutations to have been exploited extensively in yeast, the remainder of this article, except for Section VI, refers explicitly only to *ts* mutations. However, most of what follows here, and in Sections II to V, could also be applied, with only minor changes in wording, to cold-sensitive, pH-sensitive, and osmotic-remedial mutants.

ts mutations are presumed in most cases to be missense mutations in the structural genes for proteins. The resulting amino acid replacements partially destabilize the protein products of the mutant genes, with the result that these proteins can maintain their proper three-dimensional structures, and their proper functions, only at relatively low temperatures. A variety of genetic and chemical arguments supports this view, and it has been verified directly, in varying degrees of detail, in many cases (e.g., Maas and Davis, 1952; Yanofsky *et al.*, 1961; Edgar and Lielausis, 1964; Eidlic and Neidhardt, 1965; Wittmann-Liebold *et al.*, 1965; Jockusch, 1966; Hartwell and McLaughlin, 1968b; McLaughlin and Hartwell, 1969; Eiserling and Dickson, 1972; Thompson *et al.*, 1973). However, several exceptional situations must be noted. [There may also be other occasional exceptions to the simple traditional picture. See Eggertsson (1968); the discussion in Cox and Strack (1971) of mutants that are simultaneously *ts* and cold-sensitive (and cf. Hartwell *et al.*, 1970b); and Overath *et al.* (1970).] First, it seems almost certain that in some cases small deletions, or nonsense mutations near the ends of genes, give rise to temperature-sensitive protein products [Jones (1972a,b); the observation by Singh and Sherman (1974) of an osmotic-remedial nonsense mutation also supports this contention]. Second, it seems clear that in some cases the defective protein in a *ts* mutant can retain its proper structure and activity at restrictive temperatures if it has once been properly folded (or assembled into an oligomer) at permissive temperatures (Edgar and Lielausis, 1964; Sadler and Novick, 1965; Kornberg and Smith, 1966; Roodman and Greenberg, 1971; Meyer and Schweizer, 1972). In such cases the mutant is said to be *ts for synthesis* of the defective protein. (When the defective protein cannot remain active at restrictive temperatures, the mutant is said to be *ts for function*.) Third, it seems clear that in some cases *ts* mutations are located in genes coding for tRNA species (Galluci *et al.*, 1970; Smith *et al.*, 1970; Nagata and Horiuchi, 1973; Beckman and Cooper, 1973; Rasse-Messenguy and Fink, 1973). In particular, some mutants are *ts* because they harbor simultaneously nonsense mutations and tRNA species which are active in nonsense suppression only at relatively low temperatures (see Section VI,D).

Because the defective gene product in a *ts* mutant must function more or less normally during growth at the permissive temperature, while being more or less nonfunctional after exposure to the nonpermissive temperature, it usually is to some extent a compromise. That is, function is frequently not fully normal at the permissive temperature, or not fully absent at the nonpermissive temperature, or both. As regards the former point, it may be noted that Klyce and McLaughlin (1973) isolated 1000 yeast mutants which formed colonies at 23°C but not at 36°C; further testing revealed that at least one-fourth of these mutants grew substantially more slowly than the wild type at 23°C. (In some cases these reduced growth rates may have been due to mutations other than the ones conferring temperature sensitivity; see Section IV.) A further illustration is provided by the well-characterized *ts* mutant whose failure to grow in rich medium at 36°C is due to a defect in methionyl-tRNA synthetase activity (McLaughlin and Hartwell, 1969). Even at 23°C the mutant strain grows more slowly than its wild-type parent, displays a requirement for methionine which the parent does not have, and has almost no methionyl-tRNA synthetase activity detectable in crude extracts.

At the same time it can clearly be quite difficult to interpret the behavior of a *ts* mutant that does not cease growth processes rapidly after a shift to the restrictive temperature. It is possible that the mutant is *ts* only for synthesis, and not for function, of the defective gene product, and that the gene product already present at the time of the shift can support a period of continued growth. It is possible that neither synthesis nor function of the defective gene product is fully blocked by the mutation. (This is called a *leaky* mutation.) It is possible that both synthesis and function of the defective gene product are fully blocked by the mutation, but that the gene product's role in the cell is not such that its absence leads to an immediate cessation of growth. There is some hope of discriminating among these possibilities if mutants carrying several different alleles of the gene in question are compared; it is possible, for example, that an allele will be found that does give rapid cessation of growth after a shift to the restrictive temperature. It should be noted that the comparison of strains harboring different alleles also has other advantages. It may allow the selection of the strain that is most nearly normal at the permissive temperature; it reduces the danger that a phenotype actually due to the acquisition of an abnormal function by the product of a particular mutant allele will be attributed to a simple loss of gene function (see also Section III); and it reduces the danger of being misled by double mutants (see also Section IV).

A variety of genetic and biochemical studies (including those cited in Section I,A) have made it clear that most of the genes of an organism, regardless of the types of biological functions their protein or RNA products may have in the cell, are potential sites for *ts* mutations. (See also Section VI.) Although the most useful *ts* mutants are *ts* lethals, with mutations in essential genes, *ts* mutations in other genes can sometimes be more useful than ordinary (unconditional) mutations in the same genes. (See also Section V.)

II. Induction of Mutants

In a few situations (e.g., the selection of survivors or of drug-resistant mutants), it may be practical to isolate spontaneous *ts* mutants; in general, however, it is desirable to increase the frequency of mutants by using some mutagenic agent. A variety of mutagens and mutagenesis procedures has been used successfully with yeast (see Kilbey, this volume, Chapter 11; Fink, 1975, 1970; Hawthorne, 1969; Meuris *et al.*, 1967; and other references in the reviews by Mortimer and Hawthorne, 1966, 1969). Most, if not all, of these procedures yield *ts*, among other, mutants. Hence we confine ourselves to the following remarks (see also Section IV, paragraph 2):

1. Unless the goals of a particular study dictate otherwise, it is desirable for genetic and biochemical studies of *Saccharomyces* to focus on one or a few well-defined strains. Particularly recommended is the already widely disseminated S288C family of strains (S288Cα, X2180 a/α, X2180-1Aa, C276 a/α, etc.; see Fink, 1970; Duntze *et al.*, 1970; Francis and Hansche, 1972; Wilkinson and Pringle, 1974). Also in very wide use is strain A364Aa and its derivatives; this strain was the parent for the large collections of *ts* mutants isolated by Hartwell (1967) and by Klyce and McLaughlin (1973). These various strains can be obtained from the Berkeley stock collection (R. K. Mortimer, Yeast Stock Center, Donner Laboratory, University of California, Berkeley, California), or from many yeast workers, but a word of caution is in order. We have obtained strains from the S288C family from two different sources (neither of which was the Berkeley collection). Although the strains obtained should have been isogenic, they in fact differed in at least one locus, and probably two, which affected reserve carbohydrate metabolism. (Other differences probably would not have been noticed.) Whether this difference resulted from a mislabeling of strains, or from actual evolution during repeated subculturing in different laboratories, is not known. That such an event can happen with a healthy prototrophic strain emphasizes the necessity for caution in the maintenance of mutant stocks (see Section IV).

2. If clear-cut results due to single mutations are to be obtained, it is of course desirable that the parent stock used for mutagenesis grow well, and with few morphologically aberrant cells, at the restrictive temperature to be used.

3. In a *ts* mutant the defective gene product must not be too drastically altered (Section I,B). Thus mutagens that tend to induce single base exchanges are more suitable than mutagens that tend to induce frame shifts or gross structural changes in the chromosomes (see Kilbey, this volume, Chapter 11). It is worth noting that the potent mutagens *N*-methyl-*N'*-nitro-*N*-nitrosoguanidine (NG), nitrous acid (NA), ethylmethane sulfonate (EMS), and ultraviolet light have all been used successfully to induce *ts* mutations (NG: Hartwell, 1967; Klyce and McLaughlin, 1973; NA: Lacroute, 1967, using the procedure described by Meuris *et al.*, 1967; EMS: Hartwell *et al.*, 1973, using the procedure described by Hawthorne, 1969; ultraviolet light: Esposito and Esposito, 1969).

4. Against the high mutagenic potency of NG must be balanced its apparent tendency to induce multiple mutations within a short region of the chromosome (Guerola *et al.*, 1971; Botstein and Jones, 1969; Clarke *et al.*, 1973; Dawes and Carter, 1974). The presence of multiple mutations may of course appreciably obfuscate the subsequent analysis (see also Section IV).

5. In our opinion most published procedures, and actual practice in most laboratories, do not emphasize sufficiently the need for caution in dealing with potent chemical mutagens. It must be remembered that the problem is not one of easily recognized acute toxicity, but of insidious long-term carcinogenic and teratogenetic potential, and that colleagues and subordinates, as well as the investigator, may be placed at risk. Although the risk may be small, and difficult to quantitate, it is irresponsible not to lean well in the direction of caution; the necessary precautions [use of protective clothing, careful instruction of students and technicians, localization (ideally in a hood) and subsequent careful decontamination of potentially contaminated equipment and glassware (even of disposable glassware), chemical inactivation of mutagen before disposal] are not really so much trouble. Apart from any actual assessment of the risks involved, there is the empirical sociological fact that concern about a colleague's possibly careless use of mutagens has led to counterproductive and disagreeable friction in at least one laboratory of which we are aware.

6. For a variety of reasons, it is desirable to conduct mutagenesis and mutant isolation in such a way that it is clear that the various mutants isolated are of independent origin (i.e., are not multiple progeny of one or a few original mutant cells). First, it is a waste of time to isolate the same mutant twice. Moreover, any argument based on a comparison of the phenotypes of mutants presumed to harbor different alleles of a particular gene (see Sections I,B and IV), or on a statistical analysis of the frequencies of various types of mutants (see, for example, Hartwell *et al.*, 1973; Hranueli *et al.*, 1974), is clearly dependent on the mutants being of independent origin. A variety of suitable protocols has been used. Growing cells can be treated

with mutagen in multiple parallel small cultures, and only one or two mutants per culture saved for subsequent analysis (Hartwell, 1967). It is probably more convenient, and minimizes the amount of glassware contaminated with mutagen, to conduct the mutagenesis in a single tube or flask containing nonproliferating cells; following mutagen treatment and washing of the cells, aliquots can either be plated directly, or used to inoculate multiple liquid stock cultures. The latter approach is convenient for some types of subsequent screening, and also allows time for mutations to be expressed, for sickly types to be selected against, and for clumps to become genetically homogeneous (see Pringle and Mor, 1975; note that a colony derived from a clump that is part mutant and part wild type may not be detected as a mutant). The medium and temperature chosen for this period of growth must not select against the desired mutants. Nonproliferating cells can be obtained either by using a culture pregrown to stationary phase or by shifting a growing population to a medium or buffer containing no utilizable carbon source. (At least in our experience, very little further cell division follows such a shift.) In our experience with mutagenesis by EMS, cells from a population in, or approaching, stationary phase give a higher proportion of usable mutants among viable cells than do cells from an exponentially growing population that has been abruptly deprived of a carbon source. It is also possible that the abruptly arrested cells are more likely than either stationary-phase cells or actively growing cells to sustain multiple closely linked mutations during NG mutagenesis.

III.   Isolation of Mutants

In one approach to the isolation of *ts* mutants, which we may term "classic nonselective isolation," the mutant collection consists of all clones that form colonies at the permissive temperature but fail to do so at the restrictive temperature. Mutants in which some particular process of interest is affected are then identified by further screening. This approach has the great advantage that the isolation of interesting mutants does not depend on any a priori knowledge of the nature, or even of the existence, of the functions that will be affected; in principle (see, however, Section VI), mutations can be obtained in any gene whose product is essential for growth. However, this approach also has two significant limitations. First, it may be necessary to isolate and screen many mutants in order to obtain a few in which a particular process is affected. Thus it may often be profitable to use some sort of selective system to increase the frequency of mutants of a particular type. Second, the isolation is not fully nonselective; mutations are not obtained in genes whose products are necessary for other cellular processes (such as sporulation, or the maintenance of stationary-phase cell viability), but not for vegetative growth, or in genes whose products are merely useful, but not essential, to the cell. When *ts* mutations in such genes are desired, they must be sought by appropriate screening or selective procedures.

A judicious initial choice of permissive and restrictive temperatures is helpful to a study of *ts* mutants. It is clear that the restrictive temperature should be as high as possible, so as to maximize the chances of obtaining *ts* mutations in genes whose products are normally relatively thermostable. At the same time, physiological studies of the mutants will be facilitated if the wild type is healthy and physiologically "normal" at the restrictive temperature. For mutations affecting vegetative growth, a good compromise seems to be 36°C; above this temperature the generation times of most wild-type strains begin to increase rapidly with increasing temperature. (These increases are presumed to reflect unhealthiness.) The permissive temperature should be enough below the restrictive that conditional mutants are feasible, yet, ideally, not so low that cultures at the permissive temperature must always be specially cooled, or that these cultures grow inordinately slowly. The use of room temperature as permissive is very convenient, especially in laboratories with reasonably good year-round temperature control. The large existing collection of *ts* mutants was isolated using 23° and 36°C as the permissive and restrictive temperatures, respectively (Hartwell, 1967; Klyce and McLaughlin, 1973). Because these temperatures are known to work well, and because comparisons with the existing mutants will be facilitated, other isolations should probably use the same temperatures. It should be noted that some flexibility is possible in working with mutants isolated using this pair of temperatures. For example, one mutant that seemed somewhat leaky at 36°C was studied at 38°C (Hartwell, 1973a; Hereford and Hartwell, 1974; Simchen, 1974), and Simchen (1974) used 33.5°C as a restrictive temperature with most of the mutations while testing their effects on sporulation. [This was important because the wild-type strains available sporulated poorly at temperatures higher than 34°C. Presumably for the same reason, Esposito and Esposito (1969) used 34°C as the restrictive temperature in searching specifically for *ts* sporulation mutants.]

A. Classic Nonselective Isolation

1. ISOLATING THE MUTANTS

The procedure is extremely simple. Samples of the mutagen-treated culture are spread on solid medium so as to give about 250 colonies per 9-cm-diameter petri dish. When the colonies appear, two replica plates are made in the usual way. The first replica is incubated at the restrictive temperature, and the second at the permissive temperature. Clones that grow on the second plate, but not on the first, are picked from the second plate, diluted with water, and streaked on two further plates. Again, one plate is incubated at each temperature. If many colonies appear at the permissive temperature, and none at the restrictive temperature, cells from one colony on the permissive temperature plate are picked to form a stock of this newly identified *ts* mutant (see Section IV). If colonies appear also at the restrictive temperature, but in reduced number, it is probably worth restreaking a few clones from the permissive temperature plate to see if a homogeneous *ts* clone (i.e., one that gives no colonies at the restrictive temperature) is obtained. The fraction of all clones that are *ts* varies of course with the mutagenesis procedure. In three actual mutant hunts, frequencies of 0.1 to 1% were obtained (Hartwell, 1967; Lacroute, 1967; Klyce and Mc-Laughlin, 1973).

It should be noted that this isolation procedure can employ either minimal or rich medium. If minimal medium is used, some of the mutants isolated will be *ts* auxotrophs; i.e., they will grow at the restrictive temperature in appropriately supplemented medium. Lacroute (1967) found 8 such *ts* auxotrophs among the 84 *ts* mutants he isolated on minimal medium. Hartwell (1967) and Klyce and McLaughlin (1973) excluded most mutants affecting small-molecule metabolism by isolating their large collections of *ts* mutants on rich medium. Nonetheless, mutants affected in some pathways, such as glycolysis and the synthesis of many phosphorylated compounds, are probably included in these collections.

It should also be noted that the time of initial isolation is probably the time to be ruthless about discarding sickly and leaky mutants. When the presumptive mutant clones (after the replica plate test) are streaked for retesting, the resulting colonies on the permissive-temperature plate can be compared in size to those on a comparable plate of the parent strain, and the surface of the restrictive temperature plate can be examined microscopically to see approximately how much the individual mutant cells grew after the shift to the restrictive temperature. (For the best test the plates should be warmed to the restrictive temperature before the streaking is done.) Many clones (sickly) give colonies significantly smaller than those of the parent at the permissive temperature, while other clones (presumptively leaky) continue to grow for several generations at the restrictive temperature. [These points are well illustrated by the data of Klyce and McLaughlin (1973). These workers did not select strongly against sickliness and leakiness during their initial isolation of 1000 *ts* mutants. Subsequent screening revealed that at least one-fourth of these mutants grew substantially more slowly than the wild type at the permissive temperature, and that roughly one-third of the mutants that grew well at the permissive temperature could complete 3.5 or more generations of growth during the 24 hours after a shift to the restrictive temperature. See also the comments in Section I,B.] For most purposes it seems desirable simply to discard clones of both types at this point. First, presumptive mutant clones are plentiful, and testing them at this level is economical. Second, good growth at the permissive temperature appreciably facilitates subsequent analysis of a mutant strain. Third, sickliness is in most, if not all, cases an allele-specific rather than a gene-specific phenomenon. (That is, a gene for which some *ts* alleles are sickly in general also has *ts* alleles that are not sickly. Thus the risk of missing altogether some interesting gene is increased very little if *n* nonsickly clones, rather than simply *n* clones, are saved for subsequent more detailed examination.) Finally, for a great number of genes of interest (e.g., genes whose products are necessary for RNA synthesis, for protein synthesis, or for a specific step in the cell division cycle), a nonleaky mutant, *ts* for the function of the defective gene product, stops growth within one or two generations at the restrictive temperature. The analysis of such a mutant will in general be easier and more informative than the analysis of a mutant, harboring another allele of the same gene, which can continue growth for 3 to 10 generations.

It must be emphasized, however, that this ruthless approach to mutant isolation is in some respects short-sighted. First, there are undoubtedly genes, whose functions should be analyzed, for which even a nonleaky, *ts* for function allele does not give rapid cessation of growth at the restrictive temperature. A search for mutations in such genes must simply be prepared to cope with a high background of mutants that harbor leaky alleles of other genes. Second, mutants carrying nonleaky, *ts* for synthesis alleles can in principle be quite useful (e.g., in determining when gene products necessary for a particular step of the cell cycle are actually synthesized; see also Sections V and VI,D); the ruthless approach lessens the chances of isolating such mutants. However, the potential usefulness of such mutants is much limited in practice by the difficulty of distinguishing them from mutants that are simply leaky. Finally, there are almost certainly genes in which *ts* mutations are intrinsically difficult to obtain because the normal gene products are unusually thermostable (see Section VI). The chances that a mutant collection will completely lack mutants affected in such a gene may be appreciably increased if the selection against presumptive leaky mutants is too stringent. However, a mutant that is too leaky to analyze well is not much better than no mutant at all, and the danger of missing important genes cannot in any case be eliminated simply by examining leaky *ts* mutants (see Section VI).

2. SCREENING THE MUTANTS

A variety of approaches may be taken to screening a collection of "non-selected" *ts* mutants, depending on the type (s) of mutants deemed to be of interest. When Hartwell first isolated 400 *ts* mutants, he was interested primarily in studying macromolecule synthesis and its regulation. Thus his original screening procedure (Hartwell, 1967) focused on a simple pulse-labeling experiment to detect mutants defective in the synthesis of protein, RNA, or DNA. This allowed the detection of 21 mutants with possible defects in protein synthesis, 38 with presumptive defects in RNA synthesis, and 4 with an apparent preferential inhibition of DNA synthesis. [The original set of 400 mutants ultimately yielded 10 mutants with lesions in genes apparently involved in DNA replication or its initiation (Hartwell et al., 1973; Hartwell, 1973a, 1974; Hereford and Hartwell, 1974).] Lac-route (1967) observed similar frequencies after a similar screening of his smaller collection of mutants. Further screening of the mutants potentially defective in protein synthesis (Hartwell and McLaughlin, 1968a) allowed the identification of 10 that actually had specific defects in protein synthesis or in mRNA production; the other mutants seemed defective in some aspect of energy metabolism or in membrane function. The 10 mutants (which define genes *rna1*, *prt1*, *prt2*, *prt3*, *mes1*, and *ils1*) were subsequently characterized in some detail (Hartwell and McLaughlin, 1968b, 1969; Hutchison et al., 1969; McLaughlin and Hartwell, 1969; McNeill and McLaughlin, 1973; Petersen and McLaughlin, 1974). Meanwhile, 23 of the mutants with presumptive defects in total RNA synthesis (which define genes *rna2* to *11*) could also be characterized further (Hartwell et al., 1970b; Warner and Udem, 1972).

The 400 mutants in Hartwell's collection were also examined microscopically to determine cell morphology after incubation at the restrictive temperature. This revealed about 20 mutants, whose defects define 14 genes (designated *cly1* to *-14*), in which the cells lyse relatively rapidly after a shift to the restrictive temperature (Hartwell, 1967; Mortimer and Hawthorne, 1973; L. H. Hartwell, personal communication). These mutants are tentatively assumed to be defective in the synthesis of cell membrane or cell wall, but have not been studied in detail.

Morphological screening has also been used to search the Hartwell collection for mutants defective in specific steps of the cell division cycle. This screening has been based on the assumption that, when an asynchronous population of such a mutant is shifted to the restrictive temperature, the cells eventually become arrested in development as a morphologically homogeneous population; the particular terminal morphology is characteristic of the particular cell cycle step that is defective. It is possible that there are genes that should be regarded as cell cycle genes (i.e., their products function at specific steps in the cell cycle), but whose mutants cannot be detected by means of this criterion. It is at present not clear what other criterion could be used. However, it is apparent that many cell cycle mutants can be recognized by looking for such morphologically homogeneous populations. At the present time about 150 mutants (defining the genes *cdc1* to *-35*) have been identified among about 1500 *ts* lethals screened (Hartwell *et al.*, 1973, 1974; Hartwell, 1974). [Some of these mutants were actually detected by complementation screening (see pages 245–246), rather than by morphological screening.] Fifty-nine of these *cdc* mutants were from the original collection of 400 mutants, which was screened somewhat more intensively than subsequent collections. Originally, the morphological screening was done by examining each mutant with time-lapse photography (Hartwell *et al.*, 1970c). Subsequently, it was realized that a liquid-culture screening procedure (Hartwell *et al.*, 1973), in which only the terminal morphologies were examined, was both less laborious and more likely to detect all mutants that met the essential criterion. (In particular, the time-lapse screening procedure tended to miss mutants that were somewhat leaky, or perhaps *ts* for synthesis, as well as the important mutants that terminated as unbudded cells of various sizes.) Observations of external morphologies should be coupled to observations of nuclear morphologies (Hartwell *et al.*, 1973; Robinow, 1975) in such a screening procedure.

By integrating observations of terminal morphologies with observations of growth and viability patterns in a single photomicrographic procedure, and by identifying the patterns likely to be displayed by RNA synthesis or protein synthesis mutants, Klyce and McLaughlin (1973) streamlined the identification of the types of mutants discussed above.

The screening procedures that have been discussed have allowed at least the tentative identification of the primary defects in about one-third of the 400 mutants of the original Hartwell collection. To date, however, there exist few clues as to the primary defects in the remaining mutants. In some cases it is probably difficult to recognize a specific defect simply because the mutant is leaky (or *ts* for synthesis), but this is certainly not the whole problem. Some additional types of screening (as for defects in some pathways of small-molecule metabolism, or in lipid metabolism, or in uptake) would presumably allow identification of some further specific defects, but it seems likely that many of these mutants are defective in functions that we do not yet know enough to screen for.

Two further points about screening should be made. First, although very few dominant mutations have been found among the *ts* lethals screened to date (Hartwell, 1967; Klyce and McLaughlin, 1973), it is probably still worthwhile to make this test, since a dominant mutation may well be one in which a gene product has acquired an abnormal function. This information would clearly be crucial to a physiological interpretation of the mutant. (As pointed out in Sections I,B and IV, a comparison of the phenotypes of independently isolated mutants defective in the same gene is also a strong control against this possibility.) Second, the value of complementation screening should be stressed. Since mutants isolated by any screening procedure will presumably be grouped by complementation testing in any case, it becomes a simple matter to test representatives of each complementation group against the unidentified mutants in the original large collection. Frequently (Hartwell and McLaughlin, 1968a; Hartwell *et al.*, 1973), this allows the identification of other mutants, defective in the already defined genes, that were missed in the original screening because they are leaky, or *ts* for synthesis; or because their phenotypes were masked by the presence of additional mutations (see also Section IV); or because their phenotypes (or perhaps those of the prototype testers) are simply atypical. This process is useful because it minimizes the danger to physiological interpretation of the occasional mutant with an atypical phenotype (see Hartwell *et al.*, 1973), because it identifies some mutants in the collection that would otherwise remain obscure, and because the alleles obtained may even be especially suitable for physiological analysis.

B.   Enrichment and Selective Isolation Procedures

1.   MUTATIONS IN ESSENTIAL GENES

Classic nonselective isolation, with associated screening procedures, has yielded many interesting mutants and will doubtless yield many more. Yet this approach is clearly in some respects inefficient. First, it would be useful to increase the frequency of *ts* lethals among surviving cells prior to plating. Methods are available for doing this (see below), but two points must be emphasized. (1) All the available methods seem to be selective; i.e., the overall frequency of *ts* lethals is increased by increasing selectively the frequency of certain types of *ts* lethals. Other types of *ts* lethals may in fact be reduced in frequency or even eliminated from the population. (2) The isolation of *ts* lethals is not the major point of inefficiency in the nonselective approach; such mutants actually occur in rather high frequency after moderately heavy mutagenesis (Section III,A,1).

Instead, the major problem is the need to isolate and screen a relatively large number of *ts* lethal mutants in order to isolate a relatively small number of mutants affected in some particular process of interest. For example, only 9 of Hartwell's original 400 mutants could be shown to have specific defects in protein synthesis (Section III,A,2; note that 3 of these 9 mutants are defective in aminoacyl-tRNA synthetases). It is clear that to obtain a complete set of protein synthesis mutants by the nonselective approach would require the isolation and screening of thousands of *ts* lethals. The problem is exacerbated when the nonselective approach has already been pursued at length. For example, morphological screening for cell cycle mutants (Section III,A,2) has been very fruitful. Approximately 10% of the *ts* lethals screened have shown distinctive cell cycle phenotypes, and 35 *cdc* genes have been identified. However, it is now clear (see Hartwell *et al.*, 1973) that a point of diminishing returns has been reached. Since the known mutations are distributed among the known genes in a far from random fashion (10 genes are represented by a single allele apiece and 3 other genes are represented by 18, 14, and 13 independently isolated alleles), with an average of 4.6 mutations per gene, it seems (1) that there are probably many *cdc* genes that have not yet been identified because *ts* mutations in them are rare, and (2) that further screening by the method used to date will yield primarily mutants in genes already known. (Although it is impossible to predict accurately, it seems unlikely that screening an additional 1500 *ts* lethals, and finding an additional 150 *cdc* mutants, would identify more than 5 additional *cdc* genes.) Thus it seems desirable either to begin looking for *cdc* mutants among conditional lethals of some other type (see Section VI), or to use a procedure that increases the frequency of one or more types of *ts cdc* mutants.

In this section we describe briefly several methods with which it should be possible to increase the frequency of certain types of *ts* lethal mutants among the survivors of a mutagenesis procedure. Not all of these methods have been tested in practice, and the intent is more to indicate what might be possible than it is to present a catalog of well-established procedures.

In one class of rather powerful selective methods, a mutagen-treated population (generally after at least a short period of growth to allow mutations to be expressed) is shifted to the restrictive temperature and is then treated in some way that leads to a loss of viability by cells that continue to grow at this temperature. One method that has been used successfully to increase the frequency of *ts* lethals is tritium suicide (Littlewood and Davies, 1973; Littlewood, 1975). After the shift to the restrictive temperature, tritiated uridine or tritiated amino acids are supplied to the cells, and mutants that fail to incorporate these precursors then survive preferentially during a subsequent period of storage. Thus this procedure increases the overall frequency of *ts* lethals among surviving cells primarily by increasing the frequency of mutants defective in RNA or protein synthesis, or in energy metabolism. Cell cycle mutants, however, are selected against, since they continue macromolecule synthesis at the restrictive temperature (see, for example, Culotti and Hartwell, 1971; or Hartwell, 1971a,b). It is likely that this method would also work with other types of tritiated substances, with corresponding changes in the types of mutations that would lead to failure of incorporation, hence to survival. A variant of this approach, which has been used successfully with several organisms and might work with yeasts, is to supply bromodeoxyuridine after the shift to the restrictive temperature (Kessin *et al.*, 1974). Those cells that incorporate the compound are then killed selectively by subsequent exposure to ultraviolet light.

A method that has been widely used to select auxotrophic and fermentation mutants of yeasts, and which should be adaptable also to the selection of *ts* lethals, is the treatment of the mutagenized population with polyene antibiotics such as nystatin or amphotericin B (Snow, 1966; Thomulka and Moat, 1968; Strömnaes and Mortimer, 1968; Fink, 1970). It is not entirely clear what aspect(s) of growth must be blocked to give rise to resistance to the antibiotics, although various blocks of energy metabolism or macromolecule biosynthesis seem effective. Again, it appears that cell cycle mutants would be selected against, because of their continued growth.

Further possible variants of this approach are suggested by the work of Henry and Culbertson (Henry, 1973; Culbertson and Henry, 1974), who showed that certain mutants that require fatty acid or inositol supplements display an exponential loss of viability when deprived of these substances. Cycloheximide prevents this loss of viability, which suggests strongly that *ts* mutations blocking protein synthesis, and possibly also other types of *ts* mutations, could be selected by using one of the Henry and Culbertson strains as a parent. In another possible variant, mutants defective in protein synthesis are selected because they resist thymineless death; this method has been used successfully with *Escherichia coli* (Kaplan and Anderson, 1968), and might be adaptable to yeast now that strains showing thymineless death are available (Brendel and Langjahr, 1974).

A point to emphasize with respect to the types of selections discussed above is that a procedure that selects *for* protein synthesis mutants (for example) may also select *among* protein synthesis mutants; i.e., some types of mutations may be much more effective than others in promoting survival under the particular selective conditions used. Thus it is probably not safe to use any one selective system as a sole substitute for nonselective isolation. A second point to emphasize is that many *ts* lethal mutants lose viability at an appreciable rate at the restrictive temperature (Hartwell, 1967, 1971a,b; Littlewood, 1972; Klyce and McLaughlin, 1973). If such mutants are to be recovered, the selective procedure must kill the growing cells more rapidly than the nongrowing cells lose viability because of their mutations. This would probably not often be a problem with the tritium suicide and polyene antibiotic methods, since the periods of exposure to the restrictive temperature can be kept relatively short. It might be a problem with a method based on the fatty acid starvation results of Henry (1973), since an exposure of 10 hours or more would apparently be required to obtain a satisfactory degree of killing.

These considerations are especially important in the context of another type of survivor selection, in which the rapid loss of viability displayed by some *ts* lethal mutants at the restrictive temperature is used as a basis for selecting secondary *ts* mutations that prevent the loss of viability. This approach has been used by B. Reid and L. Hartwell (personal communication) to select for mutations in *cdc* genes whose products function at about the time of the "start" event (Hartwell *et al.*, 1974; Hartwell, 1974). [A very similar approach has been used by Littlewood (1972) to select for ordinary (i.e., not *ts*) antibiotic-sensitive mutants.] Because of the apparent importance of this part of the cell cycle with respect to the control of division, such mutants were of special interest. Unfortunately, the 150 *cdc* mutants obtained by nonselective isolation included only a few mutants with blocks at this part of the cycle, and many of these were too leaky to analyze in detail. Since stationary-phase yeast cells are in general arrested in this same part of the cycle, it seemed likely that at least some types of mutants that were arrested there would retain viability relatively well at the restrictive temperature. Indeed, when a *ts cdc* mutant that loses viability rapidly at the restrictive temperature was treated with mutagen and exposed to the restrictive temperature, a rather high proportion of the survivors seemed to have new *ts* mutations which led to arrest at or near the time in the cycle of the "start" event. (Presumably, when a population of such a mutant is shifted to the restrictive temperature, a portion of the cells are arrested at the point defined by the "old" *ts* mutation and die, while a portion of the cells are arrested at the point defined by the "new" *ts* mutation and survive.) It is not clear whether all types, or only some types, of "start" mutants can be isolated in this way. Other survivors were not analyzed in detail, but almost certainly included, among others, mutants defective in protein synthesis, since many such mutants retain viability well at the restrictive temperature (Hartwell, 1967; Klyce and McLaughlin, 1973). It is perhaps worth emphasizing that probably not all types of protein synthesis mutants would be selected by such a procedure, since at least some mutants that cease protein synthesis rapidly after a shift to the restrictive temperature also lose viability relatively rapidly (Hartwell, 1967).

It may be worth mentioning several other types of selections which seem to have some potential for facilitating the isolation of cell cycle mutants. First, since most cell cycle mutants continue growth without continuing division (Culotti and Hartwell, 1971; Hartwell, 1971a,b), it might be possible to enrich for such mutants by shifting a mutagen-treated population to the restrictive temperature, waiting several hours, and selecting the largest cells by a centrifugal technique. This type of approach has been used, with promising results, to select for *ts* cell cycle mutants of *Schizosaccharomyces pombe* (P. Nurse and B. L. A. Carter, personal communication). Note that this method would miss any mutants that lost viability too rapidly at the restrictive temperature, and would also miss any mutants that failed to continue growth at the restrictive temperature (such as the *cdc1* mutants; see Hartwell, 1971b, 1974; Hartwell *et al.*, 1974). Second, it is possible that at least some mutants that arrest at or near the "start" event may take on the properties of stationary-phase cells (which are also arrested at this same point in the cell cycle). These properties include a greater density than that of cells from a growing population, as well as an increased resistance to heat (Schenberg-Frascino and Moustacchi, 1972), to snail gut enzymes (Deutch and Parry, 1974), and to mutagens such as NA and EMS (J. M. Parry, personal communication). Any of these properties could clearly be the basis of an enrichment procedure for such mutants. The last-mentioned two properties may be the most suitable for this purpose, since they are known to show rather sharp transitions as wild-type populations progress from the exponential to the stationary phase (J. M. Parry, personal communication).

2. OTHER MUTATIONS

The methods available for the selection of *ts* mutations in dispensable genes are essentially identical to the methods that can be used for the selection of ordinary (i.e., unconditional) mutations in the same genes. All that is necessary is to manipulate the temperature in the appropriate fashion. In some cases, such as the Espositos' procedure for the isolation of mutants defective in sporulation (Esposito and Esposito, 1969; Esposito, 1975), the isolation of *ts* (or other conditional) mutants may be necessary to allow a subsequent analysis which would otherwise be difficult for technical reasons. (The mutants are obtained in homozygous form in a diploid strain which neither mates nor—unless the mutation is conditional—sporulates, making genetic analysis difficult.) For the same reason the method of Dawes and Hardie (1974) for detecting sporulation mutants could probably also be most profitably applied to the detection of conditional mutants. In contrast, the procedure devised by Roth and Fogel (1971) could be applied equally well to the isolation of conditional or of unconditional sporulation mutants; however, the procedure is only suitable for selecting mutants blocked in the early stages of sporulation.

IV. Handling and Preservation of Strains; The Problem of Genetic Complexity

In this section we primarily call attention to a group of related problems which complicate attempts to interpret the phenotypes of mutants. Lacroute (1975) has also emphasized these problems.

The first problem is that a mutant strain, as isolated, may have suffered more than one mutation that affects (directly or indirectly) the cellular process of interest. The probability of isolating such double (or multiple) mutants is of course higher when a vigorous mutagenesis has been used to induce mutations. Thus, when relatively common types of mutants are sought, it may be desirable to use a rather mild mutagenesis, so as to minimize the multiple-mutation problem. However, when rare mutants are sought, it is probably more effective to use a vigorous mutagenesis, and then to attend with special care to the precautions to be described. We emphasize, however, that in no case should these precautions be ignored.

Several variants of this problem must also be noted. First, although in general the two mutations in a double mutant are not closely linked [according to Mortimer and Hawthorne (1973), the chance that two mutations picked at random will show *any* linkage is about 1%], such close linkage is always a possibility, and may be a significant danger after NG mutagenesis (see Section II). Thus observation of 2:2 segregations in a limited number of tetrads (see below) is not definitive proof that a single mutation is responsible for all aspects of the mutant phenotype observed. A closely related possibility is that a single mutation is present, but that it is a multisite mutation (e.g., a deletion) which affects more than one gene. Finally, it is possible that the expression of a particular mutation in a particular strain is influenced not simply by one or two other newly induced mutations with easily definable effects, but by the "genetic background" of the parent strain. That is, another strain harboring the same mutation, but having a different genetic background, would display a somewhat different phenotype.

It may be thought that the possibilities mentioned are dangers only in theory, and not in practice. This is most emphatically not the case. The following examples, all but one drawn only from work on *ts* mutants of yeast, should leave no doubt on this point. Lacroute (1967) tested six *ts* mutants by crossing them with wild type; one of these did not show a 2:2 segregation of temperature sensitivity. Hartwell's mutant *ts296* (defective in methionyl-tRNA synthetase) apparently contained two unlinked *ts* mutations, and possibly other defects as well (McLaughlin and Hartwell, 1969). Backcrosses were a necessary prerequisite to a proper characterization of this mutant. Mutant *ts136* (defective in mRNA production) carried a second mutation which interfered with genetic analysis by influencing spore survival (Hutchison et al., 1969). Of the 10 *rna* mutants tested by tetrad analysis (Hartwell et al., 1970b), two segregated excesses of temperature-sensitive spores (suggesting the presence of two unlinked *ts* mutations), while one segregated an excess of wild-type spores (suggesting that the simultaneous presence of two independently segregating mutations was responsible for the temperature sensitivity of the original strain). The *ts* cell cycle mutants provide the most impressive data (Hartwell et al., 1973; Hartwell, 1974).

Of the 148 cell cycle mutants recognized, 9 contained 2 different *ts cdc* mutations; 11 contained a *ts cdc* mutation together with a second *ts* lethal mutation
of another type; 1 contained two different mutations, both of which were
necessary for temperature sensitivity; 1 contained a *ts cdc* mutation together
with a temperature-independent mutation leading to a modification in cell
morphology; 3 contained *ts cdc* mutations in known genes, but had slightly
atypical phenotypes (probably because of additional uncharacterized
secondary mutations); and 1 had apparently become diploid. It must be
emphasized that a rather elaborate program (Hartwell *et al.*, 1973) of tetrad
analysis, complementation analysis, and comparisons of the phenotypes of
independently isolated allelic mutants was necessary to detect all these
complications, and that there were undoubtedly other more subtle complications which were not recognized. Study of several of these mutants
without the benefit of the genetic analysis would have been quite misleading.

Since the Hartwell mutants were all isolated from a single parent strain,
the various complications described were presumably all introduced during
the mutagenesis procedure. An example of a somewhat different type is
provided by the "glycogenless" mutant isolated by Rothman-Denes and
Cabib (1970). The inability of this mutant to grow on acetate proved not to
be an obligatory consequence of its inability to accumulate glycogen.
Rather, it seems that the simultaneous presence of the mutation affecting
glycogen metabolism and another mutation (or elements of the genetic
background) apparently present in the "wild-type" strain used as parent for
the mutant induction led to the inability to grow on acetate. When the
original mutant was crossed to a different wild-type strain, all the glycogen-
accumulating segregants, and some of the glycogenless segregants, grew
well on acetate (Pringle *et al.*, 1974).

It should be clear from these few examples that any study of the physiological or developmental effects of mutations *must* incorporate controls
against the complications outlined above. A judicious use of the following
approaches is recommended. (In any actual study it is necessary to compromise somewhat between the level of proof and the time necessary to
obtain it.):

1. *Tetrad analysis and backcrossing.* Any mutant that is to be subjected
to detailed analysis should be crossed with some standard wild-type strain,
and several tetrads should be analyzed. [Note that it is usually unnecessary
to analyze in detail all the mutants yielded by the initial isolation procedure.
Ordinarily, the mutants of interest can be sorted into complementation
groups (in working with a previously unstudied type of mutant, it is necessary first to analyze a subset of the mutants in order to establish some complementation groups and to obtain testers), and one or two promising
members of each complementation group can be selected for detailed

analysis.] If the various relevant cellular properties segregate together, and in a 2:2 fashion, in each of about 10 tetrads, this is a rather strong argument that these properties are all due to a single mutation. A more detailed comparison of the phenotypes of several individual mutant segregants allows a statement about the lack of observable effects of genetic background. (To make a stronger statement, one can analyze segregants resulting from crosses of the mutant with two unrelated wild-type strains.) If the first set of tetrads reveals any irregularities, one or two mutant segregants of appropriate mating type can be picked and again crossed with the standard wild-type strain. With luck, a regular segregation will then be observed, and detailed physiological or developmental analysis can focus on these segregants, or on diploids constructed from them. Although this point may be debatable, it seems to us that a detailed physiological or developmental analysis should always focus on the segregants from a first, or even a second, cross with a standard wild-type strain, even when no irregularities are apparent in the first set of tetrads. We argue as follows. Ideally, one would like to study the effects of various mutations in a single genetic background. Although the different mutants may have been isolated from a single parent strain, the mutagenesis will certainly have destroyed their "isogenicity," even if this is not obvious from the tetrad analysis. (For example, it has been noted in Section I,B that differences in growth rate at the permissive temperature may sometimes be due to the presence of additional mutations.) Crossing each mutant twice with a wild-type strain will not restore isogenicity [except to the extent that this wild-type strain is isogenic to the parent strain of the mutants; this may be a strong argument for working with the S288C family of strains (see Section II); one could isolate mutants in strain X2180-1Aa, and then, in effect, backcross to the isogenic (except at the mating-type locus) strain X2180-1Bα (Duntze *et al.*, 1970)] but will at least tend to discard clear abnormalities (i.e., genetic background not found in typical wild-type strains) introduced by the mutagenesis, and will give, for each mutant, some closely related segregants whose properties can be compared. It should be noted that diploids constructed from such segregants have the advantage that much genetic variability is cloaked by heterozygosity.

Finally, it must be noted that tetrad analysis and backcrossing are not strong controls against the possibility of closely linked or multisite mutations.

2. *Complementation analysis.* Comprehensive complementation analysis can identify many cases in which the phenotype due to a particular mutation is masked or distorted by the presence of another mutation. (This and other advantages of complementation analysis were already noted in Section III,A,2.) Hartwell *et al.* (1973) provide several illustrations. This approach

increases in power as a system becomes better studied (i.e., as more and more complementation groups are defined). It has the potential to identify or suggest the presence of closely linked or multisite mutations, since a particular mutant may fail to complement with testers for each of two linked but distinct genes, or a particular allele (identified by complementation) may seem to give a phenotype different than that given by other alleles of the same gene (thus suggesting that something is amiss).

3. *Comparison of the phenotypes due to multiple independently isolated alleles.* This is a powerful control against genetic complications, especially when used in conjunction with complementation analysis so that the recognition of different allelic mutants is not dependent on their having the same phenotype. This approach may detect cases of closely linked or multisite mutations, or cases in which the gene product of a particular mutant allele has acquired an abnormal function. Other advantages of this approach were indicated in Section I,B; its use is well illustrated by Hartwell *et al.* (1973).

4. *Revertant analysis.* Determining that two or more phenotypic characteristics revert simultaneously is a classic and very useful argument that a single mutation is responsible for both.

Finally, it should be noted that the time of mutagenesis is not the only point at which genetic complications can appear. There is also a real danger that strains will evolve, during storage and subculturing, through the appearance of reversions, suppressor mutations, and secondary modifying mutations. This is a particular danger with a collection of original (i.e., never crossed with a wild-type strain) mutant isolates, since the mutagenesis may well have introduced various lesions which reduce growth rate or the retention of viability during storage, thus supplying selection pressure if more fit modified mutants should appear. Any *ts* mutant should be regarded as a special risk, even if other mutations have been eliminated by crosses with the wild type, since the strain is likely to be somewhat defective even at permissive temperatures because of the *ts* mutation itself (see Section I,B). From these considerations two morals can be drawn. First, as far as possible, a mutant strain should be stored as some kind of permanent stock which does not require subculturing and in which no evolution or selection is possible. Second, the characteristics of the clones on which experiments are to be done should be checked (as far as possible) from time to time against the recorded characteristics of the original mutant strain.

It is not clear that there is a perfect solution to this problem of the storage of strains. The following approach, which we have used extensively, seems reasonably satisfactory. As soon as a new mutant has been identified, cells from a single colony are used to inoculate a small culture in YM-1 medium. (YM-1 medium contains, per liter, 10 gm succinic acid, 6 gm NaOH, 7 gm

Difco Yeast Nitrogen Base without amino acids, 5 gm Difco yeast extract, 10 gm Difco Bacto peptone, and 20 gm glucose.) This culture is allowed to grow until the early stationary phase is reached. A portion is then mixed thoroughly with an equal volume of sterile glycerol solution (30 ml glycerol made up to 100 ml with water), and the mixture is stored at $-20°$ to $-80°$C. These frozen samples serve as permanent reference stocks. When necessary, a frozen stock is thawed, and a miniscule amount of cell suspension withdrawn with a sterile toothpick and streaked on solid medium. Cells from a single colony are then picked for inoculation into YM-1; this liquid culture is allowed to grow to early stationary phase, and is then stored at $0°$ to $4°$C. The characteristics of the strain are checked with a sample from this liquid stock, which then serves as a source of cells for experiments until it is used up, contaminated, or shows signs of unhealthiness (e.g., a significant number of dead or lysed cells, or an abnormally long lag period on inoculation into fresh medium). A new liquid stock is then derived from the frozen stock by the procedure described, and *not* by subculturing of the previous liquid stock. Any strain constructed from the original mutants is stored and handled in an identical fashion.

It should be emphasized that this method is not foolproof; there is a significant mortality during each freeze-thaw cycle, and this may in some cases be selective. On one occasion successive streakings from the frozen stock of a standard wild-type diploid revealed a progressively higher proportion of a small-colony variant. The nature of this variant was not investigated, and it is not clear why its frequency increased so markedly. A single colony of normal size was used to make a new frozen stock, and the problem has not recurred. Since such problems can occur with a healthy wild-type strain, one must be especially alert when working with mutants. Perhaps the optimal solution (recommended also by Lacroute, 1975) is to mate each mutant with a standard wild-type strain, and to let a frozen stock of this heterozygous diploid serve as the primary storage form for the mutation.

The use of dried stocks (either lyophilized or mixed with silica gel; see Fink, 1970) may be a satisfactory alternative to the use of frozen stocks, but we have had no personal experience with these methods.

V. Experimental Uses of Temperature-Sensitive (*ts*) Mutants

A. General and Miscellaneous

It was emphasized in Section I that the major usefulness of *ts* mutants is that they provide one of a limited number of ways to probe genetically the indispensable functions of the cell. It should be noted that the mutants can be isolated without knowing anything about the functions that will be affected; thus they may reveal the existence of previously unknown indispensable functions. Moreover, sometimes the existence of a function is known from biochemical studies, but it is not clear whether or not the function is indispensable to the cell. If mutants lacking such a function are sought, it is clearly prudent to seek *ts* (or other conditional) mutants.

It was also pointed out in Section I,B that *ts* mutations in dispensable genes may sometimes be more useful than ordinary mutations in the same genes. This is a consequence of the following possible uses of *ts* mutations (note that *ts* mutations in indispensable genes can also be used in these ways):

1. A mutation that is *ts* is probably in the structural gene for a protein and is probably a missense mutation. Note that there are some exceptions to each of these conclusions (see Section I,B), but that the former can be much strengthened if a suspected protein product can be shown to be temperature-sensitive *in vitro* (see the references cited in Section I,B). The demonstration of *in vitro* temperature sensitivity can be particularly useful in distinguishing between genes that control the structure of a protein of interest and genes that regulate the synthesis of that protein or control the production of an essential cofactor (e.g., see MacDonald and Cove, 1974).

2. Purification of the product of a mutant gene may be facilitated if it has an activity that can be followed when extracts are prepared from cells grown at permissive temperatures.

3. A *ts* mutant can be shifted at will from a wild-type to a mutant phenotype, and frequently in the reverse direction as well. This may be experimentally convenient in some situations; the use of *ts* mutations in the *cI* gene of phage λ (Herskowitz, 1973) and of *ts* behavioral mutants of *Paramecium* (Chang and Kung, 1973) provide examples, as do several of the studies of macromolecule synthesis cited in Section V,C.

4. A special case of the usefulness of such shifts is in efforts to learn about the timing of various gene-controlled steps in sequences such as those of the mitotic division cycle, or of meiosis and sporulation. It must be noted, however, that this is a tricky business. The results of shifts from the permissive to the restrictive temperature (Hartwell *et al.*, 1970c, 1973; Hartwell, 1974; Esposito *et al.*, 1970) are in principle relatively simple to interpret, although interpretation is confounded in practice by the difficulty of distinguishing among mutants that are *ts* for function, those that are *ts* for synthesis, and those that are leaky to various degrees (see further comments in Sections I,B and VI,D). Sometimes these problems can be at least tentatively circumvented by a comparison of different allelic mutants [see the argument of Hartwell *et al.* (1973, p. 284–285) and of Hartwell (1974, p. 168)]; sometimes they can be circumvented if a biochemical handle is available. For example,

biochemical (Kane and Roth, 1974) and genetic (Pringle *et al.*, 1974) arguments suggest that the accumulation and subsequent utilization of glycogen are necessary for ascospore production. If a *ts* mutant blocked in glycogen accumulation were available, it would be possible to determine, by appropriate shifts to the restrictive temperature, how much glycogen accumulation is sufficient for sporulation. The effects of the shift could be monitored by direct measurements of glycogen. The results of shifts from the restrictive to the permissive temperature (Esposito *et al.*, 1970) seem to be difficult to interpret, even in principle; they seem to define the time at which recovery (from an initial failure to perform some function) is no longer possible, rather than the time at which that function normally occurs.

A variety of other miscellaneous uses of *ts* mutants may be mentioned. Mortimer and Hawthorne (1973) made extensive use of *ts* lethals from the Hartwell (1967) and Klyce and McLaughlin (1973) collections in improving the quality of the yeast genetic map. Further efforts along these lines, using additional *ts* (or other conditional) lethals, would not be wasted. Note that for this purpose it is not necessary to know anything about the functions that are defective in the mutants used. Hartwell (1973b) simplified his analysis of the kinetics of mating considerably by performing the experiment at the restrictive temperature with an *a* and *α* pair carrying the same *cdc* mutation. *ts* mutants defective in the initiation of DNA replication have been used to prepare synchronized cultures for studies of DNA replication (Hereford and Hartwell, 1973; Newlon *et al.*, 1974). *ts* mutants that die rapidly at the restrictive temperature have been used to select other mutations (see Section III,B,1). Lacroute (1975) has described a clever use of *ts* mutants in metabolic studies.

Finally, *ts* mutants are useful in the study of sporulation for several reasons. First, as already noted in Section III,B,2, some isolation schemes for sporulation-defective mutants make it highly desirable to isolate conditional mutants, since it is otherwise difficult to recover the mutations for genetic analysis. Second, the potential usefulness of *ts* mutants in allowing the timing of events during sporulation was mentioned above. Third, only the use of *ts* (or other conditional) mutants makes possible an analysis of the roles in sporulation of genes that also function in the mitotic cell cycle (Simchen, 1974). Other uses of *ts* mutants defective in sporulation have been described by Esposito and Esposito (1974), Moens *et al.* (1974), and Esposito (1975).

B. Studies of the Cell Division Cycle

The various aspects of the yeast mitotic cell division cycle, and their study by the use of *ts* mutants, have recently been reviewed in some detail by

Hartwell (1974). Thus, we summarize here only the *types* of uses to which *ts* mutants have been put in studying this complex process. The isolation of *ts* cell cycle mutants, and some of the complications that have arisen in studying them, have been discussed elsewhere in this article (Sections III,A,2 and B,1, and IV, and the introductory portion of Section VI).

1. *Identification of previously unknown functions.* The best example to date is provided by the sequence of gene-controlled steps that precedes, and is necessary for, the initiation of DNA synthesis (Hereford and Hartwell, 1974; Hartwell *et al.*, 1974; Hartwell, 1974; Byers and Goetsch, 1974). The existence of this sequence was not suspected, and would have been quite difficult to detect, before the appropriate mutants were available. It may also be noted that many of the 14 known genes (Hartwell *et al.*, 1973) whose products seem necessary for nuclear division, but not for DNA synthesis, must control functions which are at present unknown.

2. *Investigation of the molecular mechanisms of cell cycle events.* The potential usefulness of *ts* mutants in such studies has hardly begun to be exploited, but is surely obvious—one has only to imagine the army of *Escherichia coli* biologists attempting to slog its way through the quagmire of DNA replication without the support of the various *ts* mutants (Horiuchi and Nagata, 1973; Beyersmann *et al.*, 1974; Wickner *et al.*, 1974; Tait and Smith, 1974; Konrad and Lehman, 1974; Hiraga and Saitoh, 1974; Sakai *et al.*, 1974; Paul and Inouye, 1974). Promising beginnings have also been made in exploiting the available mutants to study DNA replication in yeast (Hereford and Hartwell, 1973; Newlon *et al.*, 1974; Wintersberger *et al.*, 1974; Bandas *et al.*, 1973; Hartwell, 1974; Petes and Newlon, 1974; Petes and Williamson, 1974). Another possibility which currently seems promising is that the mutants affected in bud emergence and in cytokinesis (Hartwell, 1971b, 1974; Hartwell *et al.*, 1973) may help in testing the ideas about these processes that have been developed by Hayashibe and Katohda (1973) and by Cabib *et al.* (1974).

3. *Examination of the interdependence of cell cycle events.* Even in the absence of knowledge about the molecular defects in the various mutants, it is possible to infer from their phenotypes some aspects of the coordination of major cell cycle events (Hartwell *et al.*, 1974; Hartwell, 1974). For example, since DNA replication and nuclear division can occur in a mutant in which bud emergence fails to occur, we infer that neither of the former two processes depends on the prior occurrence of budding.

4. *Assessment of progress through the cell cycle.* Because a cell harboring a *ts* cell cycle mutation can complete one normal division cycle at the restrictive temperature if it has completed the temperature-sensitive event at the permissive temperature (Hartwell *et al.*, 1970c, 1973; Hartwell, 1974), it is possible to assess the position of a cell within the cycle by determining which of the temperature-sensitive events (defined by the various mutations)

it has completed. By using this approach it was possible to show that cells arrested by the α and a mating factors (Bücking-Throm *et al.*, 1973; Wilkinson and Pringle, 1974), and cells in stationary-phase populations (J. Pringle, R. Maddox, and L. Hartwell, in preparation), were all arrested at the very beginning of the cell cycle.

5. *Distortion of the normal relationship between growth and division.* In one possible mechanism for the coordination of growth with the division cycle, continued growth would be dependent on progress through the division cycle. The fact that the cell cycle mutants seem in general to continue growth while their division is blocked (Culotti and Hartwell, 1971; Hartwell, 1971a,b) seems to be an argument that this type of coordinating mechanism does not exist (see Hartwell, 1974, p. 190).

6. *Synchronization of cells* by means of reversible arrest at a mutational block was mentioned in Section A, above.

7. *The selection of additional cell cycle mutants* by using a cell cycle mutant as parent strain was discussed in Section III,B,1.

C. Studies of Macromolecule Synthesis

ts mutants have been used in a variety of ways to study the mechanisms, regulation, and functions of RNA and protein synthesis in yeast. It is impossible to do more here than to provide a superficial summary, with appropriate references, of this variety of uses.

The properties of *ts* mutants defective in methionyl-tRNA synthetase (McLaughlin and Hartwell, 1969) and in isoleucyl-tRNA synthetase (Hartwell and McLaughlin, 1968b) have indicated that the yeast cell normally contains a single species of each of these enzymes, which is responsible for charging all relevant tRNA species. In the former case this presumably includes the tRNA species that initiates translation. These mutants have also provided useful information about the overall coordination between protein synthesis and RNA synthesis (McLaughlin *et al.*, 1969); the conclusions drawn agree with those obtained by examining the total spectrum of *ts*-mutant types obtained (Hartwell, 1967) and by other methods not employing *ts* mutants (Roth and Dampier, 1972). The isoleucyl-tRNA synthetase mutant has also provided information about the regulation of the synthesis of the enzymes of isoleucine biosynthesis (McLaughlin *et al.*, 1969). Similar studies of mutants defective in other activating enzymes would clearly also be useful.

The various *ts* mutants defective in bulk RNA accumulation have been useful in studies of the mechanism and control of ribosome assembly (Hartwell *et al.*, 1970b; Warner and Udem, 1972), and may be expected to contribute still more in this area (see, for comparison, the similar work in bacterial systems; Davies and Nomura, 1972).

The *ts* mutants defective, respectively, in RNA (including mRNA) production (Hutchison *et al.*, 1969) and in the initiation of protein synthesis
(Hartwell and McLaughlin, 1969) have also been useful in a variety of ways.
The former mutant has helped to provide an estimate of the average half-
life of mRNA in yeast (Hutchison *et al.*, 1969; Tønnesen and Friesen, 1973)
and has, together with the latter mutant, provided information about the
factors governing polyribosome stability (Hartwell *et al.*, 1970a) and the
transport of mRNA from nucleus to cytoplasm (Shiokawa and Pogo, 1974).
The mutant defective in mRNA production has also been used to provide
a specific block of nuclear mRNA production in a study of mitochondrial
autonomy (Mahler and Dawidowicz, 1973), in a study of the dissociability
of ribosomes by salt (Martin and Hartwell, 1970), and in a study of the mode
of biosynthesis of the mating pheromone α factor (Scherer *et al.*, 1974). The
last of these studies also used the mutant defective in the initiation of protein
synthesis (as well as two mutants apparently defective in polypeptide chain
elongation) to test the effects of specific blocks in protein synthesis; the
results obtained agreed with those obtained after blocking protein synthesis
with cycloheximide or by amino acid starvation. The specific blocks in
protein synthesis provided by the mutants have also been exploited in
a study of the role of protein synthesis during sporulation (Magee and
Hopper, 1974), and in a study of the role of polyadenylate sequences in
yeast mRNA (McLaughlin *et al.*, 1973).

It is worth noting in a general way the advantages of using *ts* mutants to
provide blocks in RNA and protein synthesis in a variety of experimental
contexts. First, the number of specific inhibitors that are effective in yeasts
is fairly small and, in any case, with inhibitors there is always the danger
of side effects. The specificity of a mutant block can be assumed to be very
high. Second, it is frequently desirable to compare the effects of inhibiting
protein or RNA synthesis in several different ways; the use of various
mutants, in conjunction with such inhibitors as are available, clearly greatly
increases the possibilities. Third, in some cases the blocks imposed by the
mutations are more readily reversible than those imposed by the available
inhibitors. Against these advantages may in some cases be set the disadvantage that the obligatory shift in temperature renders interpretation
of the results difficult (Magee and Hopper, 1974).

VI.   Other Types of Conditional Mutants

In this final section we call attention to the possibility, and to the importance, of complementing studies of conventional *ts* mutants with studies of

other types of conditional mutants. Studies of other conditional mutants are important for at least two general reasons (see also Section VI,D):

1. Attempts such as those of Jarvik and Botstein (1973) and Hereford and Hartwell (1974) to deduce "order-of-function" sequences for the gene-controlled steps of pathways are dependent on having two *different* kinds of conditional blocks. Some progress can be made by using *ts* mutants in conjunction with inhibitors (Hereford and Hartwell, 1974), but a more comprehensive effort can be made if two types of conditional mutants (such as *ts* and cold-sensitive mutants; Jarvik and Botstein, 1973) are available.

2. There is a significant danger that by working only with conventional *ts* mutants we will fail to observe (i.e., fail to isolate any mutants altered in) some genes that do not readily mutate to *ts* alleles. Since different individual normal proteins differ substantially in their thermostabilities, it seems evident a priori that the different genes coding for them will differ substantially in the frequencies with which they give rise to *ts* mutations. (For example, a protein normally only marginally stable at 36°C will probably be readily converted by a single amino acid change to a protein stable at 23°C but unstable at 36°C, while a protein normally stable even at 50°C will in general be more difficult to destabilize to just the right degree.) In addition, some genes may possess mutational hot spots which tend to give temperature-sensitive products, while other genes may not. Experience seems to support these expectations. That is, although it seems possible to obtain *ts* mutations in most of the genes of an organism (see Section I, and particularly the references on the work with phages λ and T4), it is clear that the mutations in a given collection are not distributed randomly among the different possible genes. This is apparent in the T4 work (Epstein *et al.*, 1963; Edgar and Lielausis, 1964), and also in the work with *cdc* mutants of yeast (Hartwell *et al.*, 1973). (The large number of alleles obtained in a gene apparently concerned with energy metabolism seems to be another example of an especially susceptible gene; see Hartwell and McLaughlin, 1968a.) It must be emphasized that while failing to obtain mutations in some genes can be merely inconvenient (e.g., it would be useful to have mutations in the genes coding for the remaining 18 aminoacyl-tRNA synthetases; see Section V,C), it can also be seriously misleading. For example, mutants defective in gene *cdc1* were originally considered to demonstrate the typical effects of a specific defect in bud emergence (Hartwell, 1971b). Thus nuclear division, and even the continuance of RNA and protein synthesis, were thought to depend on the emergence of a bud. The subsequent discovery of mutants defective in gene *cdc24* (Hartwell *et al.*, 1973; Hartwell, 1974), which continued nuclear division and macromolecule synthesis despite their failure to bud, resulted in a reevaluation of the

significance of the *cdc1* mutants and in the construction of a cell cycle model based (in part) on the properties of the *cdc24* mutants (Hartwell *et al.*, 1974; Hartwell, 1974).

At least four additional types of conditional mutants seem likely to be useful in yeast studies. These are (1) cold-sensitive mutants, (2) pH-sensitive mutants, (3) osmotic-remedial mutants, and (4) *ts* mutants isolated in strains harboring temperature-sensitive nonsense suppressors. [A fifth type of conditional mutant, streptomycin-suppressible mutants such as those described for *E. coli* by Murgola and Adelberg (1970), may also now be obtainable in yeasts (Bayliss and Vinopal, 1971)]. Since none of these types has as yet been exploited extensively in yeast studies, the comments that follow deal with what might be, rather than what has been, done. As these comments will note in more detail, it seems likely that the usefulness of these types of mutants for revealing new genes will be in the order $4 > 1 > 2 > 3$, and that their suitability for use (in conjunction with conventional *ts* mutants) in the determination of order-of-function sequences will be in the order $1 > 2 > 3$.

A.   Cold-Sensitive Mutants

Such mutants resemble the wild type at temperatures in the middle or upper part of the normal growth range, but display a mutant phenotype at lower temperatures. They have been isolated in many different organisms, including the bacteriophages λ (Cox and Strack, 1971; Herskowitz, 1973), phiX-174 (Dowell, 1967), T4 (Scotti, 1968), and P22 (Jarvik and Botstein, 1973); the bacteria *E. coli* (O'Donovan and Ingraham, 1965; Guthrie *et al.*, 1969; Nikiforov *et al.*, 1974; Waskell and Glaser, 1974) and *Salmonella typhimurium* (Tai *et al.*, 1969; Brenchley and Ingraham, 1973; Ginther and Ingraham, 1974); *Neurospora* (Horowitz and Leupold, 1951; Roberds and DeBusk, 1973; Schlitt and Russell, 1974); and *Drosophila* (Falke and Wright, 1974). Cold-sensitive yeast mutants have been isolated by Esposito and Esposito (1969), Littlewood and Davies (1973), Weislogel and Butow (1970), Bayliss and Vinopal (1971), Singh and Manney (1974), and Hartwell *et al.* (1970b; these mutants were simultaneously *ts* and cold-sensitive; cf. Cox and Strack, 1971), but no systematic effort has been made.

It is at present not clear how much one can generalize about the molecular defects of cold-sensitive mutants. There is support for the idea that such mutants are frequently defective in genes whose protein products must have allosteric function, or participate in assembly into complex structures (see Brenchley and Ingraham, 1973, and references cited by them). At the same time, simple protein chemical considerations, both theoretical and empirical, suggest that many proteins should be convertible by partially destabilizing amino acid substitutions into forms that lose activity at low temperatures. [Since hydrophobic forces are important in keeping poly-peptide chains properly folded, and since these forces weaken as the tem-perature decreases, it is not surprising that some normal proteins exhibit cold lability (see Brown *et al.*, 1973; Huang and Cabib, 1973; and other references cited in these papers). Note that cold-labile proteins can also be denatured by heating. It thus seems likely that many proteins can be con-verted, by appropriately destabilizing amino acid substitutions, to *either* abnormally cold-labile or abnormally heat-labile forms. It even seems likely that in some cases the *same substitution* will give both abnormal cold lability and abnormal heat lability.] Empirically, the studies cited in the preceding paragraph have made it clear that (1) a large number of genes are potential sites of cold-sensitive mutations; (2) in many of these genes it is also possible to isolate *ts* mutations, and in some cases a single mutation is both *ts* and cold-sensitive; (3) cold-sensitive mutations certainly occur in some genes much more often than they do in other genes; and (4) the genes that are especially prone to cold-sensitive mutations are frequently not the ones that are especially prone to *ts* mutations.

From these considerations it is clear that searching for cold-sensitive mutants is reasonably likely to reveal genes that have been overlooked while collecting *ts* mutants, and that a combination of *ts* and cold-sensitive mutants is a reasonable approach to order-of-function tests, as already demonstrated in practice by Jarvik and Botstein (1973).

It is worth noting the problem of determining which cold-sensitive muta-tions are located in genes already defined by *ts* mutations. A standard com-plementation test does not work, and the method used by Jarvik and Botstein (1973), which depends on having a known nonsense mutation in the same gene, would be difficult to apply in yeast. However, it seems that a combination of linkage testing and order-of-function testing would allow an almost certain determination of allelism, given the great genetic length of the yeast map and the paucity of cases of clustering of functionally related genes (Mortimer and Hawthorne, 1973), and the power of order-of-function tests (Jarvik and Botstein, 1973; Hereford and Hartwell, 1974).

B. pH-Sensitive Mutants

Such mutants resemble the wild type when the environmental pH is rela-tively high, and display a mutant phenotype when the environmental pH is relatively low, or vice versa. Few attempts have been made to isolate such mutants, and it is not entirely clear that they can be isolated in sufficient number and variety to be highly useful. Theoretical considerations are moderately encouraging. Certainly protein function is sensitive to changes in pH, and the available information does suggest that the internal pH of yeast cells varies significantly with changes in the environment (Conway and Downey, 1950; Kotyk, 1963; Neal *et al.*, 1965). Thus it is reasonable to suppose that a partially destabilizing amino acid substitution could make a protein unable to function well at an extreme of intracellular pH. Moreover, the only serious attempt to isolate pH-sensitive mutants of which we are aware (Campbell, 1961, working with phage λ) was moderately successful; several mutants that were simultaneously *ts* and pH-sensitive, and two that were pH-sensitive but not *ts*, were isolated. pH-sensitive mutants did seem to occur somewhat less frequently than *ts* mutants, but the sample was small, and it is not clear that the pH values used were optimal. From simple protein chemical considerations, and from the modicum of evidence from Campbell's study, it seems likely that the set of genes prone to give pH-sensitive mutants would only partly overlap with the set of genes prone to give *ts* mutants. Thus if pH-sensitive mutants can be obtained in any frequency, they should be useful both in identifying new genes and in order-of-function studies.

C.   Osmotic-Remedial Mutants

Such mutants display mutant phenotypes in normal media, but resemble the wild type in media of increased osmolarity. Such mutants occur in high frequency in yeast (Hawthorne and Friis, 1964; Bassel and Douglas, 1968; Esposito, 1968; Jones, 1972a,b; Singh and Sherman, 1974; Venkov *et al.*, 1974) and in *E. coli* (Russell, 1972; Bilsky and Armstrong, 1973; Egan and Russell, 1973). In some cases osmotic-remedial mutants seem to be defective in the synthesis of cell wall or membrane, so that they exist as fragile forms which lyse or leak their contents in the absence of a high osmotic pressure in the environment (see Russell, 1972; Venkov *et al.*, 1974; and the references given in these articles). However, it is clear from the references cited above that the great majority of osmotic-remedial mutations occur in genes that have no obvious direct function in wall or membrane synthesis. Although it is possible that nonobvious effects at the cell membrane level are involved in the osmotic-remedial phenotype of some of these mutants (Russell, 1972; Bilsky and Armstrong, 1973; Egan and Russell, 1973), it seems that the general explanation is that many not too drastically modified mutant proteins are stabilized in their active conformations by an increase in extracellular osmolarity. Since ionic (e.g., potassium chloride) and nonionic (e.g., sucrose) solutes are in almost all cases equally effective in permitting the growth of osmotic-remedial mutants, the crucial factor is probably plasmolysis of the cell rather than a direct stabilizing effect of the solute itself. It is not surprising that most osmotic-remedial mutations seem to be missense mutations (Hawthorne and Friis, 1964), although exceptions have been noted (Jones, 1972a,b; Singh and Sherman, 1974).

Given this explanation of the osmotic-remedial phenotype, it is not surprising that many mutants isolated as *ts* mutants are osmotic-remedial, and vice versa (see the various references cited at the beginning of this paragraph). This high degree of overlap may significantly limit the usefulness of osmotic-remedial mutants. However, it should be noted that the overlap is not complete (so that order-of-function studies might be possible), and that in at least some studies (Jones, 1972a; Singh and Sherman, 1974) there are suggestions that osmotic-remedial mutants may occur in greater numbers than *ts* mutants (so that it is at least possible that some genes that do not readily give *ts* alleles would give osmotic-remedial alleles).

D. *ts* Suppressors

The isolation of *ts* nonsense suppressors in yeast (Rasse-Messenguy and Fink, 1973) is an important development. Studies of bacteriophages have relied heavily on the use of mutants that harbor nonsense mutations in essential genes (Campbell, 1961; Hershey, 1971; Herskowitz, 1973; Epstein *et al.*, 1963; Wood *et al.*, 1968; Eiserling and Dickson, 1972; Jarvik and Botstein, 1973). Such mutants are conditional because they can be propagated in host cells that possess an appropriate nonsense suppressor, and studied in host cells that lack such a suppressor. This type of conditional mutation has, however, been of almost no use in the study of essential *cellular* functions. The problem is obvious: a cell that contains both nonsense mutation and suppressor does not display the mutant phenotype, and a cell that contains nonsense but no suppressor cannot grow. This makes it difficult to isolate mutants, and impossible to study the effects of any mutations isolated except by observing the germination and subsequent limited lives of individual spores. What has been needed, clearly, is a method to terminate suppressor activity simultaneously in all the cells of a vegetative population. The *ts* nonsense suppressors, isolated first in bacteria (Galluci *et al.*, 1970; Smith *et al.*, 1970; Nagata and Horiuchi, 1973; Beckman and Cooper, 1973), and now in yeast (Rasse-Messenguy and Fink, 1973), provide this method.

Before proceeding, the following complication must be noted. A strain that harbors a nonsense mutation and a nonsense suppressor can be *ts* for either of two reasons. (Note also that these possibilities are not mutually exclusive.) First, the suppressor may function effectively only at permissive temperatures; thus, at restrictive temperatures, only "nonsense fragments" can be produced. Second, suppression may be active at all temperatures, but the protein product of the gene containing the nonsense mutation can be *ts* (either for synthesis or for function) because the suppressor has inserted an amino acid different from that present in the wild-type protein. [This second possibility can be illustrated by the data of Campbell (1961), who showed that several nonsense mutations of phage λ were *ts* when grown in the permissive host.] This complication must be kept in mind if additional *ts* suppressors are to be isolated. Several different nonsense mutations must be tested with the putative *ts* suppressor in order to establish that it is *suppression*, and not a particular protein produced by suppression, that is temperature-sensitive. Although the *ts* suppressors of Rasse-Messenguy and Fink have passed this test, the problem should also be kept in mind while using them. It is possible that the suppressor could revert to temperature insensitivity, and yet a collection of *ts* mutants isolated in a strain containing the suppressor would still contain some mutants that harbored suppressible nonsense mutations that were *ts* for the reason indicated. However, the special advantages discussed below only apply when suppression itself is temperature-sensitive.

To exploit the *ts* nonsense suppressors in a study of essential genes, one isolates *ts* lethals in a conventional way (Section III), but using a parent strain that possesses the *ts* suppressor gene. Ideally, the parent strain should also contain at least one auxotrophic mutation suppressible (at permissive temperatures) by the *ts* suppressor, and the isolation should thus employ rich medium. The mutants isolated will fall into two basic classes: (1) conventional *ts* mutations, unrelated to the *ts* suppressor (i.e., predominantly missense mutations in various essential genes); and (2) nonsense mutations, in various essential genes, which are suppressible (at permissive temperatures) by the *ts* suppressor. The class to which a particular mutant belongs can be determined by either or both of two methods. In the first method (suggested by the tests used by Beckman and Cooper, 1973), temperature-resistant "revertants" (i.e., clones once again capable of growth at the restrictive temperature) are isolated. If the mutant is in class 1, almost all such revertants will still possess the auxotrophic mutation and *ts* suppressor, and so will be auxotrophs at the restrictive temperature. In contrast, a significant proportion of the revertants of a mutant from class 2 should be strains in which the suppressor function is no longer temperature-sensitive, and which are therefore prototrophic at the restrictive temperature. In the second method, the newly isolated *ts* mutant is crossed with a wild-type strain, and tetrads are analyzed for the growth of spores on complete medium at the permissive temperature. Mutants of class 2, but not those of class 1, should give tetrads with one (tetratype) or two (nonparental ditype) inviable spores.

An effort to isolate *ts* mutants in a *ts* suppressor strain seems to present three major advantages. First, mutants of class 2 should be obtainable in *any* essential gene, without regard to the properties of its protein product. This deals directly with the danger discussed in the introductory part of Section VI. Note that, if a search for additional genes is the only goal, it is not even necessary to sort out the mutants obtained into classes 1 and 2.

For example, in the context of cell cycle studies, one could identify cell cycle mutants (which might be either in class 1 or in class 2) on morphological grounds (Section III), and then enter these mutants directly into complementation tests with representatives of the known genes. These tests would work with mutants of either type, so long as the tester strains contained no relevant suppressors.

Second, as pointed out by Beckman and Cooper (1973), class-2 mutants, at the restrictive temperature, produce only fragments of the protein coded for by the gene of interest. This may facilitate the identification of the protein product of that gene, by making possible the search for missing or displaced protein bands (after electrophoresis) that has been so fruitful in virus work [see Laemmli (1970) or any of numerous other references in Eiserling and Dickson (1972)].

Finally, mutants isolated using a *ts* nonsense suppressor should be useful in learning about the temporal aspects of gene-controlled events. This can perhaps best be made clear with an example from cell cycle studies. For each gene whose product seems to be necessary for a particular single step of the cell division cycle, one would like to know, first, when in the cycle that gene product acts, and second, when in (or before) the cycle the gene product is synthesized. In principle, one can gain a significant amount of information relevant to these questions by the study of conventional *ts* mutants. The *execution point* (Hartwell *et al.*, 1970c, 1973; Hartwell, 1974) of a mutant that is a clear-cut *ts* for function should define the point in the cycle at which the gene product completes its action relevant to that cell cycle, while the execution point of a mutant that is a clear-cut *ts* for synthesis should define the point at which enough of the gene product has been synthesized to allow completion of the division cycle. In practice, the difficulty of determining which mutants are *ts* for synthesis, which are *ts* for function, and which are leaky (and to what degree) severely limits the amount of information that can be extracted from execution point data. The combination of a nonsense mutation in a *cdc* gene with a *ts* nonsense suppressor seems more likely to be manageable. *If* suppressor action ceases abruptly when the temperature is raised, and *if* the protein made (under the influence of the suppressor) at the permissive temperature is not itself thermolabile, such a mutant should behave like a conventional *ts* mutant which was a clear-cut *ts* for synthesis. Knowing the execution point of such a mutant would not only supply the information noted above, but would also help in interpreting the execution points of conventional *ts* mutants. The crucial point is that the rapidity and completeness with which suppressor activity shuts off following a shift to the restrictive temperature can be tested directly by doing temperature-shift experiments in strains carrying the *ts* suppressor and nonsense mutations in the structural genes of known enzymes that can be assayed. Note that such experiments should ideally be done with several different nonsense mutations in each enzyme tested, to minimize the chances that in a particular case the enzyme synthesized under the influence of suppression is itself thermolabile. There is of course also some danger that the protein product of a *cdc* gene being studied would itself be thermolabile when it is synthesized (at the permissive temperature) under the influence of the suppressor. This danger could also be minimized by comparing the execution points obtained with several different alleles in combination with the same *ts* suppressor; it is unlikely that in each case suppression would by coincidence lead to a thermolabile protein product.

ACKNOWLEDGMENTS

I thank Lee Hartwell, in whose laboratory I learned about *ts* mutants, and Armin Fiechter, in whose laboratory this chapter was written, for their advice and encouragement. I also thank Joe Culotti and Brian Reid for innumerable illuminating discussions, and B. S. Mitchell for a penetrating analysis. I have been supported by Swiss National Science Foundation Grant No. 3.628.71, U. S. Public Health Service Postdoctoral Fellowship FO2-GM41910, and by a training grant to the Department of Genetics, University of Washington.

REFERENCES

Bandas, E. L., Bekker, M. L., Luchkina, L. A., Tkatchenko, V. P., and Zakharov, I. A. (1973). *Mol. Gen. Genet.* **126**, 153.

Bassel, J., and Douglas, H. C. (1968). *J. Bacteriol.* **95**, 1103.

Bayliss, F. T., and Vinopal, R. T. (1971). *Science* **174**, 1339.

Beckman, D., and Cooper, S. (1973). *J. Bacteriol.* **116**, 1336.

Beyersmann, D., Messer, W., and Schlicht, M. (1974). *J. Bacteriol.* **118**, 783.

Bilsky, A. Z., and Armstrong, J. B. (1973). *J. Bacteriol.* **113**, 76.

Botstein, D., and Jones, E. W. (1969). *J. Bacteriol.* **98**, 847.

Brenchley, J. E., and Ingraham, J. L. (1973). *J. Bacteriol.* **114**, 528.

Brendel, M., and Langjahr, U. G. (1974). *Mol. Gen. Genet.* **131**, 351.

Brown, M. S., Dana, S. E., Dietschy, J. M., and Siperstein, M. D. (1973). *J. Biol. Chem.* **248**, 4731.

Bücking-Throm, E., Duntze, W., Hartwell, L. H., and Manney, T. R. (1973). *Exp. Cell Res.* **76**, 99.

Byers, B., and Goetsch, L. (1974). *Cold Spring Harbor Symp. Quant. Biol.* **38**, 123.

Cabib, E., Ulane, R., and Bowers, B. (1974). *Curr. Top. Cell. Regul.* **8**, 1.

Campbell, A. (1961). *Virology* **14**, 22.

Chang, S.-Y., and Kung, C. (1973). *Science* **180**, 1197.

Clarke, S. J., Low, B., and Konigsberg, W. H. (1973). *J. Bacteriol.* **113**, 1091.

Conway, E. J., and Downey, M. (1950). *Biochem. J.* **47**, 355.

Cox, J. H., and Strack, H. B. (1971). *Genetics* **67**, 5.

Culbertson, M. R., and Henry, S. A. (1974). *Genetics* **77**, s15.

Culotti, J., and Hartwell, L. H. (1971). *Exp. Cell Res.* **67**, 389.

Davies, J., and Nomura, M. (1972). *Annu. Rev. Genet.* **6**, 203.

Dawes, I. W., and Carter, B. L. A. (1974). *Nature (London)* **250**, 709.

Dawes, I. W., and Hardie, I. D. (1974). *Mol. Gen. Genet.* **131**, 281.

Deutch, C. E., and Parry, J. M. (1974). *J. Gen. Microbiol.* **80**, 259.

Dowell, C. E. (1967). *Proc. Nat. Acad. Sci. U.S.* **58**, 958.

Duntze, W., MacKay, V., and Manney, T. R. (1970). *Science* **168**, 1472.

Edgar, R. S., and Lielausis, I. (1964). *Genetics* **49**, 649.

Egan, A. F., and Russel, R. R. B. (1973). *Genet. Res.* **21**, 139.

Eggertsson, G. (1968). *Genetics* **60**, 269.

Eidlic, L., and Neidhardt, F. C. (1965). *J. Bacteriol.* **89**, 706.

Eiserling, F. A., and Dickson, R. C. (1972). *Annu. Rev. Biochem.* **41**, 467.

Epstein, R. H., Bolle, A., Steinberg, C. M., Kellenberger, E., Boy de la Tour, E., Chevalley, R., Edgar, R. S., Susman, M., Denhardt, G. H., and Lielausis, A. (1963). *Cold Spring Harbor Symp. Quant. Biol.* **28**, 375.

Esposito, M. S. (1968). *Genetics* **58**, 507.

Esposito, M. S., and Esposito, R. E. (1969). *Genetics* **61**, 79.

Esposito, M. S., and Esposito, R. E. (1975). *In* "Methods in Cell Biology" (D. M. Prescott, ed.), Vol. XI, p. 303. Academic Press, New York.

Esposito, M. S., Esposito, R. E., Arnaud, M., and Halvorson, H. O. (1970). *J. Bacteriol.* **104**, 202.

Esposito, R. E., and Esposito, M. S. (1974). *Proc. Nat. Acad. Sci. U.S.* **71**, 3172.

Falke, E. V., and Wright, T. R. F. (1974). *Genetics* **77**, s21.

Fink, G. R. (1970). *In* "Methods in Enzymology" (H. Tabor and C. W. Tabor, eds.), Vol. 17A, p. 59. Academic Press, New York.

Francis, J. C., and Hansche, P. E. (1972). *Genetics* **70**, 59.

Galluci, E., Paccheti, G., and Zangrossi, S. (1970). *Mol. Gen. Genet.* **106**, 362.

Ginther, C. L., and Ingraham, J. L. (1974). *J. Bacteriol.* **118**, 1020.

Greer, H., and Fink, G. R. (1975). *In* "Methods in Cell Biology" (D. M. Prescott, ed.), Vol. XI, p. 247. Academic Press, New York.

Guerola, N., Ingraham, J. L., and Cerdá-Olmedo, E. (1971). *Nature (London), New Biol.* **230**, 122.

Guthrie, C., Nashimoto, H., and Nomura, M. (1969). *Proc. Nat. Acad. Sci. U.S.* **63**, 384.

Hartwell, L. H. (1967). *J. Bacteriol.* **93**, 1662.

Hartwell, L. H. (1971a). *J. Mol. Biol.* **59**, 183.

Hartwell, L. H. (1971b). *Exp. Cell Res.* **69**, 265.

Hartwell, L. H. (1973a). *J. Bacteriol.* **115**, 966.

Hartwell, L. H. (1973b). *Exp. Cell Res.* **76**, 111.

Hartwell, L. H. (1974). *Bacteriol. Rev.* **38**, 164.

Hartwell, L. H., and McLaughlin, C. S. (1968a). *J. Bacteriol.* **96**, 1664.

Hartwell, L. H., and McLaughlin, C. S. (1968b). *Proc. Nat. Acad. Sci. U.S.* **59**, 422.

Hartwell, L. H., and McLaughlin, C. S. (1969). *Proc. Nat. Acad. Sci. U.S.* **62**, 468.

Hartwell, L. H., Hutchison, H. T., Holland, T. M., and McLaughlin, C. S. (1970a). *Mol. Gen. Genet.* **106**, 347.

Hartwell, L. H., McLaughlin, C. S., and Warner, J. R. (1970b). *Mol. Gen. Genet.* **109**, 42.

Hartwell, L. H., Culotti, J., and Reid, B. J. (1970c). *Proc. Nat. Acad. Sci. U.S.* **66**, 352.

Hartwell, L. H., Mortimer, R. K., Culotti, J., and Culotti, M. (1973). *Genetics* **74**, 267.

Hartwell, L. H., Culotti, J., Pringle, J. R., and Reid, B. J. (1974). *Science* **183**, 46.

Hawthorne, D. C. (1969). *Mutat Res.* **7**, 187.

Hawthorne, D. C., and Friis, J. (1964). *Genetics* **50**, 829.

Hayashibe, M., and Katohda, S. (1973). *J. Gen. Appl. Microbiol.* **19**, 23.

Henry, S. A. (1973). *J. Bacteriol.* **116**, 1293.

Hereford, L. M., and Hartwell, L. H. (1973). *Nature (London), New Biol.* **244**, 129.

Hereford, L. M., and Hartwell, L. H. (1974). *J. Mol. Biol.* **84**, 445.

Hershey, A. D. ed. (1971). "The Bacteriophage Lambda." Cold Spring Harbor Lab., Cold Spring Harbor, New York.

Herskowitz, I. (1973). *Annu. Rev. Genet.* **7**, 289.

Hiraga, S., and Saitoh, T. (1974). *Mol. Gen. Genet.* **132**, 49.
Horiuchi, T., and Nagata, T. (1973). *Mol. Gen. Genet.* **123**, 89.
Horowitz, N. H. (1948). *Genetics* **33**, 612.
Horowitz, N. H. (1950). *Advan. Genet.* **3**, 33.
Horowitz, N. H., and Leupold, U. (1951). *Cold Spring Harbor Symp. Quant. Biol.* **16**, 65.
Hranueli, D., Piggot, P. J., and Mandelstam, J. (1974). *J. Bacteriol.* **119**, 684.
Huang, K.-P., and Cabib, E. (1973). *Biochim. Biophys. Acta* **302**, 240.
Hutchison, H. T., Hartwell, L. H., and McLaughlin, C. S. (1969). *J. Bacteriol.* **99**, 807.
Jarvik, J., and Botstein, D. (1973). *Proc. Nat. Acad. Sci. U.S.* **70**, 2046.
Jockusch, H. (1966). *Biochem. Biophys. Res. Commun.* **24**, 577.
Jones, E. W. (1972a). *Genetics* **70**, 233.
Jones, E. W. (1972b). *Genetics* **71**, 217.
Kane, S. M., and Roth, R. (1974). *J. Bacteriol.* **118**, 8.
Kaplan, S., and Anderson, D. (1968). *J. Bacteriol.* **95**, 991.
Kessin, R. H., Williams, K. L., and Newell, P. C. (1974). *J. Bacteriol.* **119**, 776.
Klyce, H. R., and McLaughlin, C. S. (1973). *Exp. Cell Res.* **82**, 47.
Kohiyama, M., Cousin, D., Ryter, A., and Jacob, F. (1966). *Ann. Inst. Pasteur, Paris* **110**, 465.
Konrad, E. B., and Lehman, I. R. (1974). *Proc. Nat. Acad. Sci. U.S.* **71**, 2048.
Kornberg, H. L., and Smith, J. (1966). *Biochim. Biophys. Acta* **123**, 654.
Kotyk, A. (1963). *Folia Microbiol. (Prague)* **8**, 27.
Lacroute, F. (1967). *C. R. Acad. Sci., (Ser. D)* **264**, 2226.
Lacroute, F. (1975). *In* "Methods in Cell Biology" (D. M. Prescott, ed.), Vol. XI, p. 235. Academic Press, New York.
Laemmli, U. K. (1970). *Nature (London)* **227**, 680.
Littlewood, B. S. (1972). *Genetics* **71**, 305.
Littlewood, B. S. (1975). *In* "Methods in Cell Biology" (D. M. Prescott, ed.), Vol. XI, p. 273. Academic Press, New York.
Littlewood, B. S., and Davies, J. E. (1973). *Mutat. Res.* **17**, 315.
Maas, W. K., and Davis, B. D. (1952). *Proc. Nat. Acad. Sci. U.S.* **38**, 785.
MacDonald, D. W., and Cove, D. J. (1974). *Eur. J. Biochem.* **47**, 107.
McLaughlin, C. S., and Hartwell, L. H. (1969). *Genetics* **61**, 557.
McLaughlin, C. S., Magee, P. T., and Hartwell, L. H. (1969). *J. Bacteriol.* **100**, 579.
McLaughlin, C. S., Warner, J. R., Edmonds, M., Nakazato, H., and Vaughan, M. H. (1973). *J. Biol. Chem.* **248**, 1446.
McNeill, R. G., and McLaughlin, C. S. (1973). *Genetics* **72**, s169.
Magee, P. T., and Hopper, A. K. (1974). *J. Bacteriol.* **119**, 952.
Mahler, H. R., and Dawidowicz, K. (1973). *Proc. Nat. Acad. Sci. U.S.* **70**, 111.
Martin, T. E., and Hartwell, L. H. (1970). *J. Biol. Chem.* **245**, 1504.
Meuris, P., Lacroute, F., and Slonimski, P. P. (1967). *Genetics* **56**, 149.
Meyer, K. H., and Schweizer, E. (1972). *Biochem. Biophys. Res. Commun.* **46**, 1674.
Moens, P. B., Esposito, R. E., and Esposito, M. S. (1974). *Exp. Cell Res.* **83**, 166.
Mortimer, R. K., and Hawthorne, D. C. (1966). *Annu. Rev. Microbiol.* **20**, 151.
Mortimer, R. K., and Hawthorne, D. C. (1969). *In* "The Yeasts" (A. H. Rose and J. S. Harrison, eds.), Vol. 1, p. 385. Academic Press, New York.
Mortimer, R. K., and Hawthorne, D. C. (1973). *Genetics* **74**, 33.
Murgola, E. J., and Adelberg, E. A. (1970). *J. Bacteriol.* **103**, 20.
Nagata, T., and Horiuchi, T. (1973). *Mol. Gen. Genet.* **123**, 77.
Neal, A. L., Weinstock, J. O., and Lampen, J. O. (1965). *J. Bacteriol.* **90**, 126.
Neidhardt, F. C. (1964). *Progr. Nucl. Acid Res. Mol. Biol.* **3**, 145.

Newlon, C. S., Petes. T. D., Hereford, L. M., and Fangman, W. L. (1974). *Nature (London)* **247**, 32.

Nikiforov, V. G., Kalyaeva, E. S., and Velkov, V. V. (1974). *Mol. Gen. Genet.* **130**, 1.

O'Donovan, G. A., and Ingraham, J. L. (1965). *Proc. Nat. Acad. Sci. U.S.* **54**, 451.

Overath, P., Schairer, H. U., and Stoffel, W. (1970). *Proc. Nat. Acad. Sci. U.S.* **67**, 606.

Paul, A. V., and Inouye, M. (1974). *J. Bacteriol.* **119**, 907.

Petersen, N. S., and McLaughlin, C. S. (1974). *Mol. Gen. Genet.* **129**, 189.

Petes, T. D., and Newlon, C. S. (1974). *Nature (London)* **251**, 637.

Petes, T. D., and Williamson, D. H. (1974). *Cell*, in press.

Pringle, J. R., and Mor, J.-R. (1975). *In* "Methods in Cell Biology" (D. M. Prescott, ed.), Vol. XI, p. 131. Academic Press, New York.

Pringle, J. R., Friedman, M., and Fiechter, A. (1974). *Proc. Int. Symp. Yeasts. 4th*, 1974, Part I, p. 37.

Rasse-Messenguy, F., and Fink, G. R. (1973). *Genetics* **75**, 459.

Roberds, D. R., and DeBusk, A. G. (1973). *J. Bacteriol.* **115**, 1121.

Robinow, C. F. (1975). *In* "Methods in Cell Biology" (D. M. Prescott, ed.), Vol. XI, p. 1. Academic Press, New York.

Roodman, S. T., and Greenberg, G. R. (1971). *J. Biol. Chem.* **246**, 4853.

Roth, R., and Dampier, C. (1972). *J. Bacteriol.* **109**, 773.

Roth, R., and Fogel, S. (1971). *Mol. Gen. Genet.* **112**, 295.

Rothman-Denes, L., and Cabib, E. (1970). *Proc. Nat. Acad. Sci. U.S.* **66**, 967.

Russell, R. R. B. (1972). *J. Bacteriol.* **112**, 661.

Sadler, J. R., and Novick, A. (1965). *J. Mol. Biol.* **12**, 305.

Sakai, H., Hashimoto, S., and Komano, T. (1974). *J. Bacteriol.* **119**, 811.

Schenberg-Frascino, A., and Moustacchi, E. (1972). *Mol. Gen. Genet.* **115**, 243.

Scherer, G., Haag, G., and Duntze, W. (1974). *J. Bacteriol.* **119**, 386.

Schlitt, S. C., and Russell, P. J. (1974). *J. Bacteriol.* **120**, 666.

Scotti, P. D. (1968). *Mutat. Res.* **6**, 1.

Shiokawa, K., and Pogo, A. O. (1974). *Proc. Nat. Acad. Sci. U.S.* **71**, 2658.

Simchen, G. (1974). *Genetics* **76**, 745.

Singh, A., and Manney, T. R. (1974). *Genetics* **77**, 651.

Singh, A., and Sherman, F. (1974). *Genetics* **77**, s60.

Smith, J. D., Anderson, K., Cashmore, A., Hooper, M. L., and Russell, R. L. (1970). *Cold Spring Harbor Symp. Quant. Biol.* **35**, 21.

Snow, R. (1966). *Nature (London)* **211**, 206.

Strömnaes, O., and Mortimer, R. K. (1968). *J. Bacteriol.* **95**, 197.

Suzuki, D. T. (1970). *Science* **170**, 695.

Tai, P.-C., Kessler, D., and Ingraham, J. (1969). *J. Bacteriol.* **97**, 1298.

Tait, R. C., and Smith, D. W. (1974). *Nature (London)* **249**, 116.

Tasaka, S. E., and Suzuki, D. T. (1973). *Genetics* **74**, 509.

Thompson, L. H., and Baker, R. M. (1973). *In* "Methods in Cell Biology" (D. M. Prescott, ed.), Vol. VI, p. 209. Academic Press, New York.

Thompson, L. H., Harkins, J. L., and Stanners, C. P. (1973). *Proc. Nat. Acad. Sci. U.S.* **70**, 3094.

Thomulka, K. W., and Moat, A. G. (1968). *J. Bacteriol.* **96**, 283.

Tønnesen, T., and Friesen, J. D. (1973). *J. Bacteriol.* **115**, 889.

Venkov, P. V., Hadjilov, A. A., Battaner, E., and Schlessinger, D. (1974). *Biochem. Biophys. Res. Commun.* **56**, 599.

Warner, J. R., and Udem, S. A. (1972). *J. Mol. Biol.* **65**, 243.

Waskell, L., and Glaser, D. A. (1974). *J. Bacteriol.* **118**, 1027.

Weislogel, P. O., and Butow, R. A. (1970). *Proc. Nat. Acad. Sci. U.S.* **67**, 52.

Wickner, S., Wright, M., and Hurwitz, J. (1974). *Proc. Nat. Acad. Sci. U.S.* **71**, 783.

Wilkinson, L. E., and Pringle, J. R. (1974). *Exp. Cell Res.* **89**, 175.

Wintersberger, U., Hirsch, J., and Fink, A. M. (1974). *Mol. Gen. Genet.* **131**, 291.

Wittmann-Liebold, B., Jauregi-Adell, J., and Wittmann, H. G. (1965). *Z. Naturforsch. B.* **20**, 1235.

Wood, W. B., Edgar, R. S., King, J., Lielausis, I., and Henninger, M. (1968). *Fed. Proc., Fed. Amer. Soc. Exp. Biol.* **27**, 1160.

Yanofsky, C., Helinski, D. R., and Maling, B. D. (1961). *Cold Spring Harbor Symp. Quant. Biol.* **26**, 11.

Chapter 13

In Vivo and in Vitro Synthesis of Yeast Mitochondrial DNA[1]

L. J. ZEMAN[2,3] AND C. V. LUSENA

Division of Biological Sciences,
National Research Council of Canada,
Ottawa, Ontario, Canada

I. Introduction

Rather paradoxically, most of our current knowledge about the replication of mitochondrial DNA (mtDNA) (for recent reviews, see Borst, 1972; Linnane *et al.*, 1972; Mahler, 1973) derives from experiments with animal cells, while much less is known about this important process in lower eukaryotes, e.g., in yeasts or *Neurospora*. From these organisms only hetero-

---

[1] NRCC Publication No. 14069.
[2] NRCC postdoctoral fellow 1972–74.
[3] *Present address:* Department of Biological Chemistry, Harvard Medical School, Boston, Massachusetts.

geneous collections of mtDNAs (most often fragments) have been prepared. Active nucleases have been detected in mitochondria of *Neurospora* (Linn and Lehman, 1965; Martin *et al.*, 1973) and yeasts (Wintersberger, 1968). Within the same preparation the individual DNA molecules may differ not only in size and form (linear or circular), but sometimes also in base composition. Moreover, these mtDNAs are endowed with unusual physical properties (melting, renaturation, buoyant density, elution from hydroxyapatite, etc.), which add to the difficulties of rational experimentation (Bernardi *et al.*, 1972; Christiansen *et al.*, 1974).

Nevertheless, a better biochemical and biophysical characterization of yeast mtDNA and its metabolism would be of great use. *Saccharomyces cerevisiae* has become the most important organism for the study of mitochondrial genetics (Coen *et al.*, 1970; Linnane *et al.*, 1972; Wilkie and Thomas, 1973). The availability of genetically mapped mutants and the relative ease and variety of physiological manipulations recommend it for a concerted investigation of mtDNA by both genetic and biochemical means. The occurrence of deletions, sequence reiterations, and other alterations in mtDNA from yeast petite mutants (Faye *et al.*, 1973) is of general biological interest.

In this chapter we describe the methods used in our laboratory for studying the synthesis of yeast mtDNA.

II. *In Vivo* Synthesis

In yeasts and other fungi, the absence of thymidine kinase (Grivell and Jackson, 1968) makes the specific labeling of DNA difficult. Thus radioactive thymine and thymidine are ineffective as precursors. Mutants that can incorporate thymidine-5'-monophosphate specifically into DNA have been isolated (Brendel and Haynes, 1973; Wickner, 1974), but their general usefulness has yet to be demonstrated.

The most common precursors and recommended concentrations for radioactive labeling of yeast nucleic acids (of which 96–98% is RNA) are: adenine-2-^{3}H (3–20 μCi/ml); adenine-8-^{14}C (0.5–1.5 μCi/ml); uracil-6-^{3}H (3–20 μCi/ml); uracil-2-^{14}C (0.5–1.5 μCi/ml); for phosphorus labeling orthophosphate-^{32}P (1–2 μCi/ml). According to Hatzfeld (1973), the incorporation of uracil and even of thymine is improved if adenine-supplemented minimal medium is used. The efficiency of incorporation is strain- and medium-dependent. Pulse-labeling is complicated by the fact that a large amount of RNA is labeled and may even act as an intermediate pool between the label and DNA. In an attempt to circumvent this difficulty, Hatzfeld (1973) developed a method of "pseudo-pulse-labeling" of yeast DNA. The optimum cell concentration for the *in vivo* labeling is 1×10^7–5×10^7 cells/ml. Extensive labeling (more than 1% of total bases radioactive) should be avoided to prevent complications due to autoirradiation.

Preferential labeling of mtDNA is achieved if cytoplasmic protein synthesis is arrested by cycloheximide (Grossman *et al.*, 1969), by a temperature shift with sensitive mutants (Grossman *et al.*, 1969, Cotrell *et al.*, 1973), by amino acid starvation (Grossman *et al.*, 1969), or if the initiation of nuclear DNA (nDNA) is inhibited by the α-mating factor (Petes and Fangman, 1973).

In our work the haploid strain of *S. cerevisiae* (D 273–10B) is grown with vigorous aeration in 3.5% potassium lactate, 1% yeast extract, and 2% Bacto peptone at 30°C. Labeling is done with 3 μCi/ml of adenine-^{3}H or 1 μCi/ml of adenine-^{14}C. For preferential labeling of mtDNA, cells are pretreated with 200 μg/ml of cycloheximide for 15 minutes and then labeled with radioactive adenine for 5 hours in the presence of the antibiotic. After this time about 90% of the incorporated label is in mtDNA, and the relative amount of mtDNA is doubled. The labeled DNA is extracted from the mitochondrial fraction as described in Section IV.

III. *In Vitro* Synthesis

A. Isolated Mitochondria

1. PREPARATION OF A CRUDE MITOCHONDRIAL FRACTION

Unless otherwise stated, all operations described are carried out at 4°C. Cytoplasm is freed from the cell walls either by spheroplasting or by mechanical disruption. Our best spheroplasting results (high yield and good stability) were obtained using the method of Hutchison and Hartwell (1967). All conditions specified by these investigators, especially the low culture density at the time of harvest, must be carefully followed. Spheroplasts are collected by centrifugation at 2000 g for 15 minutes, and the pellet is suspended in 5 mM Tris–Cl, 1 mM EDTA, 0.25 M sorbitol (pH 7.5) (pellet/buffer, 1:10 by volume). After 30 minutes the spheroplast lysis is complete. The pH is adjusted to 7.5 with 1 N NaOH. Heavy cellular components are removed by centrifugation at 2000 g for 20 minutes. The crude mitochondrial fraction is collected by centrifugation at 37,000 g for 15 minutes and subsequently washed once in 10 mM Tris–Cl, 2 mM EDTA, and 0.5

M sorbitol (TES) and once in 10 mM Tris–Cl, 10 mM $MgCl_2$, and 0.25 sorbitol (pH 7.5) (TMS).

When mechanical disruption of cells is preferred, 15 gm (wet weight) of cells is washed twice with distilled water and once with TES, suspended in 10 ml of TES, and shaken for 15 seconds at full speed with 50 gm of 0.5-mm glass beads in a cooled Braun homogenizer (B. Braun, Melsungen, West Germany). This operation can be scaled down to 1 gm of yeast using special bottles for small samples. The cellular material is washed off the beads with TES (final volume 100 ml), and the pH adjusted to 7.5. The rest of the procedure is the same as in the spheroplast method.

2. *In Vitro* Incorporation of Radioactive Precursors

In contrast to mitochondria from higher eukaryotes, isolated mitochondria from yeasts do not incorporate thymidine or thymidylate into mtDNA effectively (Zeman and Lusena, 1974). This reflects their lack of the respective kinase enzymes. As first observed by Wintersberger (1966), TTP and dATP are readily incorporated into the acid-precipitable DNA. Incorporation is enhanced by phosphoenol pyruvate (PEP) and pyruvate kinase (PK). As expected, the requirement for Mg^{2+} or another divalent cation is absolute. No labeling of RNA is observed.

Our incubation medium for the *in vitro* labeling of mtDNA (virtually that of Wintersberger, 1966) contains in 10 ml of TMS: mitochondria (about 80 mg of protein), 0.33 μmoles of dATP, dCTP, dGTP, 10 mg PEP, 1 mg PK, and 50 μCi TTP-^{3}H (10–20 Ci/mmole). Incubation is usually for 20 minutes at 37°C, after which time an excess of unlabeled TTP is added. The mitochondria are collected by centrifugation and twice washed in TES. During 20 minutes of incubation about 1 pmole of TTP is incorporated per milligram of mitochondrial protein. This amounts to less than 1% of the available radioactive label. Virtually the same results are obtained when dATP-^{3}H is used. Of the two bromodeoxyribonucleoside triphosphates tested, only BrdUTP was incorporated, while no incorporation was observed with BrdCTP.

3. Removal of Nuclear DNA

Of many methods tried, only treatment of the crude mitochondrial fraction with pancreatic DNase was found effective in removing nDNA contamination. Mitochondria are suspended in TMS (about 10 mg protein/ml) with 200 μg/ml of DNase I (RNase-free product of Worthington Biochemical Co.) and incubated for 30 minutes at 4°C. After this time, 0.6% (by volume) diethyl pyrocarbonate (Calbiochem) is added and the suspension incubated for 15 minutes (deactivation of DNase). The mitochondrial are then washed two times in TES and are referred to as DNase-purified mitochondria. No difference in the properties of the *in vitro* labeled mtDNA was observed when DNase treatment preceded incorporation.

B. Permeable Cells

Yeast cells, when permeabilized by the detergent Brij 58, incorporate TTP-^{3}H and carry out limited synthesis of DNA (Hereford and Hartwell, 1971; Banks, 1973). It was suggested by Banks (1973) that the incorporation is only into mtDNA. However, our unpublished experiments with Brij-58-permeabilized cells indicate that nDNA may be labeled as well, because the cesium chloride gradient profile strongly resembles the one obtained with the *in vitro* labeled DNA extracted from crude mitochondria. The mtDNA replication complex is likely to be membrane-bound, and the user of this technique should be concerned about artefacts caused by the interaction between the detergent and mitochondrial membranes.

IV. Extraction and Analysis of mtDNA

A. Extraction of DNA

The mitochondrial pellet (about 50 mg of protein or less) is gently suspended in 4 ml of 0.1 M EDTA, 0.15 M NaCl, and 2.5% sodium dodecyl sulfate (SDS) (pH 8.0) containing 30 μl of diethyl pyrocarbonate to block the detergent-activated nucleases. Lysis is carried out at 60°C for 10 minutes; the suspension is then made 1 M with respect to NaCl and kept on ice for at least 1 hour. The heavy precipitate containing most of the protein and SDS is separated by centrifugation (40,000 g for 10 minutes), and the DNA-containing supernatant is dialyzed for at least 12 hours against 500 ml of 0.15 M NaCl and 0.015 M sodium citrate (pH 7.0) (SSC) with at least three changes.

B. Isopycnic Cesium Chloride Gradients

Fixed-angle rotors effect better separation of DNA mixtures than swinging-bucket rotors (Flamm *et al.*, 1969). Alternatively, the addition of distamycin A (Calbiochem) enhances the separation of yeast mt- and nDNA (Banks, 1973). Our comparison shows that separation of the two DNAs on a fixed-angle rotor (Spinco 40) is superior to that obtained with a swinging-bucket rotor (Spinco SW 50.1), even when distamycin A is used in the latter case. The reason for improved separation in the presence of distamycin A is not understood, and it is not clear whether the DNA can be subsequently recovered in a pure form.

To prepare a cesium chloride–distamycin A gradient in a SW 50.1 rotor (as in Banks, 1973), 1.91 gm of cesium chloride is mixed with 2.00 gm of DNA extract containing 125 μg of distamycin A (the antibiotic is first dissolved in 10 μl of ethanol). The solution is overlayed in the centrifuge tube with paraffin oil, and the sample centrifuged for 60 hours at 35,000 rpm at 18°C.

To prepare a cesium chloride gradient for a Spinco 40 rotor, 6.00 gm of cesium chloride is gently mixed with 5.00 gm of DNA extract to yield 6.6 ml of solution with a density of 1.670 gm/cm^3. The solution is overlayed in the centrifuge tube with paraffin oil, and the sample is centrifuged for 60 hours at 33,000 rpm at 18°C.

The gradients are fractionated with a device similar to the one described by Flamm et al. (1969). Fractions are always collected from the bottom of the tube. Each fraction (75–200 μl) is diluted with 0.6 ml of water, the absorbancy at 260 nm is read on a Gilford 240 spectrophotometer, and the whole fraction or an aliquot is mixed with 10 ml of Aquasol (New England Nuclear). Radioactivity is measured with a Beckman LS-250 scintillation counter. Counting efficiency is about 30% for ^{3}H and 85% for ^{14}C.

C.   Sedimentation Velocity Centrifugation

To estimate the size of DNA, we use cesium chloride band centrifugation (Vinograd et al., 1963). The DNA-containing fractions from the isopycnic cesium chloride gradient are pooled and dialyzed against SSC. For each 5-ml tube, 2.88 gm of cesium chloride is mixed with 3.83 gm of SSC and overlayed with 200 μl of the DNA solution. Samples are centrifuged 5 or 10 hours in a SW 50.1 rotor at 30,000 rpm and 18°C. Fraction collection, absorbancy, and radioactivity measurements are the same as for isopycnic gradients. Escherichia coli tRNA (4.5 S) is used as a marker.

V.   Examples of Results

There are marked differences between the in vivo and the in vitro labeled mtDNA of yeasts (Figs. 1–3). The in vivo labeled DNA extracted from crude mitochondria can be partly resolved into n- and mtDNA (the lighter component) on a cesium chloride–distamycin A gradient in the SW 50.1 rotor

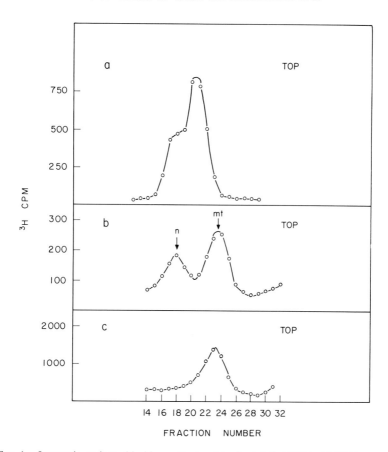

FIG. 1. Isopycnic cesium chloride gradients of *in vivo* labeled DNA; (a) DNA extracted from crude mitochondria; 50 μg/ml distamycin A, initial density 1.558 gm/cm³; Spinco SW 50.1 rotor, 35,000 rpm, 60 hours, 18°C; 75-μl fractions collected. (b) DNA extract as in Fig. 1a, initial density 1.670 gm/cm³. Spinco 40 rotor, 33,000 rpm, 60 hours, 18°C; 200-μl fractions collected; (c) DNA extracted from DNAse-purified mitochondria; conditions as in Fig. 1b. Reproduced by permission of the National Research Council of Canada from the *Can. J. Biochem.* **12**, (1974).

(Fig. 1a). Clear separation of the two peaks is achieved on a fixed-angle rotor (Fig. 1b). Contaminating nDNA is almost completely removed if mitochondria are purified by DNase treatment (Fig. 1c). The *in vitro* labeled mtDNA (Fig. 2) gives rise to broader peaks on isopycnic cesium chloride gradients than the *in vivo* labeled DNA, reflecting the lower molecular weight of the former. The peak is asymmetrical with a skew toward high density. The computer-generated resolution of the profile into gaussian peaks (Fig.

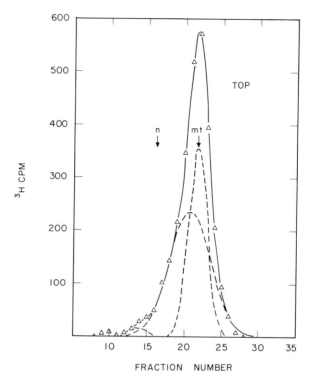

FIG. 2. Isopycnic cesium chloride gradient of *in vitro* labeled mtDNA extracted from DNase-purified mitochondria. Centrifugation conditions as in Fig. 1b. Broken line, Computer-generated resolution; Solid line, fitting a sum of gaussian distributions on the experimental points (triangles). Reproduced by permission of the National Research Council of Canada from the *Can. J. Biochem.* **12** (1974).

2, dashed line) reveals a sharp peak which coincides with the equilibrium position of the *in vivo* labeled mtDNA and a broad peak with a maximum at a position between mt- and nDNA. The result is not compatible with the idea that the skew is due to the *in vitro* labeling of the nDNA contaminant (compare Figs. 2 and 1b). This possibility is also contradicted by the appearance of the profile after a denaturation-renaturation step (results not shown). A more detailed characterization of the *in vitro* synthesized mtDNA will be presented elsewhere. Sedimentation velocity analysis (Fig. 3) confirms that the *in vitro* product is on the average much smaller than the *in vivo* synthesized product. After denaturation the difference is even more pronounced. Control experiments exclude the possibility that fragmentation is caused by the DNase treatment used to remove nDNA.

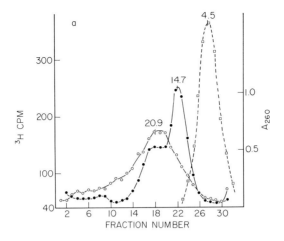

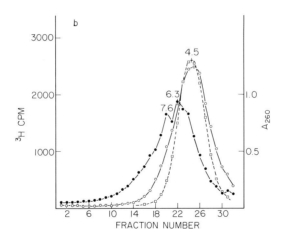

FIG. 3. Cesium chloride sedimentation velocity centrifugation of mtDNA. Solid circles, ^{3}H counts per minute of native mtDNA; open circles, ^{3}H counts per minute of mtDNA denatured by heating for 5 minutes at 100°C and rapidly cooled on ice; squares, A_{260} of *Escherichia coli* tRNA (4.5 S). Solution of 3.8 *M* cesium chloride in SSC, density 1.49 gm/cm³. Centrifugation at 30,000 rpm in a Spinco SW 50.1 rotor at 18°C. Fractions of 150 μl were collected. Numbers above the peaks refer to calculated *S* values. (a) *In vivo* labeled mtDNA (fractions 22 to 26 from Fig. 1b); centrifugation, 5 hours. (b) *In vitro* labeled mtDNA from DNase-purified mitochondria; centrifugation, 10 hours. Reproduced by permission of the National Research Council of Canada from the *Can. J. Biochem.* **12**, (1974).

VI. Concluding Remarks

The *in vitro* labeled mtDNA molecules of yeasts isolated by the same method as *in vivo* labeled molecules are on the average about 10 times shorter. The asymmetrical profiles of the *in vitro* product on cesium chloride gradients such as the one shown in Fig. 2 reflect most probably the base composition heterogeneity of fragmented yeast mDNA (Piperno *et al.*, 1972; Carnevali and Leoni, 1972). The profile is, however, different from the broad symmetrical peak obtained with sonicated *in vivo* labeled mtDNA (F. Carnevali and L. Leoni, unpublished results, 1972). This indicates that either the replication itself or the nuclease cuts accompanying the synthesis are nonrandom. The small size of the fragments recovered makes it difficult to establish (by density-labeling with BrdUTP) whether the *in vitro* synthesis is due to replication or repair (see discussion in Borst, 1972, p. 345). However, some fully hybrid molecules are synthesized, and about 65% of the radioactivity recovered from the cesium chloride gradient is displaced by more than 6 mg/cm^3, a density shift expected for 8% substitution of thymine by BrdU (unpublished results). The *in vivo* density transfer experiments with mtDNA of *S. cerevisiae* (Sena, 1972) suggest that the mode of replication of this DNA is dispersive rather than semiconservative. The dispersive mode of replication could be explained by molecular recombination occurring concomitantly with replication. The mtDNA fragmentation observed during the *in vitro* synthesis might illustrate part of this process.

ACKNOWLEDGMENTS

We thank Dr. D. H. Williamson for bringing to our attention the usefulness of distamycin A in cesium chloride gradients, Mrs. G. Dixon for technical assistance, the staff of the Biomathematics Section for the computer resolution of the cesium chloride gradient profiles and Dr. F. Sherman for providing yeast strain D273-10B.

REFERENCES

Banks, G. R. (1973). *Nature (London), New Biol.* **245**, 196–199.
Bernardi, G., Piperno, G., and Fonty, G. (1972). *J. Mol. Biol.* **65**, 173–189.
Borst, P. (1972). *Annu. Rev. Biochem.* **41**, 333–376.
Brendel, M., and Haynes, R. H. (1973). *Mol. Gen. Genet.* **126**, 337–348.
Carnevali, F., and Leoni, L. (1972). *Biochem. Biophys. Res. Commun.* **47**, 1322–1331.
Christiansen, C., Christiansen, G., and Bak, A. L. (1974). *J. Mol. Biol.* **84**, 65–82.
Coen, D., Deutsch, J., Netter, P., Petrochilo, E., and Slonimski, P. P. (1970). *Symp. Soc. Exp. Biol.* **24**, 449–496.
Cottrell, S., Rabinowitz, M., and Getz, G. S. (1973). *Biochemistry* **12**, 4374–4377.
Faye, G., Fukuhara, H., Grandchamp, C., Lazowska, J., Michel, F., Casey, J., Getz, G. S., Locker, J., Rabinowitz, M., Bolotin-Fukuhara, M., Coen, D., Deutsch, J., Dujon, B., Netter, P., and Slonimski, P. P. (1973). *Biochimie* **55**, 779–792.

Flamm, W. G., Birnstiel, M. L., and Walker, P. M. B. (1969). *In* "Subcellular Components, Preparation and Fractionation" (G. D. Birnie, and S. M. Fox, eds.), pp. 125–154. Butterworth, London.

Grivell, A. R., and Jackson, J. F. (1968). *J. Gen. Microbiol.* **54**, 307–317.

Grossman, L. I., Goldring, E. S., and Marmur, J. (1969). *J. Mol. Biol.* **46**, 367–376.

Hatzfeld, J. (1973). *Biochim. Biophys. Acta* **299**, 34–42.

Hereford, L. M., and Hartwell, L. H. (1971). *Nature (London), New Biol.* **234**, 171–172.

Hutchison, H. T., and Hartwell, L. H. (1967). *J. Bacteriol.* **94**, 1697–1705.

Linn, S., and Lehman, I. R. (1965). *J. Biol. Chem.* **241**, 2694–2699.

Linnane, A. W., Haslam, J. M., Lukins, H. B., and Nagley, P. (1972). *Annu. Rev. Microbiol.* **26**, 163–198.

Mahler, H. R. (1973). *CRC Crit. Rev. Biochem.* **1**, 381–460.

Martin, C. E., Lin, D., and Wagner, R. P. (1973). *J. Cell Biol.* **59**, 216a.

Petes, T. D., and Fangman, W. L. (1973). *Biochem. Biophys. Res. Commun.* **55**, 603–609.

Piperno, G., Fonty, G., and Bernardi, G. (1972). *J. Mol. Biol.* **65**, 191–205.

Sena, E. P. (1972). Ph.D. Thesis, University of Wisconsin, Madison.

Vinograd, J., Bruner, R., Kent, R., and Weigle, J. (1963). *Proc. Nat. Acad. Sci. U.S.* **49**, 902–910.

Wickner, R. B. (1974). *J. Bacteriol.* **117**, 252–260.

Wilkie, D., and Thomas, D. Y. (1973). *Genetics* **73**, 367–377.

Wintersberger, E. (1966). *Biochem. Biophys. Res. Commun.* **25**, 1–7.

Wintersberger, E. (1968). *In* "Biochemical Aspects of the Biogenesis of Mitochondria," pp. 189–201. Adriatica Editrice, Bari, Italy.

Zeman, L. J., and Lusena, C. V. (1974). *FEBS (Fed. Eur. Biochem. Soc.) Lett.* **40**, 84–87.

Chapter 14

Isolation of Mitochondria and Techniques for Studying Mitochondrial Biogenesis in Yeasts

ANTHONY W. LINNANE AND H. B. LUKINS

Department of Biochemistry, Monash University,
Clayton, Victoria,
Australia

I. Introduction

During the past decade the yeast cell has become the favored organism for studies of mitochondrial biogenesis, as a result of the variety of experimental treatments that can be exploited to modify the structure, function, and development of the yeast organelle. These treatments fall into two categories: manipulation by genetic means of mutations affecting mitochondria, for which yeasts are particularly suitable, and modification by a variety of physiological procedures among which the effects of oxygen limitation and catabolite repression on mitochondrial development are of paramount importance. It is outside the scope of this single chapter to describe the experimental details relevant to all the genetic and physiological procedures employed in the study of yeast mitochondriogenesis, and only a brief discussion surveying these various experimental approaches is given. More detailed treatments of various aspects will be found in other chapters of this volume, for instance, in chapters discussing the nucleic acid aspects of mitochondria or the genetic manipulation of mitochondrial development. We devote this chapter primarily to a consideration of the effects of cultural conditions on the structure and activity of mitochondria in yeast and to the technology for the isolation and purification of yeast mitochondria.

Several comprehensive articles on the theoretical aspects of employing yeast cells in the study of mitochondria genesis have been published in recent years (Linnane and Haslam, 1970; Schatz, 1970; Boardman *et al.*, 1971; Borst, 1972; Linnane *et al.*, 1972; Mahler, 1973; Kroon and Saccone, 1974).

II.   Growth Determinants of
Mitochondrial Structure and Activity

A.   Catabolite Repression

As a result of the facultative anaerobic nature of the yeast *Saccharomyces cerevisiae*, the developmental state of mitochondria in this organism can be varied progressively by cultural conditions from a nonfunctional state with respect to respiratory activity, such as occurs in anaerobic cells, to a state such as that occurring in aerobic cells displaying high respiration comparable to mitochondria from mammalian sources. Under aerobic conditions alone the development of mitochondria can be extensively varied by catabolite repression, which under common cultural conditions results from the presence of a rapidly fermentable substrate such as glucose. Cells grown on nonfermentable substrates such as lactate, acetate, glycerol, or ethanol, however, are completely dependent on mitochondrial function and show maximal derepression of mitochondrial function and formation. Such cells provide a basis of comparison when assessing the effects of catabolite repression in cells grown on other media, and are the cells of choice for the isolation of mitochondria of high activity and quality. Intermediate degrees of repression are seen in cells grown on more slowly fermentable sugars such as galactose, raffinose, or melibiose (Reilly and Sherman, 1965), or by growing cells in a chemostat with a limitation of glucose supply and consequent slowing of growth rate, it is possible to achieve extensive derepression even with this substrate (Bowers et al., 1967; Kellerman et al., 1971). Cells cultured under derepressed conditions elaborate very well-formed organelles with clearly demonstrable cristae and a high content of cytochromes, of the order of 3–4 nmoles/mg protein (Lukins et al., 1968). Such mitochondria are comparatively resistant to mechanical damage, are easily isolated in a relatively pure form, and have a high respiration rate. The mitochondria of repressed cells have underdeveloped cristae, a lower cytochrome content, especially cytochromes aa_3, and a low oxidative capacity. They probably also display a lower protein synthetic capacity, along with less mitochondrial RNA (mtRNA) (for review, see Linnane and Haslam, 1970). These changes can be so extensive that cells grown on media of high glucose concentration can show an almost negligible content of cytochromes aa_3.

It is important to recognize that growth conditions established with one strain to give a certain degree of catabolite repression cannot be applied either to other strains of the same species or to yeasts of another genus, since there exists a wide range of susceptibility to catabolite repression. *Candida* yeast strains, for instance, show little if any response to catabolite repression (Bulder, 1964; Kellerman et al., 1969).

B. Anaerobiosis

Cells grown under anaerobic conditions are unable to form several of the normal components of mitochondria, notably the holocytochromes of the respiratory chain; hence considerable changes result in the structure and activity of the mitochondria. Furthermore, the cells are forced to obtain their total energy yield from the fermentation of carbohydrate, so catabolite repression is always present to some degree. The actual magnitude of catabolite repression, even anaerobically, depends on the rate at which the carbohydrate can be fermented; thus galactose-grown anaerobic cells are less repressed than those grown on glucose media (Wallace et al., 1968).

The lipids ergosterol and unsaturated fatty acids are also components of the mitochondria not formed under anaerobic conditions, and these lipids are essential nutritional supplements to anaerobic culture media (Andreason and Stier, 1953). This is of particular significance from the experimental standpoint, since it is possible to achieve a wide range in mitochondrial properties under anaerobic conditions by variations in both the nature and magnitude of lipid supplements and in the carbon substrate determining the degree of catabolite repression. At one extreme end of the range are highly glucose-repressed anaerobic cells cultured on growth-limiting lipid supplements. Such cells can be grown to a stage at which the total lipid contains only about 5–10% unsaturated fatty acid instead of the normal 75–80% and the total cellular ergosterol is reduced to about 0.2 mg/gm dry weight cells instead of about 5 mg/gm cells in derepressed aerobically grown cells. The exact degree of essential lipid depletion achieved is dependent on the sugar supplied, and unsaturated fatty acid levels as low as 5% are generally obtained only on glucose medium. Cells grown to maximal unsaturated fatty acid and sterol depletion are in a precarious balance, and death of the majority will occur if they are held in stationary phase for more than a few hours (judged by dye staining or counting of viable colonies); consideration must also be given to a probable increase in the proportion of petite mutants in such cultures. However, if care is taken, cells can be obtained and studied at a time of extreme depletion when they are essentially all viable and nonmutant. Mitochondria from such cells are extremely fragile and require special precautions to be taken during their isolation; the ATPase activity of these organelles displays a greatly reduced sensitivity to oligomycin, and they have a very low mitochondrial protein synthesis activity (Davey *et al.*, 1969; Forrester *et al.*, 1971; Watson *et al.*, 1971). At the other end of the range are relatively derepressed cells, grown on galactose medium liberally supplemented with unsaturated fatty acid and sterol. Such cells contain easily obtainable and reasonably robust mitochondrial organelles, with cristae, readily demonstrable citric acid cycle enzymes, oligomycin-sensitive ATPase, and protein synthesis; they still lack the terminal electron-transport holocytochromes, but so far no other well-defined absence has been recorded (Watson *et al.*, 1971; Plattner *et al.*, 1971).

III. Yeast Culture Conditions for Mitochondrial Isolation

For the growth of batch cultures of yeasts for the general purpose of mitochondrial isolation, we have routinely employed in our laboratory a simple yeast extract–salts medium supplemented with a carbon source appropriate to the culture conditions (i.e., aerobic or anaerobic) or type of mitochondria required. The composition of the basic yeast extract–salts medium in grams per liter of distilled water is 5 gm Difco yeast extract, 0.5 gm NaCl, 0.7 gm $MgCl_2$ (hydrated), 1.2 gm $(NH_4)_2SO_4$, 0.1 gm $CaCl_2$,

1.0 gm KH_2PO_4, and 0.005 gm $FeCl_3$. The yeast extract present in this medium contains a sufficient amount of most of the amino acids (except tryptophan) and purine and pyrimidine bases to allow growth of auxotrophic strains to about 2 mg cells (dry weight/ml medium.) Media for the growth of auxotrophic strains must therefore be supplemented with either 0.5% peptone or individual amino acids (25–100 mg per liter) and purine and pyrimidine bases (25–50 mg per liter) to obtain higher cell densities.

A. Selection of Carbon Substrate

For the majority of studies, mitochondrial suspensions with high metabolic activity and of good quality are desired. These are most readily obtained from cells grown in the absence of catabolite repression on non-fermentable substrates such as ethanol, lactate, glycerol, or acetate. In our experience ethanol at an initial concentration of 2% (v/v) has proved the carbon source of choice for the growth of fully derepressed cells in liquid media. Such cells have a high respiratory activity throughout the log growth phase and are usually harvested in the later part of this phase at a density of between 3 and 4 mg cells (dry weight)/ml of media. Minor disadvantages in the use of ethanol arise from its volatility; thus it is necessary to add the ethanol after autoclaving the medium, and some loss of substrate may occur in highly aerated liquid cultures. Glycerol is a convenient substrate, as it is nonvolatile, but yeast strains are higly variable in their capacity for glycerol utilization. Sodium lactate– or sodium acetate-grown cells yield excellent mitochondria, but many strains show poor growth on these substrates, and media containing these carbon sources require strong buffering or constant-pH control since their metabolism leads to the formation of $NaHCO_3$ and a consequent increase in pH. Yeasts grow poorly when the pH rises above 6.

Where mitochondria are to be isolated from anaerobic cells, mutant cells defective in mitochondrial function, or cells cultured in the presence of an inhibitor of mitochondrial activity, a fermentable substrate must be employed and catabolite repression of mitochondrial development will result. Glucose causes the most powerful repression and sucrose slightly less, while galactose or melibiose causes markedly less repression (Wallace et al., 1968; Reilly and Sherman, 1965). However, some yeast strains grow very poorly on galactose or melibiose under anaerobic conditions, even though they show satisfactory growth aerobically.

When grown aerobically on a fermentable substrate, respiratory-competent cells will eventually undergo derepression on exhaustion of the fermentable substrate, provided other nutrients in the medium are in excess and the cells are permitted to grow for a further one to two generations. For instance, batch cultures of yeasts supplied with 1% glucose exhaust this substrate at a cells density of about 1.2 mg (dry weight)/ml, and after further growth on the products of fermentation to a cell density of about 2 mg/ml have a well-established respiratory metabolism. If the objective is to isolate organelles from repressed cells, it is necessary to harvest cells while an appreciable concentration of the fermentable substrate still remains in the medium. Near complete derepression of aerobic cells on glucose can be achieved by chemostat culture, a technique that has the additional advantage of maintenance of constancy of medium composition and the ability to derepress cells that are respiratory-incompetent as a result of petite mutation or as a result of inhibition of mitochondria by antibiotics. By elimination of the effects of catabolite repression, it is possible to identify the primary effects of petite mutation or of antibiotic inhibition on mitochondrial development. In setting up a continous culture, all nutrients are supplied in excess, with the exception of glucose which is the growth-limiting factor. We have employed the yeast extract–salts medium as described supplemented with a glucose concentration of 2% for respiratory-incompetent cells and 1.0% for respiratory-competent cells (Marzuki et al., 1974a). By appropriate adjustment of the dilution rate, a glucose concentration at an equilibrium of 0.007% can be readily maintained. This steady-state level can be compared to a level of 0.01% which is required for appreciable glucose repression, or values of the order of 1–5% which are usual in batch cultures. A detailed discussion of continuous culture is given in Chapter 6 of Volume XI in this series.

B.   Aeration and Temperature Control

A factor of considerable importance for growth and mitochondrial development in yeasts, which is often underestimated, is the need for adequate oxygenation of aerobic cultures, particularly cultures of high cell density or of large volume.

Cultivation of yeasts for mitochondrial isolation can be achieved by the use of conical flasks incubated on a rotary shaker, in carboy flasks of 5- to 15-liter capacity aerated with sintered-glass aerators, or in fermenter vessels fitted with aerators and motor-driven stirrers. Some guidelines for obtaining adequate aeration are now given, although every culture system must be tested for efficiency of aeration by some measure such as growth rates of cells, Q_{O_2} measurements, or determination of cytochrome content. In conical flasks aeration is most easily accomplished with a rotary shaker operating at 250 rpm; a loose-fitting metal cap is best for closing the flask, and the volume of liquid should be no more than 20% of the nominal capacity of the flask. With flasks of 1 liter or greater capacity, it is generally necessary to increase the mixing efficiency by having several indentations in the side of the flask to increase turbulence. Cultures in large carboys fitted with sintered-glass aerators should be operated at about 1 liter of air per liter of media per minute, although the requirement for dispersion of the air into fine bubbles is more important than the total volume of air passed through the culture. The air must be well humidified before entering the culture to avoid rapid evaporation and a possible drop in the temperature of the culture. In fermenter vessels rapidly stirred by motor-driven paddles, the rate of aeration may be dropped to between 200 and 500 ml of air per liter of medium per minute.

For most yeast strains a temperature of between 28° and 30°C is optimal; growth at even a few degrees above this can considerably increase the frequency of the petite mutation and give misleading results with regard to mitochondrial activity. In large cultures at a density in excess of about 2 mg (dry weight)/ml, it is necessary to ensure the dissipation of metabolic heat.

C.   Anaerobic Culture

The extent of mitochondrial development in anaerobic cells can be regulated by the amounts of lipid supplements added to the growth medium, as well as by the carbon substrate. For well-formed organelles the cells are supplied with an excess of unsaturated fatty acids, conveniently supplied by the water-soluble Tween 80 (5 gm per liter), and ergosterol (20 mg per liter). Lipid limitation of anaerobic cells is obtained at levels of Tween 80 and ergosterol of the order of 1 mg per liter and 0.2 mg per liter, respectively; in the presence of excess carbon source, cells cultured at this level of supplementation grow to a cell density of about 1 mg (dry weight)/ml, and the organelles in these cells show highly abnormal membrane development and a deficiency in the products of the mitochondrial protein-synthesizing system (see Section V,A).

IV.   Isolation of Mitochondria

The isolation of mitochondria from yeast cells is complicated by the presence of a refractory polysaccharide-peptide wall surrounding the cells. The rupture of this cell wall for the release of mitochondria can be achieved by mechanical means (Vitols and Linnane, 1961; Schatz, 1967; Mattoon and Balcavage, 1967), or the cell wall can be removed by digestion with the enzymes present in snail gut juice to form yeast protoplasts (Duell et al.,

1964). The release of mitochondria from the protoplasts can be achieved by gentle rupture procedures, thus avoiding exposure of mitochondria to the high shearing forces required to break yeast cell walls by mechanical means. By the protoplast procedure high yields of mitochondria of excellent quality can be obtained; the isolation of mitochondria from protoplasts is now commonly employed in most laboratories working with yeast mitochondria.

In the original description of the protoplast procedure (Duell *et al.*, 1964), mitochondria were released from the protoplasts by osmotic shock. This treatment results in a low yield of mitochondria, considerable damage to mitochondrial membranes, and loss of respiratory control activity (Lamb *et al.*, 1968). Subsequent developments of the original procedure, for the most part, have been directed at procedures for rupture of the protoplasts, although other modifications have concerned the efficiency of protoplast formation and their stability. In the procedure found most suitable in our laboratory, the mitochondria are released from the protoplasts by low-pressure treatment in a French pressure cell. Other laboratories employ homogenization of protoplasts in a Waring Blendor run at low speeds to effect protoplast rupture and the release of intact mitochondria in good yield (Schatz *et al.*, 1967; Grivell *et al.*, 1971).

The two major disadvantages of the snail enzyme procedure are the cost of the commercial snail enzyme preparations and the protracted nature of the preparative procedure (a minimum of 7 hours from commencement of protoplast formation to collection of gradient-purified mitochondria). In view of these factors, other methods involving direct rupture of whole cells have been developed, all based on some mechanical means for severely agitating yeast cells in the presence of fine glass beads. Disintegrators of the Nossal type (e.g., Braun MSK Homogenizer), which operate on the shaking-arm principle, are suitable for small quantities of cells (Schatz, 1967), and for larger-scale preparations a homogenizer bowl fitted with a high-speed overhead stirrer has been described (Vitols and Linnane, 1961). These procedures give rise to a high proportion of fragmented mitochondria, and are best suited for the preparation of sub-mitochondrial particles, although some intact organelles can be obtained that show coupled phosphorylation but no respiratory control. More recently, Mattoon and Balcavage (1967) have described the use of a small mill under carefully controlled conditions by which it is possible to obtain mitochondria showing both coupled phosphorylation and reasonable respiratory control ratios. These investigators suggest that the speed and simplicity of the preparation outweighs the small yield, which is about one-tenth or less of that available from the snail enzyme method.

Precautions must be taken during the isolation of mitochondria from yeasts against the possibility of bacterial contamination, since the regime of centrifugations involved in obtaining the mitochondrial pellet also concentrates any bacterial contamination derived either from the yeast suspension or from solutions and reagents employed in the isolation. For this reason the yeast culture should be checked for the presence of bacteria before starting the preparation, and all reagents and solutions utilized must have been sterilized either by filtration through a 0.22-nm filter or by autoclaving, unless they are by nature sterile (Lamb *et al.*, 1968). The hexitol solutions employed as osmotic supports for protoplasts or mitochondria must be buffered before autoclaving to prevent charring.

The isolation procedure described below has been developed for the isolation of mitochondria from the yeast *S. cerevisiae*. Mitochondria or submitochondrial particles have also been prepared from several other yeast species, including *Saccharomyces carlsbergensis* (Ohnishi *et al.*, 1966), *Candida parapsilosus* (Kellerman *et al.*, 1969), *Candida utilis* (Ohnishi *et al.*, 1966), and *Schizosaccharomyces pombe* (Goffeau *et al.*, 1973). It is generally desirable to use freshly cultured cells for the isolation, but satisfactory mitochondria can also be obtained from commercially available pressed cakes of bakers' yeast. In the latter case, cells from the pressed cake should be aerated in an ethanol–salts–yeast extract medium for between 2 and 4 hours to ensure derepression of mitochondrial activity. Bacterial contamination is frequently encountered in preparations of mitochondrial from pressed yeast cakes.

A. Digestion of Cell Wall with Snail Gut Enzymes

1. SOLUTIONS AND REAGENTS

Tris–EDTA: 0.1 M Tris–HCl (pH 9.3), 0.02 M EDTA (potassium).

Citrate buffer: 0.6 M mannitol, 0.3 M sorbitol, 0.2 M K_2HPO_4, 0.01 M citric acid, 0.001 M EDTA (potassium); adjust to pH 5.8 with 40% KOH. The total hexitol concentration of 0.9 M may be made up of any combination of the two components up to the solubility limit of mannitol (about 0.6 M at 0°C); mannitol is cheaper than sorbitol. Some workers have used sorbitol concentrations up to 1.5 M (Grivell *et al.*, 1971).

Resuspension buffer: 0.5 M mannitol, 0.4 M sorbitol, 0.2 M Tris, 0.001 M EDTA (potassium); adjust to pH 7.4 with concentrated HCl. The hexitol concentration of this buffer is isotonic with protoplasts.

Dilute resuspension buffer: 300 ml of resuspension buffer diluted to 450 ml with water. The hexitol concentration of this buffer (0.6 M) is isotonic with mitochondria.

β-Mercaptoethanol.

Snail-gut juice: Suc *d'Helix pomatia* from Industrie Biologique Française

S. A., Gennevilliers, France, or Glusulase, Endo Laboratories, Inc., Garden City, N.Y.

Unless otherwise specified, all solutions are maintained and manipulations performed at 0°–2°C.

The cells from which mitochondria are to be prepared are harvested from the culture by centrifugation and washed once with Tris–EDTA to remove spent medium. A measured volume of Tris–EDTA, roughly equivalent to the volume of the packed cells, is used to resuspend the cells, and the total volume of the suspension is noted. The volume of the packed cells can then be obtained by difference. Additional Tris–EDTA is added to bring the final buffer volume/cell volume ratio to between 2:1 and 2.5:1. The temperature of the suspension is brought to 30°C by running warm water over the outside of the container, and β-mercaptoethanol is added at 0.15–0.3 ml per 100-ml cell suspension. The suspension is incubated with occasional shaking at 28°–30°C for 10–15 minutes. The mercaptoethanol serves to reduce disulfide bonds which stabilize the cell wall against snail enzyme digestion; the high pH increases the effectiveness of this reduction. However, β-mercaptoethanol is harmful to cell constituents, and only the minimum treatment necessary should be used. Thus glucose-repressed cells or log-phase ethanol-grown cells require only the lower amount of β-mercaptoethanol and the shorter incubation time, while anaerobically grown cells or late-stationary-phase cells require the higher concentration and longer incubation period. The suspension is then centrifuged (1000 g_{av} for 5 minutes), the supernatant decanted, and the cells washed once with 2 vol of citrate buffer and finally resuspended in 1.3 vol of citrate buffer.

For digestion of the cell wall, the temperature of the cell suspension is again raised to 28°–30°C, and a snail gut enzyme preparation is added in the amount of 1 ml of enzyme preparation per 10 ml of packed cell volume as determined previously. The suspension is incubated at 28°–30°C with gentle shaking. The progress of wall digestion and protoplast formation is followed by measuring the fall in turbidity (absorbance at 740 nm) of an aliquot of the suspension diluted 1:500. A zero-time reading is taken just after addition of the snail enzyme, and further samples are measured at 15-minute intervals until the absorbance has decreased to about 25% of the zero-time value. This normally takes about 45–60 minutes, but the rate and extent of protoplast formation varies considerably with the physiological state of the cells. Thus digestion of the cell wall is most rapid in aerobically grown cells harvested in the log phase of growth; as stationary phase is approached, the susceptibility to the enzyme decreases. Anaerobically grown cells are especially difficult to digest, particularly those grown on limiting lipid supplements, and in these cases it may be necessary to double the amount of snail enzyme used. When snail enzyme digestion is complete, the suspension is centrifuged (2000 g_{av} for 5 minutes) to pack the cells and protoplasts, and the supernatant is decanted. If desired, this supernatant containing snail enzyme can be collected and worked up by ammonium sulfate fractionation to recover the digestive enzymes. The protoplasts are resuspended by gentle mixing in 3 vol of citrate buffer (care being taken to avoid premature rupture of the protoplasts), and are centrifuged again at the same speed. The protoplasts are washed once more with citrate buffer, and then once with 3 vol of resuspension buffer. The final resuspension of the protoplasts and residual cells is made in 1–2 vol of *dilute* resuspension buffer. If mitochondria having a high respiratory control ratio or high P/O ratios are required, bovine serum albumin (BSA) (fraction V) should be added to the diluted resuspension buffer at this stage to a concentration of 2 mg/ml, and maintained throughout the remaining stages of the preparation.

B.   Release of Mitochondria from Protoplasts

The protoplasts are ruptured by passage through a French pressure cell operated between 3000 and 5000 lb/in². Pressures of this order achieve breakage of protoplasts without major disintegration of the released mitochondria or any effect on residual whole cells present in the suspension (pressures of the order of 8 tons/in² are required to rupture normal cells). For mitochondria of the highest quality, a pressure at the lower end of the range (i.e., 3000 lb/in²) should be used, but this will be at the expense of a somewhat lower yield. After passage of each sample of suspension through the press, the pH of the effluent should be quickly adjusted to about 7.2 by the dropwise addition of 1 M KOH. The total volume of the combined effluents is noted, and then diluted with half this volume of diluted resuspension buffer. The pH of the protoplast lysate is checked and, if necessary, adjusted to pH 7.4.

C.   Differential Centrifugation for Collection of Mitochondria

The homogenate from the French press is centrifuged at 700 g_{av} for 10 minutes, and the supernatant collected. The pellet containing cell debris and unbroken protoplasts can be reextracted with an equal volume of dilute resuspension buffer to increase the yield of mitochondria, and the supernatants pooled. The combined supernatants are centrifuged again at 700 g_{av} for 10 minutes, and the pellet discarded. Further centrifugations of the supernatants at the same speed are carried out to give a total of three to four centrifugations. This series of centrifugations is required to remove residual protoplasts, particularly those derived from small buds. Any appreciable carryover of protoplasts into the mitochondrial fraction can lead to spurious results in the determination of certain mitochondrial activities, for instance, mitochondrial protein synthesis. The number of centrifugations required for the removal of residual protoplasts can be judged either from the absence of a pellet in the centrifuge tube, or thorough microscopic examination of the suspensions under phase-contrast. If a high yield of mitochondria is not a consideration, fewer centrifugations at a higher g_{av} value ($1000 g_{av}$) can be employed, but some mitochondria will be sedimented in this case. The mitochondria are recovered from the supernatant by centrifugation at $8000 g_{av}$ for 10 minutes. The pellet of mitochondria is washed by gently suspending it in dilute resuspension buffer with the aid of a glass-Teflon homogenizer with a loose-fitting pestle. The suspension is first centrifuged at $700 g_{av}$ for 10 minutes to remove debris, and then the mitochondria are pelleted from the supernatant by centrifugation at $8000 g_{av}$ for 10 minutes. This washing procedure is then repeated twice more, or the mitochondria are further purified by gradient centrifugation. Mitochondria washed three times and taken at this stage are adequately purified for many experimental purposes, such a measurement of respiratory activity, respiratory control, oxidative phosphorylation, and protein synthesis. The yield of mitochondria at this stage from derepressed cells is of the order of 300 mg protein/100 ml packed cell volume.

D.   Gradient Purification of Mitochondria

Further purification of the mitochondria from contaminating membrane material can be achived by centrifuging the washed mitochondrial pellet on a discontinuous sorbitol gradient made of four equal steps of 50, 60, 70, and 80% (w/v) sorbitol (containing 1 mM EDTA preadjusted to pH 7.0). Alternatively, a continuous sucrose gradient from 20 to 70% (w/v) can be used. Centrifugation is performed in the SW 25.1 head of a Spinco Model L, or its equivalent, at 24,000 rpm ($58,000 g_{av}$); for purification of mitochondria from derepressed cells, centrifugation for 90–120 minutes is usually sufficient, but for preparations from repressed cells or anaerobic cells centrifugation for at least 3 hours is required for adequate banding. For certain purposes it is permissible to centrifuge overnight. The mitochondria from derepressed cells (i.e., ethanol-grown) band in the gradient at the 60–70% sorbitol interface, or in the continuous sucrose gradient which is very suitable for these mitochondria, the band is located about two-thirds of the way down the gradient. Mitochondria obtained from repressed cells, anaerobic cells, or petite mutants appear denser, perhaps because of greater solute penetration, so these mitochondria band at the 70–80% sorbitol interface. The mitochondrial band from derepressed cells can be readily recognized by its brown color, while repressed mitochondria are paler as a result of the lower cytochrome content, and anaerobic mitochondria paler still because the normal respiratory holocytochromes are absent. The mitochondrial fraction is collected from the gradient by aspirating off the upper layers and taking off the mitochondrial band with a large-bore Pasteur pipette; alternatively, the band can be collected by dripping from the bottom of the tube. The mitochondrial fraction is diluted slowly with 2 vol of dilute resuspension buffer (or 0.6 M sucrose) and collected by centrifugation for 15 minutes at about 12,000 g. The yield of gradient-purified mitochondria is of the order of 250 mg protein/100 ml packed cell volume in the case of derepressed cells, but only about 50–100 mg/100 ml packed cell volume from repressed cells.

E.   Stabilization of Mitochondria from Anaerobic Cells

The conditions of cultivation of anaerobic cells with respect to lipid supplementation markedly affect the fragility of mitochondrial organelles, hence the yield of intact particles obtained in the isolation procedure just described. Thus mitochondria from anaerobic cells grown with excess lipid supplements are obtained in satisfactory yield and are for the most part intact, whereas lipid-depleted anaerobic cells give very poor yields of highly fragmented mitochondrial structures. The low yield in the latter case is contributed to by incomplete digestion of the cell wall during protoplasts formation. Gradient purification is generally essential for mitochondria from anaerobic cells, yet the appropriate band from a gradient of lipid-depleted mitochondria is still found to be largely composed of disrupted mitochondria together with a few intact organelles. However, it is possible to isolate intact mitochondrial structures in reasonable yield from lipid-limited anaerobic cells, if the protoplasts are first subjected to a mild prefixation with glutaraldehyde prior to rupture in the press (Damsky et al., 1969; Watson et al., 1970). The fixative has a stabilizing effect due to cross-linkage of proteins, but the process also leads to progressive enzyme inactivation and the entrapment of RNA and DNA in the protein meshwork. Provided only mild treatment is used, the preservation of appreciable activity for most of the common marker enzymes such as malate and succinate dehydrogenases can be obtained. The prefixation technique involves suspending the washed protoplasts for 10 minutes at 0°C in resuspension buffer containing 2% glutaraldehyde, washing three times with resuspension buffer, and then proceeding with the breaking process as already described. Again, gradient purification of the final mitochondrial suspension is essential. If prefixation is considered undesirable, then an appreciable improvement in the yield of intact organelles can be achieved by increasing the hexitol concentrations described in the standard procedure to 1.3 M for the preparation of protoplasts and to 0.9 M for all subsequent stages of the isolation procedure.

F. Assessment of Mitochondrial Quality

Several different criteria can be examined to ascertain the biochemical integrity of mitochondria in a preparation. Respiratory studies with a polarograph are most commonly employed because of their speed and simplicity; these studies include measurements of respiratory rates with different substrates, ADP/O (P/O) ratios, or respiratory control ratios generally considered to be the most significant indices of mitochondrial quality. A brief outline of a reaction mixture for the determination of respiratory control with yeast mitochondria is given below. Good-quality mitochondria should exhibit respiratory control ratios in excess of 2 with succinate and 3 with α-ketogluterate, and respiratory rates with pyruvate (plus malate) should be comparable to that obtained with succinate. Respiratory control and pyruvate oxidation seem to decline together in damaged organelles. As mentioned previously, all buffer solutions involved in stages of the isolation procedure subsequent to protoplast rupture must include 0.2% BSA if high respiratory control ratios are desired. The P/O ratios obtained with mitochondria from *S. cerevisiae* should be in the range 1.2–1.6 with succinate and 1.5–1.8 with either pyruvate (plus malate) or ethanol. The P/O ratios observed with mitochondria from this yeast, or other faculative anaerobes such as *S. carlsbergensis*, do not exceed values of 2 for NADH-linked substrates because of the absence of site-1 phosphorylation (Vitols and Linnane, 1961; Ohnishi *et al.*, 1966; Kormancikova *et al.*, 1969). However, it has been reported that induction of site-1 phosphorylation, giving P/O ratios greater than 2 with pyruvate (plus malate) as substrate, can be achieved in these organisms by the areration of cells for several hours in phosphate buffer (Ohnishi, 1970; Schuurmans-Stekhoven, 1966). In the case of obligate aerobic yeasts, such as *C. utilis*, P/O ratios approaching 3 are attainable with NADH-linked substrates. These aerobic yeasts normally retain site-1 phosphorylation, but this activity, together with sensitivity to piericidin A and rotenone, is lost in chemostat culture of the organisms under conditions of iron or sulfate limitation (Light *et al.*, 1968; Ragan and Garland, 1971).

The respiratory and phosphorylative capacity of mitochondrial preparations can be measured polarographically in a reaction mixture of the following composition. In a polarographic cell of 3.0- to 3.5-ml capacity, add 2.8 ml of buffered sucrose [0.6 M sucrose, 10 mM K_2HPO_4, 12 mM EDTA, 10 mM Tris–HCl (pH 7.4)], 0.1 ml of BSA (fraction V, 50 mg/ml),

0.05 ml of substrate (300 mM Sodium succinate, 300 mM Sodium pyruvate + 300 mM Sodium malate, etc.), 5 μl ADP (50 mM) and 0.05 to 0.1 ml mitochondrial suspension (20 mg mitochondrial protein/ml). Respiratory control ratios are determined from rates of oxygen uptake after successive additions of mitochondria, substrate, and ADP to initiate state 3; the second state-4 oxidation rate follows ADP depletion.

Other indices of the mitochondrial quality of a different nature which may be examined include the rate of mitochondrial protein synthesis and the base composition and molecular weight of the mtRNA. With mitochondria from derepressed aerobic cells, the rate of amino acid incorporation (measured, for example, as leucine-^{14}C incorporation) should be of the order of 3–5 pmoles per minute per milligram of protein; the activity should show a requirement for an external ATP source in the presence of oligomycin and be free of contamination with a functional cytoplasmic ribosomal system as shown by insensitivity to cycloheximide (Lamb *et al.*, 1968). The extent of contamination with cytoplasmic ribosomes can be estimated by determination of the guanosine plus cytosine (GC) content of RNA extracted from the mitochondrial preparation. mtRNA has a GC content of about 28%, and GC values above this level indicate contamination with cytoplasmic ribosomes which have a GC value of about 47% (see Borst and Grivell, 1971, for review). On polyacrylamide gels the bulk of the RNA should be found in only three bands, corresponding to 21 and 15 S rRNA and a 4 S tRNA (Forrester *et al.*, 1970; Borst and Grivell, 1971).

Finally, electron microscope inspection of fixed mitochondrial preparations provides a direct means for assessing the intactness of mitochondrial organelles and the extent of contamination with membranous material. Preparations are fixed in a solution of 2% glutaraldehyde in 0.5 M sucrose–0.1 M sodium cacodylate for 1 hour, followed by 1 hour in a 1% osmic acid–0.1 M sodium cacodylate solution. The fixed samples are dehydrated through an acetone series and embedded in Epon–Araldite. An example of a gradient-purified mitochondrial preparation isolated from cells grown on ethanol is shown in Fig. 1.

V. General Techniques for the Study of Mitochondrial Biogenesis

In this section we briefly summarize the principles involved in experimental approaches to mitochondriogenesis that are applicable to the organism *S. cerevisiae* and which make this organism a most attractive one for a multi-

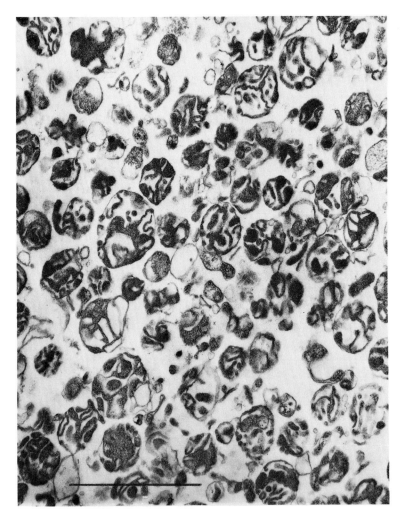

FIG. 1. Electron micrograph of gradient-purified mitochondria obtained by standard procedure from cells of *S. cerevisiae* grown on a yeast extract–salts medium supplemented with 2% (v/v) ethanol. The bar represents 1 μ

disciplinary attack on the problem. The capacity of yeasts to obtain energy for growth by fermentation processes allows cells to be cultured under a variety of conditions which lead to the elaboration of incompletely formed mitochondria or mitochondria with altered activity. Various cultural conditions that affect mitochondrial development include anaerobiosis, level of fermentable substrate present, the presence of certain drugs and antibiotics that inhibit the synthesis of nucleic acids or protein by the organelle, and the occurrence of mutations of either nuclear or mitochondrial origin which lead to changes in or loss of essential mitochondrial components. Various combinations of these cultural conditions can be employed to obtain a wide range of effects on mitochondrial development, which can then be studied at the intact-cell level as well as with isolated mitochondria *in vitro*.

A. Anaerobiosis and Oxygen Induction of Respiration

An experimental approach that forms the basis of a large proportion of studies on mitochondrial biogenesis in yeasts is the analysis of the changes in morphology and physiology that accompany the oxygen-induced transition of cells from the anaerobic state to the aerobic condition. Similar in principle to this procedure is the use of catabolite repression-derepression cycles to effect changes in mitochondrial properties. When utilized in combination with specific antibiotics, these procedures provide a dynamic means for following the sequence of production and integration of mitochondrial components into the organelle.

The synthesis of cytochromes and the development of a functional respiratory system occurs when anaerobically grown cells are exposed to oxygen; the time required for induction and the sequence of events can be modified to some extent by the conditions of growth of the anaerobic cells. Cells grown anaerobically on galactose and excess lipid supplements adapt rapidly on oxygenation, the process appearing to involve essentially the formation of cytochromes and their incorporation into the well-formed inner membranes of the promitochondrial structures preexisting in these anaerobic cells. These promitochondria, which can be readily isolated (Schatz, 1965; Criddle and Schatz, 1969; Watson *et al.*, 1970, 1971), contain normal levels of mitochondrial DNA (mtDNA) (Fukuhara, 1969; Nagley and Linnane, 1972), mtRNA (Forrester *et al.*, 1971), and an active mitochondrial protein-synthesizing system (Davey *et al.*, 1969; Schatz and Saltzgaber, 1969), as well as an energy transfer system as indicated by the presence of P_i–ATP exchange activity (Groot *et al.*, 1971).

Cells grown anaerobically on glucose and to growth-limiting levels of ergosterol and unsaturated fatty acids contain mitochondrial structures that lack a folded inner membrane system, have reduced levels of mitochondrial nucleic acids (Nagley and Linnane, 1972), and have very little or no mitochondrial protein-synthesizing activity (Watson *et al.*, 1971). These cells induce only slowly after exposure to oxygen, because of the necessity for extensive membrane reorganization involving the synthesis of unsaturated fatty acids and ergosterol and their membrane insertion, the synthesis of many components of the translational system, and subsequent acquisition of the products of the system prior to maturation of a functional respiratory membrane.

Respiratory induction experiments are usually performed by harvesting anaerobic cells, resuspending them in fresh medium containing an energy source and a source of nitrogen, and aerating them at normal growth temperature (30°C). For rapid adaptation galactose is best employed as the energy source, whereas varying glucose concentrations can be employed to slow the process by catabolite repression. It is essential to realize that the initial stages of induction can occur very rapidly on exposure to even traces of oxygen (or air), hence to observe the early stages of induction the anaerobic culture must be poured over an excess of ice immediately on opening the anaerobic culture vessel and the cells maintained at 0°C throughout the harvesting procedure. Respiratory induction accompanying cell growth can be observed using a dilute suspension [namely, 0.5 mg (dry weight)/ ml] of the harvested anaerobic cells, or induction essentially devoid of cell division is obtained by resuspending the anaerobic cells at a high cell density [in excess of 6 mg (dry weight)/ml] (Vary et al., 1969). For the rigorous control of induction experiments involving the use of inhibitors, these should be added to the anaerobic culture before exposure of the cells to oxygen and should be present during the harvesting procedure. The alternate procedure of catabolite derepression under aerobic conditions has been employed in a series of studies from the laboratory of Mahler, to which readers may refer for details (Henson et al., 1968; Mahler et al., 1971).

B.   Use of Inhibitors of Mitochondrial Development

Antibiotics that have selective inhibitor action against the protein-synthesizing system of either the mitochondria or the cytoplasmic ribosomes are extensively used to differentiate between the contributions of these two systems to the development of the organelles. Chloramphenicol or erythromycin (or several related macrolide antibiotics) is commonly used to inhibit mitochondrial protein synthesis either *in vitro* or *in vivo* (Clark-Walker and Linnane, 1967; Lamb et al., 1968), while cycloheximide or emetine is used as an inhibitor of the cytoplasmic ribosomes. Cells grown in the presence of chloramphenicol lack the holocytochromes a, a_3, b, and c_1 (Huang et al., 1966). In experiments in which the proteins formed in the presence of a selective antibiotic are labeled with radioactive amino acids, it can be shown that chloramphenicol inhibits incorporation into mitochondrial protein by about 13% in derepressed cells and 4% in repressed cells, while complementary experiments with cycloheximide yield similar estimates of the contribution of the mitochondrial protein-synthesizing system (Kellerman et al., 1971; Mahler et al., 1971).

Specific antibiotic are frequently used in conjunction with experiments on respiratory induction. By this means it has been found that the products of one protein-synthesizing system may accumulate to a limited degree under conditions in which the other system is rendered nonfunctional by inhibition; e.g., some accumulation of mitochondrial products can occur during induction in the presence of cycloheximide (Rouslin and Schatz, 1969). When induction or derepression experiments of this type are performed in combination with radioactive amino acids, it is possible to label preferentially the individual components of mitochondrial enzyme complexes formed by the two protein-synthesizing systems of the cell. The enzyme complexes of the inner mitochondrial membrane which are being studied in this way, which are formed by the cooperative action of both the mitochondrial and cytoplasmic protein-synthesizing systems, include oligomycin-sensitive ATPase and the cytochrome oxidase complex.

This approach is exemplified by the studies of Tzagoloff (Tzagoloff *et al.*, 1973, 1974) on oligomycin-sensitive ATPase. This complex consists of 10 proteins organized in three functional parts, F_1 ATPase, the oligomycin-sensitivity-conferring protein (OSCP), and an inner membrane fraction. During derepression the induction of this complex is prevented by chloramphenicol, but the soluble F_1 ATPase and the OSCP are formed on cytoplasmic ribosomes and accumulate in the cell in loose association with the mitochondria. However, when cells are incubated with radioactive amino acids in the presence of cycloheximide, four protein components of the membrane fraction become labeled and can be recovered from the solubilized complex by precipitation with antiserum to the whole ATPase complex. These four mitochondrially synthesized components, which are of low molecular weight and strongly hydrophobic, represent about 15% by weight of the oligomycin-sensitive ATP (Senior, 1973). In studies of a similar nature, yeast cytochrome oxidase was shown to be composed of about seven proteins, of which three high-molecular-weight proteins are contributed by the mitochondria (Mason and Schatz, 1973; Tzagoloff *et al.*, 1973, 1974; Sebald *et al.*, 1974; Ross *et al.*, 1974).

As well as a means for preferentially labeling mitochondrial proteins, antibiotics can be employed to label preferentially mitochondrial DNA. On aeration of lipid-supplemented anaerobic cells, a small burst of mtDNA synthesis (10–35% increase) occurs (Grossman *et al.*, 1969; Rabinowitz *et al.*, 1969); when this takes place in the presence of cycloheximide, the replication of nuclear DNA is rapidly inhibited, while mtDNA replication continues unaffected for some time (Grossman *et al.*, 1969).

C. Applications of Some Yeast Mutants

There is now a considerable variety of yeast mutants of both nuclear or mitochondrial origin which have applications in specific aspects of the study of mitochondriogenesis. The properties of many of these are considered in Chapter 17, dealing with mitochondrial genetics, and have been reviewed elsewhere (Linnane *et al.*, 1972, 1974; Sager, 1972). In the following paragraphs we discuss only two classes of mutants which are of general application to the problem of mitochondriogenesis.

1. CYTOPLASMIC PETITE MUTANTS (ρ^-)

During about 25 years of investigation the respiratory-deficient, cytoplasmic petite mutants described by Ephrussi (see Sager, 1972, for review) have played a very significant role in the development of our understanding of mitochondrial biogenesis. In recent years biochemical investigations of ρ^- mutants have resulted in two observations which form the basis for the general application of ρ^- mutants in studies of mitochondrial biosynthesis. These are:

1. The ρ^- mutation is associated with deletions of mtDNA, and in some petite mutants deletion can result in the complete absence of mtDNA from the cells (Nagley and Linnane, 1970; Goldring *et al.*, 1970). These mutants devoid of mtDNA are designated ρ^0; they are also consequently devoid of any RNA species transcribed from mtDNA, such as mitochondrial rRNA and mitochondrial specific tRNAs, and any protein encoded in mtDNA.

2. All ρ^- mutants so far described, regardless of the nature of any mtDNA they may retain, are found not to contain a functional protein-synthesizing system (Wintersberger, 1967; Kuzela and Grecna, 1969; Schatz and Salzgaber, 1969; Kellerman *et al.*, 1971).

The lack of mitochondrial protein synthesis in petites and of mtDNA in ρ^0 cells provides a means of definitive identification of the sites of synthesis of mitochondrial proteins and the location of the genes determining these proteins. Thus any component present in a ρ^0 cell must be genetically determined in the nucleus and synthesized on the cytoplasmic ribosomes. The elongation factors G and T for mitochondrial protein synthesis (Richter, 1971; Scragg *et al.*, 1971), F_1 ATPase and OSCP (Schatz, 1968; Kovac and Weissova, 1968), and some components of the cytochrome oxidase apoprotein (Tuppy and Birkmayer, 1969; Kraml and Mahler, 1967) are only a few examples of mitochondrial components confirmed as being coded for in the nucleus and synthesized on cytoplasmic ribosomes by their detection in ρ^0 cells.

Cytoplasmic petite mutants also have a wide application to the study of the structure, organization, and biosynthesis of mitochondrial nucleic acids. Of particular importance is the ability to obtain particular fragments of the mtDNA in petite cells in an enriched state, the multiple copies of fragments occurring as a result of reiteration of conserved sequences. The selection of petite mutants retaining particular fragments of the mitochondrial genome is achieved by continued selection, during a series of subcloning steps, of a particular mitochondrial antibiotic resistance marker, for instance, the gene for erythromycin resistance (Gingold *et al.*, 1969; Saunders *et al.*, 1971; Nagley *et al.*, 1973). The mtDNA from petites of this type can then be employed in a variety of hybridization-type experiments. For instance, petites have been obtained that contain mtDNA showing an increased level of hybridization, relative to ρ^+ mtDNA, for 21 S mitochondrial rRNA (Nagley *et al.*, 1974) and certain mitochondrial tRNA molecules (Cohen *et al.*, 1972). Studies of this nature involving petite DNA have only recently begun, but offer considerable potential for determining the structure of mtDNA by DNA–DNA and DNA–RNA hybridization and by heteroduplex mapping.

2. LIPID MUTANTS

The manipulation of mitochondrial properties by lipid depletion under anaerobic conditions was discussed in Section II,B. The study of membrane synthesis under anaerobic conditions is complicated by the interaction of several cellular phenomena, including the direct effect of anaerobiosis on holocytochrome synthesis and the synthesis of unsaturated fatty acids and ergosterol. Superimposed on these effects is catabolite repression. However, these complications of anaerobic growth can be avoided, and the regulation of mitochondrial membrane formation by lipid nutrition achieved under aerobic conditions, through the use of an unsaturated fatty acid auxotroph (Haslam *et al.*, 1971; Proudlock *et al.*, 1969, 1971). Further, by culturing this auxotroph in a glucose-limited chemostat, the effects of catabolite repression can also be eliminated (Marzuki *et al.*, 1974b). Thus the mutant provides a means for identifying effects arising solely from variation in the unsaturated fatty acid content of the cells. It has emerged that the mutant can be used for the study of a variety of aspects of mitochondrial biogenesis, including oxidative phosphorylation, mitochondrial protein synthesis, and mtDNA replication.

When an unsaturated fatty acid auxotroph is grown in batch culture on a limiting amount of unsaturated fatty acid, the cells will cease growing when the unsaturated fatty acid content of the cell membranes reaches a critical low level. This level may vary in different strains from 8 to 15% of unsaturated fatty acid in the cell lipids, as compared with a value of 70 to 80% for the wild-type organism. As the unsaturated fatty acid content decreases below 40%, the P/O ratios of mitochondria fall, and at a value of about 25% the mitochondria are fully uncoupled even though their respiratory activity remains essentially unchanged. When cells are cultured in a glucose-limited chemostat at several different unsaturated fatty acid levels, the P/O ratios determined *in vitro* were found to correlate with the extent of mitochondrial unsaturated fatty acid depletion (Marzuki *et al.*, 1975b). For example, P/O ratios fell from about 1.2 at normal unsaturated fatty acid levels to less than 0.1 when only about 30% of the mitochondrial fatty acids were unsaturated. The ATPase activity of unsaturated fatty acid–depleted mitochondria is normal, but P_i–ATP exchange activity is absent (Haslam *et al.*, 1971).

The loss of oxidative phosphorylation with unsaturated fatty acid limitation is a direct consequence of the changed lipids, since phosphorylation can be reconstituted by incubation of cells in medium supplemented with unsaturated fatty acids and containing both chloramphenicol and cycloheximide (Proudlock *et al.*, 1971). This experiment demonstrates that all the proteins of the organelle required for coupled oxidative phosphorylation preexisted in the cells prior to the addition of unsaturated fatty acids and the two antibiotics to the medium. A consequence of the unsaturated fatty acid depletion of the mitochondria was a large increase in the permeability of the mitochondrial membrane to protons (Linnane *et al.*, 1973). From these observations it was concluded that the decrease in the unsaturated fatty acid content of the mitochondrial membranes results in the loss of the membrane proton gradient and collapse of the membrane potential required for oxidative phosphorylation.

Previous studies with cells grown anaerobically had suggested that a functional mitochondrial protein-synthesizing system may be absent from unsaturated fatty acid–depleted cells. This conclusion is supported by recent experiments with an unsaturated fatty acid auxotroph grown in a glucose-limited continuous culture at a steady state of 20% unsaturated fatty acid in the cell lipids. These cells contain near normal levels of mitochondrial ATPase, but this ATPase activity is only slightly inhibited by oligomycin (Marzuki *et al.*, 1975a). In addition, there is a considerable reduction in cytochrome content and respiratory activity in these cells. These observations are consistent with the presence of only trace amounts of mitochondrial protein synthesis in these cells, since the oligomycin-sensitive ATPase and the cytochrome complexes aa_3 and b each contain components synthesized by the protein-synthesizing system of the mitochondria.

One further observation made with chemostat cultures maintained at a low unsaturated fatty acid content (about 15%) for an extended period of growth was a marked increase in the frequency of petite mutants (Marzuki *et al.*, 1974). This reached a constant value of about 70% petite mutants and was shown to be the result of an increase in the rate of mutation of mtDNA, rather than a selective growth advantage of ρ^- cells under these chemostat conditions. It appears that the decreased fluidity of the mitochondrial membrane that is present under these conditions affects mtDNA replication in such a way that defective replication occurs and gives rise to petite mutants.

Studies of this nature, in which the unsaturated fatty acid auxotroph is used to regulate mitochondrial development through changes in the lipid composition of the membranes, have yet to be fully exploited. It is also likely that similar studies using other mutants of lipid metabolism, e.g., mutants defective in the formation of phospholipid or ergosterol, will yield valuable information on the process of mitochondrial biogenesis.

REFERENCES

Andreason, A. A., and Stier, T. J. B. (1953). *J. Cell. Comp. Physiol.* **41**, 23–36.

Boardman, N. K., Linnane, A. W., and Smillie, R. M., eds. (1971). "Autonomy and Biogenesis of Mitochondria and Chloroplasts." Amer. Elsevier, New York.

Borst, P. (1972). *Annu. Rev. Biochem.* **41**, 334–376.

Borst, P., and Grivell, L. A. (1971). *FEBS (Fed. Eur. Biochem. Soc.) Lett.* **13**, 73–88.

Bowers, W. D., McClary, D. O., and Ogur, M. (1967). *J. Bacteriol.* **94**, 482–484.

Bulder, C. J. E. A. (1964). *Antonie van Leeuwenhoek, J. Microbiol. Serol.* **30**, 1–9.

Clark-Walker, G. D., and Linnane, A. W. (1967). *J. Cell Biol.* **34**, 1–14.

Cohen, M., Casey, J., Rabinowitz, M., and Getz, G. S. (1972). *J. Mol. Biol.* **63**, 441–451.

Criddle, R. S., and Schatz, G. (1969). *Biochemistry* **8**, 322–334.

Damsky, C. H., Nelson, W. H., and Claude, A. (1969). *J. Cell. Biol.* **43**, 174–179.

Davey, P. J., Yu, R., and Linnane, A. W. (1969). *Biochem. Biophys. Res. Commun.* **36**, 30–34.

Duell, E. A., Inoue, S., and Utter, M. F. (1964). *J. Bacteriol* **88**, 1762–1773.

Forrester, I. T., Nagley, P., and Linnane, A. W. (1970). *FEBS (Fed. Eur. Biochem. Soc.) Lett.* **11**, 59–61.

Forrester, I. T., Watson, K., and Linnane, A. W. (1971). *Biochem. Biophys. Res. Commun.* **43**, 409–415.

Fukuhara, H. (1969). *Eur. J. Biochem.* **11**, 135–139.

Gingold, E. B., Saunders, G. W., Lukins, H. B., and Linnane, A. W. (1969). *Genetics* **62**, 735–744.

Goffeau, A., Landry, Y., Foury, F., and Briquet, M. (1973). *J. Biol. Chem.* **248**, 7097–7105.

Goldring, E. S., Grossman, L. I., Krupnick, D., Cryer, D. R., and Marmur, J. (1970). *J. Mol. Biol.* **52**, 323–335.

Grivell, L. A., Reijnders, L., and Borst, P. (1971). *Biochim. Biophys. Acta* **247**, 91–103.

Groot, G. S. P., Kovac, L., and Schatz, G. (1971). *Proc. Nat. Acad. Sci. U.S.* **68**, 308–311.

Grossman, L. I., Goldring, E. S., and Marmur, J. (1969). *J. Mol. Biol.* **46**, 367–376.

Haslam, J. M., Proudlock, J. W., and Linnane, A. W. (1971). *Bioenergetics* **2**, 351–370.

Henson, C. P., Weber, C. N., and Mahler, H. R. (1968). *Biochemistry* **7**, 4431–4444.

Huang, M., Biggs, D. R., Clark-Walker, G. D., and Linnane, A. W. (1966). *Biochim. Biophys. Acta* **114**, 434–436.

Kellerman, G. M., Biggs, R., and Linnane, A. W. (1969). *J. Cell Biol.* **42**, 378–391.

Kellerman, G. M., Griffiths, D. E., Hansby, J. E., Lamb, A. J., and Linnane, A. W. (1971). *In* "Autonomy and Biogenesis of Mitochondria and Chloroplasts" (N. K. Boardman, A. W. Linnane, and R. M. Smillie, eds.), pp. 346–359. Amer. Elsevier, New York.

Kormancikova, V., Kovac, L., and Vidova, M. (1969). *Biochim. Biophys. Acta* **180**, 9–17.

Kovac, L., and Weissova, K. (1968). *Biochim. Biophys. Acta* **153**, 55–59.

Kraml, J., and Mahler, H. R. (1967). *Immunochemistry* **4**, 213–217.

Kroon, A. M., and Saccone, C., eds. (1974). "Biogenesis of Mitochondria." Academic Press, New York.

Kuzela, S., and Grecna, E. (1969). *Experientia* **25**, 776–777.

Lamb, A. J., Clark-Walker, G. D., and Linnane, A. W. (1968). *Biochim. Biophys. Acta* **161**, 415–427.

Light, P. A., Ragan, C. I., Clegg, R. A., and Garland, P. B. (1968). *FEBS (Fed. Eur. Biochem. Soc) Lett.* **1**, 4–8.

Linnane, A. W., and Haslam, J. M. (1970). *Curr. Top. Cell. Regul.* **2**, 101–172.

Linnane, A. W., Haslam, J. M., Lukins, H. B., and Nagley, P. (1972). *Annu. Rev. Microbiol.* **26**, 163–198.

Linnane, A. W., Ward, K. A., Forrester, I. T., Haslam, J. M., and Plummer, D. T. (1973). *In* "Yeast, Mould and Plant Protoplasts" (J. R. Villanueva *et al.*, eds.), pp. 345–348. Academic Press, New York.

Linnane, A. W., Kellerman, G. M., and Lukins, H. B. (1974). *In* "Techniques of Biochemical and Biophysical Morphology" (D. Glick and R. M. Rosenbaum, eds.), Vol. 2, pp. 1–98. Wiley (Interscience), New York.

Lukins, H. B., Jollow, D., Wallace, P. G., and Linnane, A. W. (1968). *Aust. J. Exp. Biol.* **46**, 651–665.

Mahler, H. R. (1973). *CRC Crit. Rev. Biochem.* **1**, 381–460.

Mahler, H. R., Perlman, P. S., and Mehrotra, B. D. (1971). *In* "Autonomy and Biogenesis of Mitochondria and Chloroplasts" (N. K. Boardman, A. W. Linnane, and R. M. Smillie, eds.), pp. 492–511. Amer. Elsevier, New York.

Marzuki, S., Hall, R. M., and Linnane, A. W. (1974). *Biochem. Biophys. Res. Commun.* **57**, 372–378.

Marzuki, S., Cobon, G., Crowfoot, P., and Linnane, A. W. (1975a). *Arch. Biochem. Biophys.* (in press).

Marzuki, S., Cobon, G., Haslam, J. M., and Linnane, A. W. (1975b). *Arch. Biochem. Biophys.* (in press).

Mason, T. L., and Schatz, G. (1973). *J. Biol. Chem.* **248**, 1355–1360.

Mattoon, J. R., and Balcavage, W. X. (1967). *In* "Methods in Enzymology" (R. W. Estabrook and M. E. Pullman, eds.), Vol. 10, pp. 135–142. Academic Press, New York.

Nagley, P., and Linnane, A. W. (1970). *Biochem. Biophys. Res. Commun.* **39**, 989–996.

Nagley, P., and Linnane, A. W. (1972). *Cell Differentiation* **1**, 143–148.

Nagley, P., Gingold, E. B., Lukins, H. B., and Linnane, A. W. (1973). *J. Mol. Biol.* **78**, 335–350.

Nagley, P., Molloy, P. L., Lukins, H. B., and Linnane, A. W. (1974). *Biochem. Biophys. Res. Commun.* **57**, 232–239.

Ohnishi, T. (1970). *Biochem. Biophys. Res. Commun.* **41**, 344–352.

Ohnishi, T., Kawaguchi, K., and Hagihara, B. (1966). *J. Biol. Chem.* **241**, 1797–1806.

Plattner, H., Salpeter, M. M., Saltzgaber, J., Rouslin, W., and Schatz, G. (1971). *In* "Autonomy and Biogenesis of Mitochondria and Chloroplasts" (N. K. Boardman, A. W. Linnane, and R. M. Smillie, eds.), pp. 175–184. Amer. Elsevier, New York.

Proudlock, J. W., Haslam, J. M., and Linnane, A. W. (1969). *Biochem. Biophys. Res. Commun.* **37**, 847–852.

Proudlock, J. W., Haslam, J. M., and Linnane, A. W. (1971). *Bioenergetics* **2**, 327–349.

Rabinowitz, M., Getz, G. S., Casey, J., and Swift, H. (1969). *J. Mol. Biol.* **41**, 381–400.

Ragan, C. I., and Garland, P. B. (1971). *Biochem. J.* **124**, 171–187.

Reilly, C., and Sherman, F. (1965). *Biochim. Biophys. Acta* **95**, 640–651.

Richter, D. (1971). *Biochemistry* **10**, 4422–4425.

Ross, E., Ebner, E., Poyton, R. O., Mason, T. L., Ono, B., and Schatz, G. (1974). *In* "The Biogenesis of Mitochondria" (A. M. Kroon and C. Saccone, eds.), pp. 477–489. Academic Press, New York.

Rouslin, W., and Schatz, G. (1969). *Biochem. Biophys. Res. Commun.* **37**, 1002–1007.

Sager, R. (1972). "Cytoplasmic Genes and Organelles." Academic Press, New York.

Saunders, G. W., Gingold, E. B., Trembath, M. K., Lukins, H. B., and Linnane, A. W. (1971). *In* "Autonomy and Biogenesis of Mitochondria and Chloroplasts" (N. K. Boardman, A. W. Linnane, and R. M. Smillie, eds.), pp. 185–193. Amer. Elsevier, New York.

Schatz, G. (1965). *Biochim. Biophys. Acta* **96**, 342–345.

Schatz, G. (1967). *In* "Methods in Enzymology" (R. W. Estabrook and M. E. Pullman, eds.), Vol. 10, pp. 197–202. Academic Press, New York.

Schatz, G. (1968). *J. Biol. Chem.* **243**, 2192–2199.

Schatz, G. (1970). *In* "Membranes and Mitochondria and Chloroplasts" (E. Racker, ed.), pp. 251–314. Van Nostrand-Reinhold, Princeton, New Jersey.

Schatz, G., and Saltzgaber, J. (1969). *Biochem. Biophys. Res. Commun.* **37**, 996–1001.

Schatz, G., Penefsky, H. F., and Racker, E. (1967). *J. Biol. Chem.* **242**, 2553–2560.

Schuurmans-Stekhoven, F. M. A. H. (1966). *Arch. Biochem. Biophys.* **115**, 555–568.

Scragg, A. H., Morimoto, H., Villa, U., Nekhorocheff, J., and Halvorson, H. O. (1971). *Science* **171**, 908–910.

Sebald, W., Machleidt, W., and Otto, J. (1974). *In* "The Biogenesis of Mitochondria" (A. M. Kroon and C. Saccone, eds.), pp. 453–463. Academic Press, New York.

Senior, A. E. (1973). *Biochim. Biophys. Acta* **301**, 249–257.

Tuppy, H., and Birkmayer, G. D. (1969). *Eur. J. Biochem.* **8**, 237–243.

Tzagoloff, A., Rubin, M. S., and Sierra, M. F. (1973). *Biochim. Biophys. Acta* **301**, 71–104.

Tzagoloff, A., Akai, A., and Rubin, M. S. (1974). *In* "The Biogenesis of Mitochondria" (A. M. Kroon and C. Saccone, eds.), pp. 405–421. Academic Press, New York.

Vary, M. J., Edwards, C. L., and Stewart, P. R. (1969). *Arch. Biochem. Biophys.* **130**, 235–244.

Vitols, E., and Linnane, A. W. (1961). *J. Biochem. Biophys. Cytol.* **9**, 701–710.

Wallace, P. G., Huang, M., and Linnane, A. W. (1968). *J. Cell Biol.* **37**, 207–220.

Watson, K., Haslam, J. M., and Linnane, A. W. (1970). *J. Cell Biol.* **46**, 88–96.

Watson, K., Haslam, J. M., Veitch, B., and Linnane, A. W. (1971). *In* "Autonomy and Biogenesis of Mitochondria and Chloroplasts" (N. K. Boardman, A. W. Linnane, and R. M. Smillie, eds.), pp. 162–174. Amer. Elsevier, New York.

Wintersberger, E. (1967). *Hoppe-Seyler's Z. Physiol. Chem.* **348**, 1701–1704.

Chapter 15

Separation and Some Properties of the Inner and Outer Membranes of Yeast Mitochondria

W. BANDLOW AND P. BAUER

Institute for Genetics,
University of Munich,
München, Germany

1. Introduction

One of the characteristic morphological properties of mitochondria is their unique arrangement in a double-membrane system which provides a variety of advantages for the cell's energy production. The most important of these are (1) provision of a structural environment for the integration of the inner membrane enzyme complexes of the respiratory chain and energy conservation; (2) buildup of an effective compartmentation for protons delivered by the electron-transport chain during substrate oxidation and for other charged low-molecular-weight substances; the charge separation results in an electrostatic potential which, according to the chemiosmotic hypothesis of energy conservation (Mitchell, 1966, 1968) may be the driving force for the production of ATP; and (3) storage of the intermediates of the citric acid cycle and the beta-oxidation of fatty acids, especially of the NADH produced, in close proximity to the dehydrogenases of the respiratory chain for regeneration of NAD^+.

The properties mentioned above are those of the inner mitochondrial membrane. But what is the function of the outer and how is it built up?

In contrast to the inner, the outer membrane is freely permeable to low-molecular-weight substances (Klingenberg and Pfaff, 1966; Pfaff et al., 1968). It contains only a few enzymes of minor importance for cell metabolism, i.e., cytochrome-b_5-dependent oxidoreductases (Sottocasa et al., 1967), and oxidases and oxygenases of aromatic systems (Greenawalt and Schnaitman, 1970; Okamoto et al., 1967; Cassady and Wagner, 1968) and of long-chain fatty acid activation (Haddock et al., 1970; Lippel and Beattie, 1970; Skrede and Bremer, 1970; van Tol, 1970; van Tol and Hülsman, 1970). Some of these have, in addition, a dual localization in the microsomal fraction. However, their structure and catalytic properties may be different in the two cell fractions (Schnaitman, 1969; Kuylenstierna et al., 1970). So it appears that, because of the lack of other essential functions, the most logical explanation for the function of the outer mitochondrial membrane is to make the inner membrane system more effective, i.e. (1) storing the protons released during the oxidation of substrates and transported through the membrane (Mitchell and Moyle, 1967) close to the outer surface of the inner membrane, and (2) keeping adenylate kinase and nucleoside mono- and diphosphokinases, enzymes of the intermembrane space (Schnaitman and Greenawalt, 1968; Lima et al., 1968) close to the inner mitochondrial compartment. These enzymes effectively prevent leakage of ADP from the matrix space by converting it to AMP and, finally, to ATP, the latter being then available for energy-dependent processes in the cytoplasm.

Mitochondria contain a system of genetic information of their own (for review, see Borst, 1972). Mitochondrial ribosomes contribute 13–15% of the total protein to the genesis of mitochondria (Hawley and Greenawalt, 1970; Schweyen and Kaudewitz, 1970). After disc electrophoretic resolution of these proteins labeled specifically in vivo in the presence of cycloheximide— an antibiotic blocking cytoplasmic ribosomal contribution (Siegel and Sisler, 1965)—seven to nine protein bands are found to be radioactive (Sebald et al., 1968; Thomas and Williamson, 1971; Groot et al., 1972). However, the function of most of them has not yet been established.

In order to investigate the function of the outer mitochondrial membrane in more detail and to test whether the mitochondrial protein-synthesizing machinery contributes to the outer membrane in addition to the inner, separation of the two membrane systems is useful. Membrane separation has confirmed the localization of the respiratory chain complexes in the inner membrane, and of the enzymes of the citric acid cycle and of fatty acid catabolism in the matrix space of mammalian mitochondria. Monoamine oxidase, kynurenine hydroxylase, and long-chain acyl-CoA synthetase were demonstrated to be localized in the outer membrane (Parsons *et al.*, 1966; Sottocasa *et al.*, 1967; Schnaitman and Pedersen, 1968; contrast: Green *et al.*, 1968; Greenawalt and Schaitman, 1970; Okamoto *et al.*, 1967; Skrede and Bremer, 1970). Membrane separation has thus contributed fundamentally to the present knowledge of the localization and environmental organization of these enzymes.

The intention of this article is to describe procedures of membrane separation for yeast mitochondria, and to include some data about the characterization of the inner and outer mitochondrial membrane.

II. Membrane Separation

Detachment of the outer membrane has been achieved with mammalian mitochondria either by swelling and shrinking, sometimes combined with gentle sonication (Parsons *et al.*, 1966; Sottocasa *et al.*, 1967), or by incubation with digitonin (Schnaitman and Greenawalt, 1968; Greenawalt and Schnaitman, 1970). Yeast mitochondria, however, are rather fragile and are sensitive to osmotic shock and strong mechanical treatment. A large degree of swelling in hypotonic medium or mild detergent treatment leads to extensive structural damage of yeast mitochondria (Bandlow, 1972). So care must be taken to maintain the intactness of the organellae during the preparation.

A. Swelling and Shrinking and Detachment of the Outer Membrane

Rapidly growing yeast strains such as *Saccharomyces carlsbergensis* CB7α *trp5*, *S. cerevisiae* M12a *ura3 ilv5 trp2*, or the diploid A 1327 A × 18I have been used and grown as described previously (Bandlow, 1972). Mitochondria were prepared after protoplastating and lysis of logarithmically growing cells as described by Kovač *et al.* (1972). Lysis was supported by homogenization in a Potter-Elvehjem homogenizer (five to six strokes). After lysis all operations were carried out at 0°C. The mitochondrial pellet was washed thoroughly with a buffer containing 0.6 *M* mannitol (pH 7.4).

These mitochondria exhibited respiratory control ratios of 2.0 to 2.9 and P/O ratios of 1.75 to 1.93 with NADH as substrate. Difference spectra taken from dithionite-reduced samples measured against an oxidized reference at 77°K as described by Bandlow *et al.* (1974) show a sharp absorption maximum for cytochrome c at 547.5 nm, exceeding the height of the cytochrome-b absorptions (see Fig. 13, trace A). Essentially no loss of cytochrome c was detectable, indicating the integrity of the inner mitochondrial compartment.

In order to preserve the intactness of the inner membrane, isotonic swelling is recommended for the disruptance of the outer membrane. This procedure employs a swelling phase in the presence of ionophors and 0.25 M mannitol. In contrast to inner membranes derived from digitonin-treated mitochondria from rat liver (Schnaitman and Greenawalt, 1968), no respiratory control was preserved after the application of ionophors, but most of the inner membranes were left intact.

As a control for the extent of swelling and shrinking, changes in turbidity were recorded at 410 nm and 25°C with an aliquot of the mitochondrial suspension. For swelling, mitochondria were suspended in a buffer containing 0.25 M mannitol, 20 M Tris–maleate, 5 mM KH_2PO_4 1 mM EDTA (final pH 7.4), and about 0.5–1 mg protein/ml. The degree of swelling is not very great under these conditions (as can be seen from Fig. 1, trace A) and is even less at pH 6.0. It can be increased more than 2-fold when additional NaH_2PO_4 (20 mM) and succinate (10 mM) are present (trace B), each of which has a separate effect (not shown). However, the degree of swelling obtained under these conditions is obviously not sufficient for the disruptance and detachment of the outer membrane. Addition of ionophors, such as gramicidin D (2 μg/ml, trace C) or valinomycin (0.5 μg/ml, trace D) initiates a phase of additional swelling, the latter being slightly more effective. The low-amplitude swelling in the absence of ionophors is almost completely reversed by the addition of 0.5 vol of 1.8 M sucrose, 12 mM ATP, 12 mM $MgSO_4$, and 0.1% bovine serum albumin. A correction for volume change has been applied in Fig. 1. The shrinking buffer had, however, little effect on the swollen state in the presence of valinomycin, while mitochondria swollen in the presence of gramicidin showed a large increase in turbidity (Fig. 1, trace C), indicating excessive shrinking. The contraction phase was even more pronounced when the previous swelling was at pH 6.0 instead of 7.4. But since at pH 7.4 the degree of swelling was much higher and the disruptance of outer membranes obviously more effective, swelling at this pH in the presence of gramicidin D was routinely used for membrane separation.

For the preparation of outer and inner membranes, the same conditions were followed as described above, except that the protein content was 10–20 mg/ml and the temperature 0°C. After a swelling period of 15 minutes at 0°C, mitochondria were sonicated (Branson B12 sonifier, microtip, 3 A, 15 seconds, ice-cooled) in order to disrupt the outer membranes. Then $\frac{1}{2}$ vol of ice-cold shrinking buffer (as above, in the cuvette experiment) was added with careful mixing. Vigorous homogenization by shearing was avoided

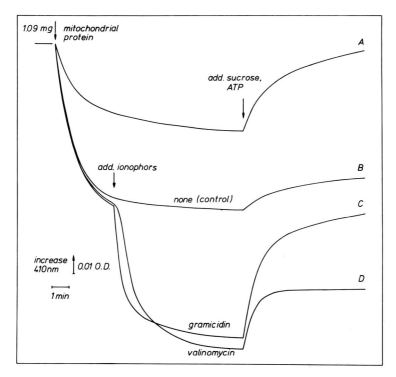

FIG. 1. Time course of swelling and shrinking. The swelling phase was started by adding 0.05 ml of a suspension of mitochondria (1.09 mg of protein) in 0.6 M mannitol, 2 mM EDTA (pH 7.4), to a cuvette containing 1.95 ml of 0.25 M mannitol, 5 mM KH$_2$PO$_4$, and 20 mM Tris–maleate (pH 7.4) (trace A). In experiments B, C, and D, 10 mM succinate and 20 mM NaH$_2$PO$_4$ (pH 7.4) were present in addition. Changes in turbidity were followed at 410 nm in an Aminco DW 2 spectrophotometer at 25°C. Additions were made at the arrows: 4 μg of gramicidin D (trace C) and 1 μg of valinomycin (trace D) in 1 μl of absolute ethanol each, and 1 ml of 1.8 M sucrose, 12 mM ATP, and 12 mM MgSO$_4$ (pH 7.4). A correction has been applied for volume change by the sucrose shrinking buffer.

during this phase. After 5 minutes the suspension was homogenized by hand with three to five strokes in a Teflon-glass homogenizer. Subsequent centrifugations (5 minutes at 5000 g and 15 minutes at 12,000 g) yielded fractions containing mainly intact mitochondria and inner membranes plus ruptured mitochondria, respectively. The brownish-yellow supernatant was diluted 1:2 with a buffer containing 0.25 M mannitol, 20 mM Tris–HCl, and 1 mM EDTA (pH 7.4) (isolation buffer) and recentrifuged at 165,000 g for 30 minutes. The resulting pellet contained outer membranes plus ruptured mitochondria.

The yield of inner membranes was increased at the expense of the outer by incubating the mitochondria with 3 mg lipase A/ml for 10 minutes at

0°C prior to the swelling, and then centrifuging and washing them free of the enzyme. Higher yields of the outer membranes were obtained by treating shrunken mitochondria with sonic oscillation, as above.

The pellets obtained after centrifugation at 12,000 g for 15 minutes, and at 165,000 g for 30 minutes, were homogenized in isolation buffer, layered on top of linear gradients of 2.0–1.1 M and 1.3–0.5 M buffered sucrose, respectively, and centrifuged at 27,000 rpm for 135 minutes at 4°C in a SW 27 rotor in a Spinco ultracentrifuge. The tubes were punctured at the bottom. Elution diagrams were prepared by recording the absorbance at 423 nm with a flow cell of 0.4-ml vol and a 1-cm light path. Fractions of 20 drops were collected. Densities were determined pycnometrically.

B. Enzymatic and Electron Optical Characterization of Inner and Outer Membranes

After centrifugation the gradient for the above-mentioned 12,000 g pellet contained five bands, and that of the 165,000 g pellet another three, one of them appearing at identical densities in both gradients. Because of the great number of different bands, the application of a linear gradient is more advisable than that of a discontinuous gradient.

The various mitochondrial subfractions have been identified and characterized by means of the following marker enzymes. Malate and glutamate (NADH) dehydrogenases were regarded as being specific markers for the soluble matrix space. Cytochrome-c oxidase, succinate: and NADH-cytochrome-c oxidoreductase and oligomycin-sensitive ATPase were taken as markers for the inner membrane, and both kynurenine hydroxylase and antimycin-insensitive NADH-cytochrome-c oxidoreductase for the outer membrane. Adenylate kinase was taken as being specific for the intermembrane space. NADPH-cytochrome-c oxidoreductase indicated microsomal contamination. Succinate: and NADH-cytochrome-c oxidoreductase were tested as described by Lang et al. (1974). All the other enzymes were assayed as summarized by Bandlow (1972).

The distribution of marker enzymes in the various fractions is shown in Figs. 2 and 3. They reveal the following. The band found at a density of 1.19 gm/cm^3 contains all the mitochondrial marker enzymes. With S. carlsbergensis a second band appears at 1.18 gm/cm^3, exhibiting the same properties. No significant differences were detectable in the two fractions. The same densities are found after gradient centrifugation of untreated mitochondria. Electron microscope examination of positively stained thin sections of these two fractions shows grossly swollen and distorted cristae encircled by an intact double membrane, as shown in Fig. 4. It is concluded that both of them contain mitochondria swollen from dilution of the sucrose-containing suspension. The reason for the dual density appearance is not yet clear, but

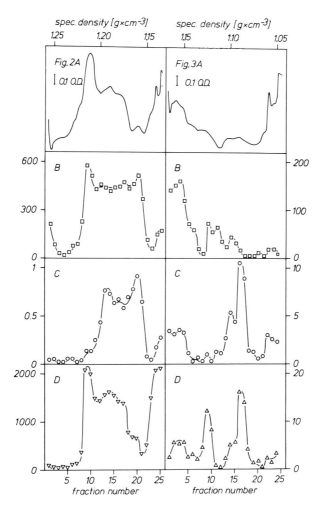

FIG. 2. Elution diagram of a sucrose gradient (2.0–1.1 *M*) of the 12 000 *g* pellet and distribution of marker enzymes. A, Tracing of optical density during elution of the gradient at 423 nm; B, succinate-cytochrome-c oxidoreductase; C, kynurenine hydroxylase; D, malate dehydrogenase. Enzyme activities are expressed as nanomoles oxidized per minute per milligram of protein.

FIG. 3. Elution diagram of a sucrose gradient (1.3–0.5 *M*) of the 165 000 *g* pellet and distribution of marker enzymes. A, Absorbance at 423 nm; B, succinate-cytochrome-c oxidoreductase; C, kynurenine hydroxylase; D, antimycin-insensitive NADH-cytochrome-c oxidoreductase.

is also found with other fractions and may reflect two states of the growth cycle of the cell. For comparison, an electron micrograph of the starting mitochondria is given in Fig. 5. It shows well-structured mitochondrial cristae. A double-membrane system is discernible.

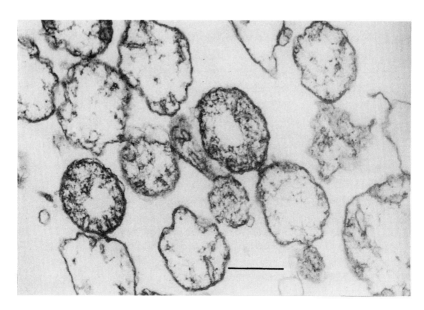

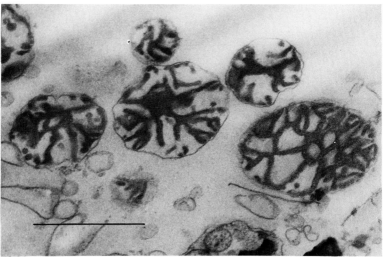

FIG. 4.   Electron micrograph of mitochondria swollen and distorted from sucrose gradient centrifugation. Thin-sectioned, fixed, and positively stained; bar, 1 μm. Electron micrographs were prepared as described earlier (Bandlow, 1972). From Bandlow (1972). Reproduced with the permission of Elsevier Scientific Publishing Company, Amsterdam.

FIG. 5.   Electron micrograph of intact mitochondria prepared as described in Section II,A (starting material). Thin-sectioned, fixed, and positively stained; bar 1 μm.

Fractions banding at 1.16 and 1.14 gm/cm³ exhibit severe loss of soluble enzymes but have retained activities of enzymes bound to both the inner and outer membrane, respectively. Electron micrographs of each fraction reveal profiles of ruptured mitochondria in both fractions (not shown).

Particles found at a density of 1.21 gm/cm³ have lost the enzymes of the outer membrane and of the intermembrane space but have retained matrix- and inner membrane–bound enzymes. A thin section of this fraction is shown in Fig. 6. It reveals profiles enclosed by a single intact membrane. Cristae structures are distorted from the osmotic shock but are still visible. It is concluded that this fraction contains complete inner membranes.

A light fraction of pale-yellow appearance is found at a density of 1.09 gm/cm³. In some preparations this band is found split, but both densities are found very close together. It contains the maxima of both kynurenine hydroxylase and antimycin-insensitive NADH-cytochrome-c oxidoreductase activities, but no cytochrome-c oxidase, adenylate kinase, or malate dehydrogenase. Very little succinate- and NADPH-cytochrome-c oxidoreductase is found to be associated with this fraction. Electron micrographs of negatively stained specimens spread on a protein surface film as in Fig. 7 show empty vesicles about 0.3 µm in diameter, in contrast to intact mitochondria which have diameters of 0.8–1.5 µm. Thin sections of this fraction show closed vesicles, indicating that these membranes fuse again after disruptance (not shown). It is concluded that this fraction contains outer membranes. The yield amounts to less than 1 mg from about 30 gm of wet weight cells.

Two additional fractions can be isolated from the gradients at densities of 1.24 and 1.12 gm/cm³. The yield of both of them is low and varies. The heavier one exhibits high oligomycin-insensitive ATPase and some adenylate kinase activity. It shows lamellar structure of 1–2 µm in diameter in electron micrographs after negative staining and spreading. It is regarded as plasmalemma contamination (Fig. 8). The lighter fraction is associated with both NADPH- and antimycin-insensitive NADH-cytochrome-c oxidoreductase activities and is regarded as microsomal contamination. Negatively stained and spread specimens show vesicular structures 0.03–0.1 µm in diameter (Fig. 9).

Figure 10 summarizes the distribution of the various marker enzymes tested. The ordinate gives specific activities, while the abscissa gives the contribution of the inner membrane, outer membrane, and soluble fraction to the total. The activities of the microsomal fraction are given for comparison. It is clear from this figure that malate, glutamate, and isocitrate dehydrogenases, ferrochelatase, succinate- and antimycin-sensitive cytochrome-c oxidoreductases, cytochrome-c oxidase, and ATPase are associ-

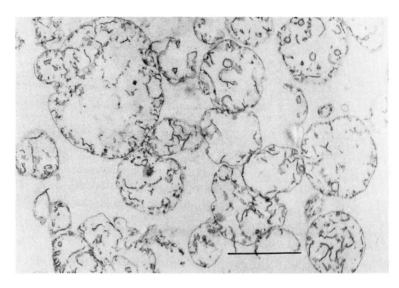

FIG. 6. Thin-section of inner membranes. Fixed and positively stained; bar, 1 μm. From Bandlow (1972). Reproduced with the permission of Elsevier Scientific Publishing Company, Amsterdam.

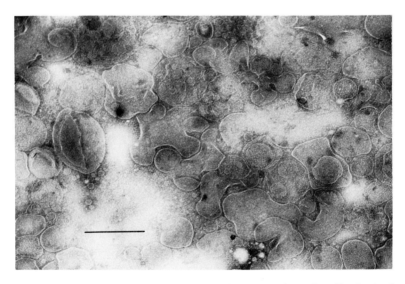

FIG. 7. Micrograph of outer membranes spread on a protein surface film. Prefixed and negatively stained as described by Bandlow (1972); bar, 0.5 μm. From Bandlow (1972). Reproduced with the permission of Elsevier Scientific Publishing Company, Amsterdam.

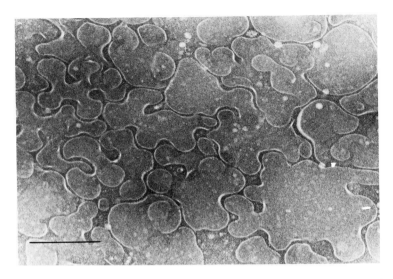

FIG. 8. Micrograph of a negatively stained fraction banding at a sucrose density of 1.24 gm/cm³ spread on a protein surface film; bar, 1 μm.

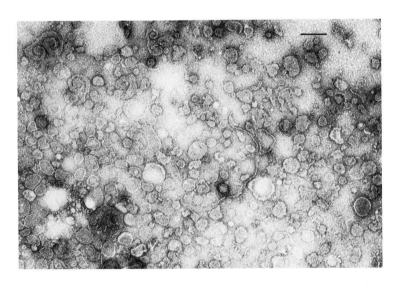

FIG. 9. Micrograph of negatively stained microsomal fraction spread on a protein surface film; bar, 0.2 μm.

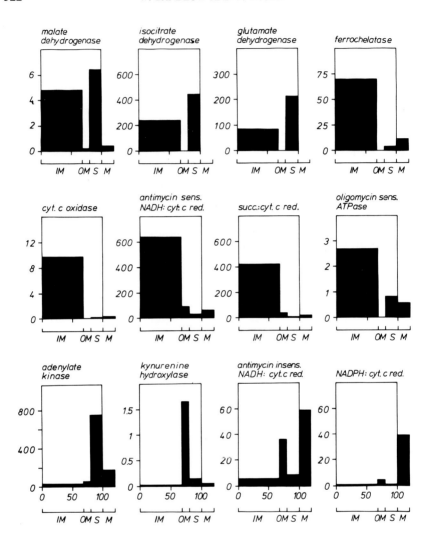

FIG. 10. Distribution of some marker enzymes in inner membrane (IM), outer membrane (OM), soluble (S), and microsomal (M) fractions. IM, OM, and S were taken as 100%. Ferrocholatase was assayed according to Jones and Jones (1970). ATPase was followed by measuring the H^+ production during hydrolysis of ATP according to Chance and Nishimura (1967). Succinate- and NADH-cytochrome-c oxidoreductase were measured as described by Lang *et al.* (1974). All the other enzymes were determined as summarized by Bandlow (1972). Cytochrome-c oxidase activity is expressed as the first-order rate constant, malate dehydrogenase and oligomycin-sensitive ATPase as micromoles per milligram per minute, and all the others as nanomoles per milligram per minute. All enzyme assays were carried out at 25°C.

ated with the inner mitochondrial compartment. Kynurenine hydroxylase and antimycin-insensitive NADH-cytochrome-c oxidoreductase are enriched with the outer membrane fraction, while adenylate kinase is released into the soluble supernatant during membrane separation. The highest activity of NADPH-cytochrome-c oxidoreductase is found associated with microsomes, and only a little with outer membranes.

C.   Spectral Properties of the Outer Membrane

Yeast mitochondrial outer membranes are devoid of monoamine oxidase (Bandlow, 1972), widely used as a specific marker in mammalian mitochondria. However, some oxidative enzymes, dependent on either NADH or NADPH are also found in yeast mitochondria. The acceptors appear to be b_5- or b_2-type cytochromes, the activity being insensitive to antimycin. So an experiment was performed in order to characterize the b cytochromes of the outer membrane. Figure 11 shows a spectrum taken at 77°K of total cytochromes present in the outer membrane by reducing the sample cuvette with dithionite using oxidized membranes as the reference (trace A). The conditions were the same as described earlier (Bandlow et al., 1974). Only the alpha and beta region is shown.

This spectrum exhibits two maxima, at 546.5 and at 557.5 nm. It resembles closely that of the microsomal fraction (shown in Fig. 12), but is clearly distinct from the corresponding mitochondrial spectrum (Fig. 13). Essentially no cytochrome-a contamination is detectable, indicating a very low contamination by cytochromes localized in the inner mitochondrial compartment. The peak absorbing at 546.5 nm is clearly not cytochrome c, which absorbs at 547.5 nm at 77°K. It is not removed by sonication and is reducible by NADH without any additional cytochrome-b reduction (Fig. 11, trace C).

The peaks absorbing at 546.5 and 557.5 nm are apparently composite. In the absence of the peak at 557.5 nm (trace C), also a long-wavelength shoulder with a maximum at about 550 nm at 77°K, has disappeared. The absorption band reduced by NADH is thus perfectly symmetrical, and its maximum is shifted to 546.0 nm. Since the bandwidth of the spectrophotometer under the experimental conditions used is 0.5 nm and the relative reproducibility is less than 1 Å, the blue shift of the peak by 0.5 nm is significant and indicates the absence of a contribution of the absorption at about 550 nm. It appears that the peak found at 546.0 nm may be the cytochrome b_2 described by Appleby and Morton (1960) in yeasts. From dual wavelength measurements at room temperature (Aminco DW 2 spectrophotometer), a content of 0.15 nmoles/mg of dithionite-reduced minus oxidized outer membranes may be calculated for this cytochrome using the

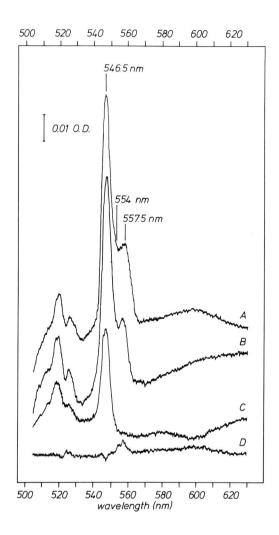

FIG. 11. Difference spectra of outer mitochondrial membranes. Spectra were recorded with a Cary Model 14 spectrophotometer equipped with an RIIC low-temperature device and a 0–0.1 OD slide wire. The temperature was 77°K and the light path 2 mm. The cuvettes received 0.6 ml of suspension (11.4 mg of protein). The buffer contained 1 M sucrose, 0.1 M KH_2PO_4 1 mM EDTA (pH 6.8). Mitochondrial fractions were preincubated in the cuvettes under exclusion of oxygen for at least 20 minutes. Oxidized references were oxidized by 0.5 mM ferricyanide. Trace A shows a dithionite-reduced minus oxidized spectrum. In trace B the sample received 3.2 mM NADPH measured against the same reference as for trace A. Trace C was obtained by recording a spectrum of NADH-reduced (3.2 mM) outer membranes against the same reference as for trace A. Spectrum D shows a spectrum reduced by NADPH in the presence of kynurenine minus an NADPH-reduced reference.

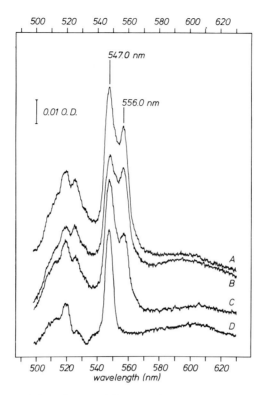

FIG. 12. Spectra of the microsomal fraction. Conditions were the same as described for Fig. 11, except that no ferricyanide was necessary to oxidize the microsomal suspension in the reference cuvette (4.9 mg of protein was used per cuvette). Trace A was obtained from dithionite-reduced minus oxidized microsomes, B from 3.2 mM NADH-reduced minus the same reference as for A, C from 3.2 mM NADPH-reduced minus the same reference as for A, and D from DL-lactate reduced (12 mM) in the presence of NADP (2 mM) minus the same reference as for A.

wavelength pair 551 minus 575 nm and the absorption coefficient of 20.0 mM^{-1} cm^{-1} reported by Klingenberg (1958).

The other cytochrome is obviously identical to cytochrome b$_5$. Its spectral properties resemble closely those published for cytochrome b$_5$ of outer membranes of mammalian mitochondria by Haddock et al. (1970) and Wohlrab and Degn (1972). However, the broad peak appearing at 557.5 nm after reduction with dithionite may be composite, too. The peak sharpens considerably after reduction with NADH, and the maximum is blue-shifted to 556.5 nm (trace B). Thus it appears that in yeast, also cytochrome b$_5$ exhibits a dual absorption band with maxima at 556.5 and 550 nm at 77° K, as in mammalian outer membranes. From the wavelength pair 557 minus 575 nm at room temperature, 0.06 nmoles/mg protein of cytochrome b$_5$

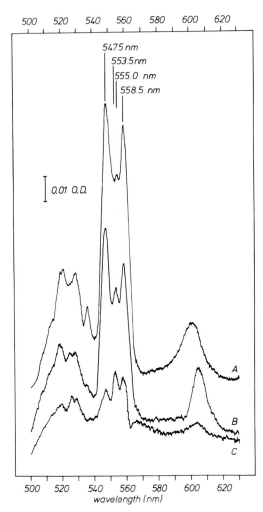

FIG. 13. Difference spectra of intact mitochondria. Conditions were the same as described for Fig. 11, except that no ferricyanide was necessary to reduce aerobic mitochondria in the reference cuvette and 2 μM FCCP was present in the sample in experiment B and in both sample and reference in experiment C. Each cuvette received 3.6 mg of protein. Trace A shows a difference spectrum of dithionite-reduced mitochondria measured against an oxidized reference; B, NADH-reduced (3.2 mM) minus oxidized; C, NADH (3.2 mM) plus ascorbate (12 mM) plus TMPD (0.5 mM) minus ascorbate plus TMPD (as in the sample).

can be determined using the same absorption coefficient as above. The nature of the long-wavelength contribution at about 558 nm to the dithionite-reduced spectrum (trace A) could not be established. Both cytochrome b$_5$ and the b cytochrome absorbing at 546.0 nm, are reduced by NADPH, while

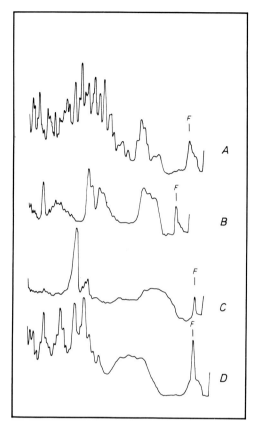

FIG. 14. Densitometric traces of polyacrylamide gels run with submitochondrial part-icles (A), inner membrane fraction (B), outer membranes (C), and microsomal fraction (d). Disc electrophoresis was performed as described by Laemmli (1970); 200 μg of submitochon-drial particles, inner membranes and the microsomal fraction, and about 80 μg of outer membranes was applied to each gel (15% polyacrylamide). Gels were stained with Coomassie brilliant blue R 250 after the front indicator (F, bromphenol blue) had been permanently marked by the injection of India ink. The gels were scanned with a Joyce Loebl Chromoscan modified for high resolution.

only the latter is reduced by NADH. This is the most pronounced difference from microsomal b cytochromes (Fig. 12).

Trace D in Fig. 11 shows a spectrum of outer membranes reduced by NADPH in the presence of kynurenine with NADPH-reduced particles in the reference cuvette. This spectrum makes it very likely that cytochrome b_5 is the acceptor of kynurenine hydroxylase. The small negative peak at 546 nm is due to a slightly better reduction of this cytochrome in the reference cell.

Microsomal spectra are given in Fig. 12. They reveal that the cyto-chrome-b content is about four times higher than in outer membranes, and also that the relative proportion of cytochrome b_5 is higher in this frac-tion. Dithionite reduces peaks at 547.0 and 556.0 nm (Fig. 12, trace A). Both NADH and NADPH reduce the same cytochromes as in trace A (traces B and C), while DL-lactate in the presence of NADP reduces a symmetric band with a maximum at 546.5 nm (trace D). A qualitatively identical spec-trum is obtained after reduction with ascorbate in the presence of TMPD. So, the acceptor for both lactate and ascorbate dehydrogenases is not cyto-chrome b_5 in yeasts but the cytochrome absorbing at 546.5 nm at 77°K.

Spectra of intact mitochondria are given for comparison in Fig. 13. They are clearly distinct from both microsomal and outer membrane spectra. They exhibit an alpha-absorption band at 600–601 nm for cytochrome a and at 547.5 nm for cytochrome c, after reduction with dithionite (trace A) or NADH (trace B). The b cytochromes show absorption maxima at 558.5 and 553.5 nm, as revealed by trace C for NADH–ascorbate–TMPD-reduced mitochondria measured against an ascorbate–TMPD-reduced reference. FCCP was present as an uncoupler both in spectra B and C.

D.   Biogenesis of the Inner and Outer Membranes

General agreement exists that both the extrinsic cytoplasmic and the intrinsic mitochondrial protein-synthesizing machineries contribute to the genesis of mitochondria (for review, see Beattie, 1971). The contribution of both systems can be elucidated *in vivo* by blocking specifically either cyto-plasmic ribosomal function by cycloheximide, or the mitochondrially governed translation by chloramphenicol or erythromycin. After the appli-cation of label, the integration of the products of the unpoisoned protein-synthesizing system into mitochondrial membrane protein can thus be studied. Subsequent membrane separation leads to additional information about the biogenesis of each individual membrane.

Neupert *et al.* (1967) and Beattie *et al.* (1967) showed with rat liver, and Bandlow (1972) with yeast mitochondria, that essentially no radioactivity was incorporated into outer mitochondrial membrane protein after labeling *in vitro* (rat) or *in vivo* in the presence of cycloheximide (yeast). These investigators observed that radioactive label was almost exclusively associ-ated with the inner membrane fraction under these conditions.

It is of some interest to know the number of individual proteins that build up the outer membrane. In contrast to results obtained with rat liver (Schnaitman, 1969), Neupert and Ludwig (1971) found that outer mem-branes of mitochondria from *Neurospora crassa* could be resolved into only a single band after polyacrylamide gel electrophoresis. Figure 14 shows the densitometric trace of an electrophoretic run of outer membranes of yeast mitochondria (trace C). We find also that in yeasts one major protein fraction of 48,000 apparent molecular weight comprises about 90% of the outer membrane protein. [For the determination of molecular weight, the following marker proteins had been used: cytochrome c (MW 12,500), myoglobin (MW 17,500), chymotrypsinogen A (MW 25,000), carbanhydrase (MW 29,000), ovalbumin (MW 44,000), and bovine serum albumin (MW 67,500). Molecular-weight determinations have been performed alternatively in 10 and 15% acrylamide gels with little deviation.] In addition, a minor second band is found with a molecular weight of 40,000 daltons. Whether the two bands reflect two individual proteins or several proteins of identical mobility in each band, or whether they are formed by an aggregation of low-molecular-weight material which might have occurred under the conditions of electrophoresis because of the highly lipophilic nature of these proteins, cannot be determined at present. At least, the minor band appears to be inhomogeneous. The two bands are obviously absent from both the inner membrane (trace B) and microsomal fractions (trace D), but are contained in submitochondrial particles obtained from whole mitochondria after sonication and centrifugation. In addition, the minor band of 40,000 molecular weight has been found in the same proportion in various preparations. Thus it apparently is not a contaminant. It is surprising that, despite the great number of enzymatic functions present in the outer membrane, this fraction is composed only of the two components that can be resolved electrophoretically.

Earlier work (Neupert et al., 1967; Neupert and Ludwig, 1971; Beattie et al.; 1967; Bandlow, 1972) revealed that the mitochondrial protein-synthesizing machinery apparently does not contribute to the formation of the outer mitochondrial membrane, because of the low level of radioactivity found associated with this fraction after the cytoplasmic ribosomal contribution had been blocked. We were interested in knowing whether this is true for both protein bands or whether the residual radioactivity is specifically associated with the minor component.

Figure 15 (trace C) shows that, after inhibition by cycloheximide, the residual radioactivity is equally distributed along the total gel. After labeling in the presence of erythromycin, the two bands of the outer membrane are found to be radioactive (Fig. 16, trace C). This strongly suggests that none of the two bands is composed of mitochondrially synthesized proteins.

After labeling in the presence of cycloheximide, the same four major, and, at least, three minor components are found in gels run with sonicated mitochondrial particles and inner membranes, respectively. From this result it may also be concluded that none of the mitochondrially synthesized proteins are separated from the inner membrane compartment simultaneously with the outer membrane.

Qualitative differences in the protein-labeling patterns of particles from

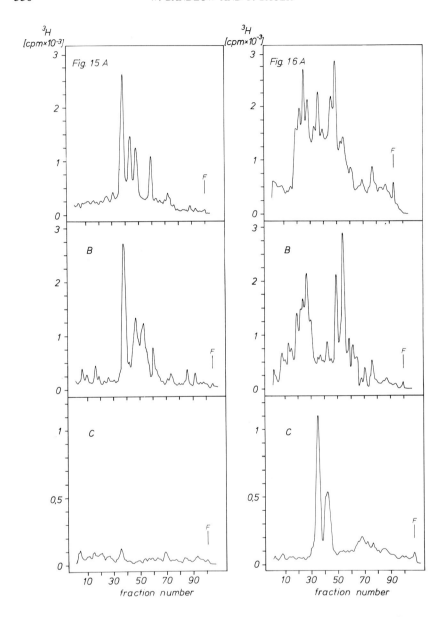

FIG. 15. A, labeling pattern of mitochondrial particles; B, inner membranes; C, outer membranes; labeled with tritium-labeled amino acids in the presence of cycloheximide (200 μg/ml). The labeling conditions were the same as described recently (Bandlow, 1972), except that cells were grown and labeled in synthetic medium containing 3% glycerol. After application of the label, cells were washed free of the antibiotic (six washes) and incubated in complete medium in the presence of the respective unlabeled amino acids (1 m*M*) for

whole mitochondria and from the inner membrane fraction, which was labeled in the presence of erythromycin, may be attributed to specific losses, during the separation and sonication procedures, of material loosely attached to the inner membrane.

III. Discussion

The most critical point in membrane separation is the preservation of the intactness of the inner membrane–matrix compartment during the swelling and shrinking and successive separation procedure. Accordingly, membrane separation was successful only with mitochondria prepared carefully from protoplasts obtained from logarithmically growing, fully derepressed cells. The quality of mitochondria was judged according to the following criteria: (1) a high respiratory control ratio and a P/O ratio close to 2 [determined according to the method of Lindberg and Ernster (1956) with NADH as substrate]; (2) a high cytochrome-c content, since this cytochrome leaks out readily from mitochondria after damage; and (3) morphological control by electron microscopy.

Membrane separation with yeast mitochondria was achieved according to the following criteria. (1) The inner membrane–matrix compartment contained both the enzymes of the respiratory chain and oxidative phosphorylation and of the soluble matrix space. This fraction was essentially free of kynurenine hydroxylase, adenylate kinase, and NADPH-cytochrome-c oxidoreductase. Electron micrographs show particles enclosed by a single membrane. (2) The outer membrane contained kynurenine hydroxylase at least 10 times enriched compared to the original mitochondria, antimycin-insensitive NADH-cytochrome-c oxidoreductase, and cytochromes b_2 and b_5. This fraction is essentially devoid of the enzyme activities of the inner membrane–matrix compartment. In low-temperature spectra no cytochrome aa_3 was detectable. Electron micrographs show that this fraction mainly contains empty envelopes.

FIG. 15 (*cont.*)

three more generations in order to reach log growth. Electrophoretic conditions were the same as described for Fig. 14; 12.5% polyacrylamide gels were sliced under freezing conditions with liquid carbon dioxide into 1-mm slices which were transferred to a Soluene (0.7 ml, Packard Instruments, Ltd.)–water (0.03 ml) mixture for swelling (5 hours at 70°C with shaking). The scintillation medium consisted of 10 ml of toluene containing 5 gm of PPO per liter and 0.3 gm of dimethyl-POPOP per liter.

FIG. 16. A, labeling pattern of mitochondrial particles; B, inner membranes; C, outer membranes; labeled in the presence of erythromycin (4 mg/ml). Procedures and conditions the same as described for Fig. 15.

Mutual contamination of outer and inner membranes is low, as may be calculated from specific marker enzymes and from the spectroscopic data of the outer membrane. Contamination of the outer membrane fraction by microsomes may be as high as 15%, as judged from the distribution of NADPH-cytochrome-c reductase in this fraction, which is regarded as a specific marker of microsomes. Microsomal contamination is very difficult to separate from outer membranes. Sonication and successive sucrose gradient centrifugation was only partially successful.

Cytochromes b_2 (functioning as an electron acceptor of lactate and ascorbate dehydrogenase) and b_5 are associated with both the outer mitochondrial membrane and the microsomal fraction of yeasts. Spectra of the two fractions are qualitatively almost identical.

Outer membranes of yeast mitochondria can be separated by SDS disc electrophoresis into two protein bands of 48,000 and 40,000 molecular weight. In accordance with previous work with rat liver mitochondria (Neupert et al., 1967; Beattie et al., 1967), Neurospora crassa (Neupert and Ludwig, 1971), and yeast (Bandlow, 1972), no significant radioactive label was found associated with the outer membrane fraction. Neither of the two protein bands of pure outer membrane is found significantly labeled after incorporation of radioactive precursors in the presence of cycloheximide, while both of them contain radioactivity after labeling in the presence of erythromycin. Outer membranes are thus definitely proven to be cytoplasmically synthesized in yeast too.

Since after the administration of the label the cells were washed free of the antibiotics and incubated in complete medium for several hours in the presence of the respective unlabeled amino acids, it appears highly unlikely that, after labeling in the presence of cycloheximide, a possibly labeled mitochondrially synthesized protein fails to be integrated because of the exhaustion of the pool of a cytoplasmically synthesized integration protein having a small pool size.

REFERENCES

Appleby, C. A., and Morton, R. K. (1960). *Biochem. J.* **75**, 258–269.
Bandlow, W. (1972). *Biochim. Biophys. Acta* **282**, 105–122.
Bandlow, W., Wolf, K., Kaudewitz, F., and Slater, E. C. (1974). *Biochim. Biophys. Acta* **333**, 446–459.
Beattie, D. S. (1971). *Sub-Cell. Biochem.* **1**, 1–23.
Beattie, D. S., Basford, R. E., and Koritz, S. B. (1967). *Biochemistry* **6**, 3099–3106.
Borst, P. (1972). *Annu. Rev. Biochem.* **41**, 333–376.
Cassady, N. E., and Wagner, R. P. (1968). *Genetics* **60**, 168.
Chance, B., and Nishimura, M. (1967). In "Methods in Enzymology" (R. W. Estabrook and M. E. Pullman, eds.), Vol. 10, pp. 641–650. Academic Press, New York.
Green, D. E., Allman, D. W., Harris, R. A., and Tan, W. C. (1968). *Biochem. Biophys. Res. Commun.* **31**, 368–378.

Greenawalt, J. W., and Schnaitman, C. (1970). *J. Cell Biol.* **46**, 173–179.
Groot, G. S. P., Rouslin, W., and Schatz, G. (1972). *J. Biol. Chem.* **247**, 1735–1742.
Haddock, B. A., Yates, D. W., and Garland, P. B. (1970). *Biochem. J.* **119**, 565–573.
Hawley, E. S., and Greenawalt, J. W. (1970). *J. Biol. Chem.* **245**, 3574–3583.
Jones, M. S., and Jones, O. T. G. (1968). *Biochem. Biophys. Res. Commun.* **31**, 977–982.
Klingenberg, M. (1958). *Arch. Biochem. Biophys.* **75**, 376–386.
Klingenberg, M., and Pfaff, E. (1966). *In* "Regulation of Metabolic Processes in Mitochondria" (J. M. Tager *et al.*, eds.), pp. 180–201. Elsevier, Amsterdam.
Kovač, L., Groot, G. S. P., and Racker, E. (1972). *Biochim. Biophys. Acta* **252**, 55–65.
Kuylenstierna, B., Nicholls, D. G., Hovmöller, S., and Ernster, L. (1970). *Eur. J. Biochem.* **12**, 419–426.
Laemmli, U. K. (1970). *Nature (London)* **227**, 680–685.
Lang, B., Burger, G., and Bandlow, W. (1974). *Biochim. Biophys. Acta* **368**, 71–85.
Lima, M. S., Nachbaur, G., and Vignais, P. (1968). *C. R. Acad. Sci., Ser. D* **266**, 739–742.
Lindberg, O., and Ernster, L. (1955). *Methods Biochem. Anal.* **3**, 1–22.
Lippel, K., and Beattie, D. S. (1970). *Biochim. Biophys. Acta* **218**, 227–232.
Mitchell, P. (1966). "Chemiosmotic Coupling in Oxydative and Photosynthetic Phosphorylation." Glynn Res., Bodmin, Cornwall, England.
Mitchell, P. (1968). "Chemiosmotic Coupling and Energy Transduction." Glynn Res., Bodmin, Cornwall, England.
Mitchell, P., and Moyle, J. (1967). *In* "Biochemistry of Mitochondria" (E. C. Slater, Z. Kaniuga, and L. Wojtczak, eds.), pp. 53–74. Academic Press, New York.
Neupert, W., and Ludwig, G. D. (1971). *Eur. J. Biochem.* **19**, 523–532.
Neupert, W., Brdiczka, D., and Bücher, T. (1967). *Biochem. Biophys. Res. Commun.* **27**, 488–493.
Okamoto, H., Yamamoto, S., Nozaki, M., and Hayaishi, O. (1967). *Biochem. Biophys. Res. Commun.* **26**, 309–319.
Parsons, D. F., Williams, G. R., and Chance, B. (1966). *Ann. N. Y. Acad. Sci.* **137**, 643–666.
Pfaff, E., Klingenberg, M., Ritt, E., and Vogell, W. (1968). *Eur. J. Biochem.* **5**, 222–232.
Schnaitman, C. A. (1969). *Proc. Nat. Acad. Sci. U.S.* **63**, 412–419.
Schnaitman, C., and Greenawalt, J. W. (1968). *J. Cell Biol.* **38**, 158–175.
Schnaitman, C. A., and Pedersen, P. L. (1968). *Biochem. Biophys. Res. Commun.* **38**, 728–735.
Schweyen, R., and Kaudewitz, F. (1970). *Biochem. Biophys. Res. Commun.* **38**, 728–735.
Sebald, W., Bücher, T., Olbrich, B., and Kaudewitz, F. (1968). *FEBS (Fed. Eur. Biochem. Soc.) Lett.* **1**, 235–240.
Siegel, M. R., and Sisler, H. D. (1965). *Biochim. Biophys. Acta* **103**, 558–567.
Skrede, S., and Bremer, J. (1970). *Eur. J. Biochem.* **14**, 465–472.
Sottocasa, G. L., Kuylenstierna, B., Ernster, L., and Bergstrand, A. (1967). *J. Cell Biol.* **32**, 415–438.
Thomas, D. Y., and Williamson, D. H. (1971). *Nature (London), New Biol.* **233**, 196–199.
Van Tol, A. (1970). *Biochim. Biophys. Acta* **219**, 227–230.
Van Tol, A., and Hülsman, W. C. (1970). *Biochim. Biophys. Acta* **223**, 416–428.
Wohlrab, H., and Degn, H. (1972). *Biochim. Biophys. Acta* **256**, 216–222.

Chapter 16

The Use of Fluorescent DNA-Binding Agent for Detecting and Separating Yeast Mitochondrial DNA

D. H. WILLIAMSON AND D. J. FENNELL

National Institute for Medical Research,
Mill Hill, London, England

I. Introduction

The advantages of *Saccharomyces cerevisiae* as an experimental system for studying the molecular genetics of mitochondria are too well known to need reiteration here. They are, however, slightly offset by the practical difficulties of isolating in reasonable yield and quantity the relatively small amount of mitochondrial DNA (mtDNA) normally carried by this species. Even under fully derepressed conditions a haploid cell may contain only 0.005 pg mtDNA, and under certain circumstances the level may be reduced still further (Williamson, 1970; Bleeg *et al.*, 1972). Some types of the respiratory-deficient petite mutant also carry very small amounts of mtDNA, and this brings us to another facet of the problem, which is the need in certain genetic studies for an easy method of distinguishing between petites that contain mtDNA (mtDNA$_+$) and others (mtDNA$_0$) that lack it altogether. In this case the problem becomes one of sensitive detection of mtDNA rather than its isolation, and present techniques leave much to be desired.

There are in fact three main routes to the isolation of pure mtDNA. The first approach involves chromatographic separation of the nuclear and mitochondrial components from extracts of broken cells or protoplasts. Examples are the hydroxyapatite column procedure developed by Bernardi *et al.* (1972) and the polylysine–kieselguhr system described by Finkelstein *et al.* (1972). Both these techniques are potentially capable of delivering large yields, but they necessarily involve considerable mechanical degradation (to which yeast mtDNA seems unusually sensitive), and in any case their scale makes them unsuitable for the routine detection of small amounts.

Similar criticisms may be leveled at the second approach to this problem, which involves the extraction of mtDNA from isolated mitochondria. Isolation of the organelles is a difficult enough task in itself, and they are almost always heavily contaminated with nuclear DNA whose elimination is a serious problem. Yeast mitochondria are frequently damaged during isolation, and even the most careful attempts to remove such contamination by treatment with DNAse are likely to lead to degradation or loss of the mtDNA molecules. Substitution of any other secondary purification robs this approach of its attraction as a potential one-step isolation procedure, and in any case it has little to recommend it as a routine for detecting small amounts of mtDNA.

This leaves the third basic strategy, which relies on the difference in base composition between the nuclear and mitochondrial components. mtDNA has a molar AT content of about 83%, as compared with only 61% for the main nuclear component, and this difference makes the mtDNA about 16 mg/cm^3 less dense than nuclear DNA in isopycnic cesium chloride gradients. The degree of separation thus achieved gives adequate resolution of the two components in analytical gradients, but preparative cesium chloride isolation of mtDNA on even a modest scale usually demands repeated centrifugation in angle rotors. The selective binding of heavy-metal ions such as Hg^{2+} or Ag^+ has been exploited with some success (e.g., Bernardi *et al.*, 1972), but this technique requires highly purified DNA preparations and is somewhat tedious in application. Modification of the gradients by inclusion of the intercalating dye ethidium bromide also fails to resolve the problem, since bulk preparations of the mtDNA of *Saccharomyces* species in practice never comprise closed circular molecules which might otherwise be amenable to this approach.

Despite these drawbacks, the cesium chloride gradient procedure offers one outstanding advantage in dealing with what appears to be a remarkably labile species of DNA, namely, the potential for minimizing both mechanical and enzymatic degradation. To make the best use of the procedure, it would be desirable to improve the resolving power of the gradient to the point where it becomes a one-step procedure. At the same time this would increase the capacity of the system by allowing heavier loading of material onto the gradients. For this reason we have explored the possibility of including in the gradient organic compounds that bind to DNA and show a preference for AT-rich species. The rationale of this approach is that DNA complexed with organic compounds almost always undergoes a decrease in buoyant density in cesium chloride. Thus banding in the presence of a compound with a preference for AT base pairs should enhance the separation between DNAs of different base composition.

We have reported elsewhere on the use of distamycin for this purpose (Williamson and Fennell, 1974). We now wish to report the use of a second compound which has some considerable advantages over distamycin. This compound is 4′, 6-diamidino-2-phenylindole (DAPI), which was first synthesized in Otto Dann's laboratory at Erlangen (Dann *et al.*, 1971) and is one of the many diamidine compounds synthesized by Dann's group in its search for trypanocides related to the trypanocidal drug berenil (Fig. 1). The latter substance is known to bind preferentially to AT-rich DNA (Newton and Le Page, 1967; Newton, 1967). We found that berenil slightly enhanced the separation of yeast nuclear DNA and mtDNA in cesium chloride gradients, but not to a useful extent. DAPI, however, gave a striking separation of the two components and, in addition, the complex it formed with DNA was

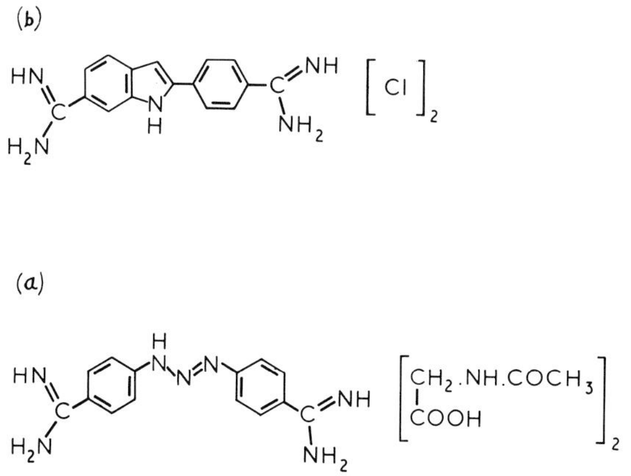

FIG. 1.   Structural formulas of (a) berenil diaceturate and (b) DAPI.

highly fluorescent. The latter property not only adds enormously to the value of the agent in the context of separating and detecting DNA in gradients, but also opens up an entirely different field of application, since it turns out that DAPI can be used as a highly sensitive and specific fluorescent probe for DNA at the level of the individual cell. In fact, the cytological applications of DAPI include several types of cell other than yeasts. In this article, however, attention is limited to the practical use of DAPI for isolating and detecting low levels of mtDNA in yeasts, and in distinguishing between mtDNA$_+$ and mtDNA$_0$ classes of petite mutants. More general cytological features of mtDNA staining of yeasts will be described elsewhere.

II.   The Use of DAPI in Cesium Chloride Gradients

A.   General Procedure

DAPI is simple to use, being added at a final concentration of about 200 μg/ml to a conventional cesium chloride gradient preparation just before centrifugation. Although pure DNA solutions may be used, the dye works just as well in crude preparations, even, for instance, in gradients containing total protoplast lysates. Routinely, however, we examine extracts prepared from 2 to 3 $\times$ 10^9 cells by crushing and Sarkosyl extraction as described elsewhere (Williamson *et al.*, 1971). Exactly 3.9 ml of the DNA-containing extract is gently pipetted onto 3.82 gm solid cesium chloride (Analar), and the latter is allowed to dissolve. This solution is mixed with 0.1 ml of an aqueous solution of DAPI (10 mg/ml), and the mixture is centrifuged at 33,000 rpm for 48–60 hours at 20°C in a Beckman SW 50.1 or SW 50L rotor. At 4°C the DAPI stock solution is stable for several weeks. The separation of the nuclear and mitochondrial components varies with the final concentration of DAPI, as illustrated in Fig. 2; for many purposes 100 μg/ml or less gives an adequate separation and, should detection of mtDNA rather than its isolation be the desired aim, even greater economy can be achieved by growing cells in YEP medium (Wickerham, 1951) containing 1 μg/ml of the dye. Ten milliliters of a stationary-phase culture prepared in this way provides enough material for analysis, and no further dye need be added to the gradients. The resulting separation is less marked than when the higher concentration is used in the gradient, but it is still adequate for detecting the mitochondrial component.

Visualization of the stained DNA bands in the gradients is achieved by illumination with a "dark-light" source such as those used for visualizing ethidium bromide–stained bands. We find a Phillips TL 20 W/08 fluorescent tube, with an emission largely in the long-ultraviolet region, satisfactory.

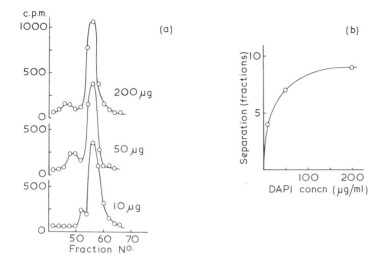

FIG. 2. Effect of DAPI concentration on the separation of yeast nuclear DNA and mtDNA in cesium chloride gradients. Crude extracts of total DNA from adenine-^{3}H-labeled cells were run in 5-ml swing-out cesium chloride gradients containing 10, 50, or 200 µg DAPI/ml. Fractions (6 drops) collected by displacement with 1,2-dibromoethane were screened for radioactive DNA by a conventional alkaline digestion–acid precipitation routine. (a) Radioactive profiles of the gradients; the mtDNA is the minor band. (b) Plot of the number of fractions separating the two bands versus the concentration of DAPI.

The intense bluish white fluorescence of the stained bands, which is much stronger than that obtained with ethidium, is only poorly illustrated by the monochrome photographs in Fig. 3. This easy visualization of the DNA bands means that for many purposes it may not be necessary to fractionate the gradients. However, when fractionation is desirable, the DNA-containing fractions may be located by fluorimetric examination, as illustrated in Fig. 4. The excitation frequency is about 365 nm, and the main emission occurs at about 450 nm. The free dye is about as fluorescent as the dye–DNA complex, and there seems to be little change in the emission spectrum when the complex is formed. This makes the use of the dye as a quantitative measure of DNA rather unreliable, and we have not investigated this aspect of its use in detail. It is in any case obvious from Fig. 4. that the fluorescence of the AT-rich mitochondrial band is disproportionately high compared with that of the nuclear DNA, reflecting the presumed AT affinity of the dye. This would further complicate its use as a quantitative reagent.

B. Identity of the Stained Bands

In our experience only bands containing DNA (i.e., sensitive to DNase) are stained with DAPI in the type of preparation described above. Bands

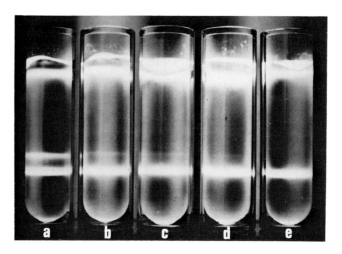

FIG. 3. Fluorescence photographs of DAPI–cesium chloride gradients run in order to assess the sensitivity of the system for detecting mtDNA. The gradients were prepared from crude cell extracts and contained 200 μgDAPI/ml. (a) DNA of a wild-type strain containing about 10% mtDNA. (e) DNA from a mtDNA$_0$ petite mutant. (b), (c), and (d) were prepared from 1:10, 1:20, and 1:50 dilutions, respectively, of the wild-type DNA extract with that of the mutant; the mitochondrial band is just detectable in (d). A non-fluorescent polysaccharide band lying immediately above the nuclear DNA band in (a) reflected light from the latter, giving the spurious appearance of a second component.

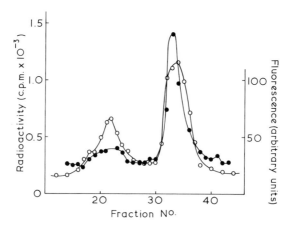

FIG. 4. Radioactive and fluorescent profiles of a DAPI–cesium chloride gradient prepared with a crude total DNA extract of adenine-^{3}H-labeled cells. Solid circles, Radioactivity; open circles, fluorescence measured with a Farrand fluorimeter. The mtDNA is the minor component.

of polysaccharide are often present in the region of the DNA. These appear pale brown in daylight and do not normally fluoresce although, if located very close to a strongly fluorescing band, they may reflect light emitted from the latter (e.g., Fig. 3). One artefact that should be mentioned is that Sarkosyl fluoresces quite strongly with an emission remarkably similar to that of DAPI and, if present in the gradient, forms a strongly fluorescent band at the surface.

It is not known whether or not DAPI can bind to any species of RNA, although this seems unlikely in view of the cytological observations described in Section III,E. RNA normally pellets in cesium chloride gradients, and gradients prepared from crude extracts often show some fluorescence of the pellet, but its significance is not known.

Single-stranded DNA prepared by heat denaturation of native molecules also binds DAPI to form fluorescent complexes, and mixtures of denatured nuclear and mitochondrial molecules can be resolved about as effectively as those of native molecules. It is conceivable that DAPI provides a means of strand separation in cases in which the strands of a duplex have different base compositions.

The separate identities of the two fluorescent components in yeast DNA preparations (as in Figs. 2–4) are readily established. First, as shown in one of the control gradients illustrated in Fig. 3, the presumed mitochondrial band has never been observed in petite mutants known for other reasons (see Section III,E) to lack mtDNA. More importantly, however, after treatment to remove the bound dye (as described in Section II,E) and analytical ultracentrifugation in cesium chloride, both the presumed nuclear and mitochondrial components appear as bands with the buoyant densities and related characteristics of these two types of DNA (Fig. 5). The heavy (γ) nuclear satellite which is about 6 mg/cc denser than the main nuclear component is not usually resolved in swing-out gradients, but may be separated in DAPI gradients in fixed angle rotors. The details of the application for this purpose will be reported elsewhere.

The high affinity of DAPI for AT-rich mtDNA prompted us to examine the mitochondrial band in these preparations for contamination with AT-rich fragments of nuclear molecules. For this purpose ^{3}H-labeled nuclear DNA from a mtDNA$_0$ petite mutant was mixed with an equal amount of unlabeled DNA from a wild-type strain and run to equilibrium in a DAPI–cesium chloride gradient. As shown in Fig. 6, no radioactivity was found in the region of the gradient where the mtDNA banded. It was concluded therefore that, in preparations of this type, in which the nuclear DNA is of moderate size (routinely 30–40S), the risk of this sort of contamination may be discounted.

Finally, it should be mentioned that petite mtDNA, when present, usually

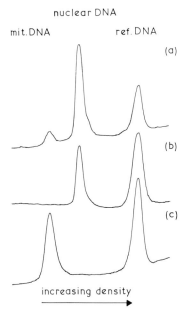

nuclear DNA

mit.DNA   ref.DNA

(a)

(b)

(c)

increasing density

FIG. 5. Densitometer tracing of ultraviolet absorption photographs of analytical isopycnic gradients. (a) Total DNA; (b) nuclear DNA isolated on a preparative DAPI–
cesium chloride gradient and treated to remove DAPI as described in the text; (c) mitochondrial DNA prepared in the same way. The reference DNA was from *Micrococcus lysodeik-
ticus* (density 1.731 gm/cm³.

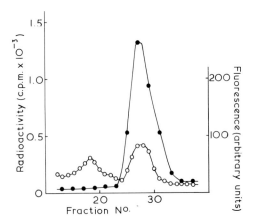

FIG. 6. DAPI–cesium chloride gradient prepared from an equal mixture of crude extracts
of total DNA from an adenine-³H-labeled mtDNA$_0$ peptite mutant and an unlabeled wild-type
strain. Solid circles, radioactivity; open circles, fluorescence.

bands at least as well as the mtDNA of the wild type. In fact, the separation is usually better, since the mtDNA of most petites has an increased AT content.

C.   Sensitivity of Detection of mtDNA

The sensitivity of this procedure for the detection of mtDNA is very high; we have assessed it by a reconstruction experiment in which an extract of total DNA from a wild-type strain containing a ratio of mtDNA to nuclear DNA of about 1:10 was diluted with increasing amounts of nuclear DNA from a mtDNA$_0$ petite mutant. An amount of total DNA equivalent to about 3×10^9 cells was loaded onto each gradient, with the result shown in Fig. 3. Although barely visible in the photograph, the mitochondrial band was clearly discernible in the 1:50 dilution of the wild-type DNA. Since in a normal wild-type diploid cell there may be about 50 mtDNA molecules, this result means that this technique should permit detection of mtDNA at the level of only a single copy per cell. The total amount of mtDNA in the most dilute of the gradients shown in Fig. 3 was about 0.1 μg.

D.   Preparative Isolation of mtDNA

For preparative use DAPI–cesium chloride–gradients may be made exactly as described above. Alternatively, an angle rotor (e.g., the Beckman type 50) may be used to still further enhance the separation. In this case the amount of solid cesium chloride used should be increased to about 4.0–4.5 g in order to position the bands correctly in the gradient, and the cesium chloride solution should be overlaid with mineral oil to prevent tube collapse.

The crushing–Sarkosyl extraction procedure referred to above (Williamson et al., 1971) in conjunction with DAPI–cesium chloride gradients gives mtDNA molecules about 5–10 μ in length. Longer molecules may be obtained by running protoplast lysates. For this purpose we have successfully used a routine in which well-washed protoplasts are pelleted and resuspended in 0.2 M tris buffer (pH 8.0) containing 0.1 M EDTA and 1–2% Sarkosyl (NL 35, Geigy) at 60°C. After extraction for 15 minutes, the solution is rapidly cooled to 4°C and centrifuged at 12,000 rpm for 15 minutes. The supernatant is then mixed with cesium chloride and DAPI as described above, ready for centrifugation.

At equilibrium any conventional means of removing the two bands separately may be used. Side puncture with a wide gauge needle is most convenient, while if long molecules are desired, it may be helpful to use a wide aperture Pasteur pipette with a bent tip.

This procedure does not lend itself to large-scale preparations, since the capacity of a cesium chloride gradient is limited. However, by heavily loading an SW 50.1 rotor, we have prepared batches of about 50 μg of mtDNA in a single operation, and the process could probably be scaled up by the use of bigger rotors.

E.   Removal of Dye from the DAPI–DNA Complex

DAPI is less easily removed from its complex with DNA than is ethidium bromide and, for reasons not fully understood, we find it necessary to use a two-stage process, an initial extraction with amyl alcohol being followed by passage through an ion-exchange resin. A high-salt environment is essential for both steps and, starting with the cesium chloride solution containing the desired DNA, the procedure we currently use is as follows.

The cesium chloride solution is diluted with an equal volume of distilled water. About half this total volume of amyl alcohol saturated with water is layered over the cesium chloride solution and gently swirled for a few minutes. It is then left to stand at 4°C for 15–30 minutes, a step that aids the extraction considerably. The amyl alcohol layer is then removed, and the process repeated four or five times. Removal of the dye is monitored at each step by visual examination for fluorescence; usually these extractions reach a stage at which the aqueous layer still visibly fluoresces, but no more dye is found to enter the amyl alcohol layer. At this stage the cesium chloride solution is removed and dialyzed for a few hours at 4°C against 0.05 M citrate phosphate buffer (pH 7.0, containing 1.0 M NaCl). The solution is then allowed to percolate slowly through a short (about 5 × 0.5 cm) column of Dowex resin (Bio Rad, Ag50W-X4) in the ammonium form. This step should remove all residual fluorescence, and the buoyant density of the DNA in analytical cesium chloride gradients should then return to normal (Fig. 5). Both nuclear DNA and mtDNA prepared in this manner undergo the usual increase in buoyant density in cesium chloride following heat denaturation.

III.   Fluorescent Staining of Cells with DAPI

The above results using DAPI in cesium chloride gradients led us to explore the possibility of employing it as a fluorescent probe for DNA at the cellular level. Both vital and postvital staining procedures may be used.

A.   Vital Staining

Like berenil (Mahler and Perlman, 1973), DAPI at moderate concentrations is selectively active against the yeast mitochondrial system, growth of most strains on glycerol media being inhibited by concentrations of about 5 μg/ml, while on glucose solid media the same dye concentration usually gives rise to 100% petite mutants. Nevertheless, concentrations of about 1.0 μg/ml in glucose YEP medium seem to have little effect on overall growth, and cells cultured in this way show brilliant fluorescent staining. No petite mutants are formed under these conditions.

B.   Postvital Staining

We have used two simple fixation procedures; although neither may be ideal, both serve the present purpose well. In the first a culture sample is mixed directly with 2 vol. of ethanol, or pelleted cells are resuspended in 70% (v/v) ethanol. After about 30 minutes they are washed once in distilled water and resuspended in a solution of 0.1–0.5 μg DAPI/ml. This procedure is recommended when immediate examination is desired. Maximal staining is achieved almost at once, but the stained cells may not withstand prolonged storage. The alternative procedure is to suspend the cells in formaldehyde (0.3%, w/v) in 0.1 M phosphate buffer (pH 7.0) for 30 minutes at room temperature. This probably gives a more meaningful fixation, but often requires a longer period of staining, e.g., overnight at 4°C, to attain the best results. The same range of dye concentration is applicable in this case also; with both fixations it may be desirable to test various levels of DAPI. Fading with this dye is minimal, but can be almost completely eliminated if the cells are washed and examined in distilled water.

C.   Microscopy

Any suitable high-resolution fluorescent microscope can be used. The preparations illustrated in Fig. 7 were photographed with a microscope fitted with dark-ground illumination from a 250-W, high-pressure mercury vapor lamp, a Chance 365-nm OXL primary excitation filter, and eyepiece ultraviolet barrier filters. The other photographs in Figs. 8–10 were taken using a Vickers M41 Photoplan microscope equipped with comparable filters, but employing incident illumination.

D.   Appearance of Stained Cells

Vitally stained cells of the wild type and of a mtDNA$_0$ cytoplasmic petite mutant are shown in Fig. 7. The important feature to note is that, whereas the petite mutant displays only a single conspicuous body identified as the nucleus, the wild-type cells possess in addition numerous smaller cytoplasmic bodies. It should be noted that the appearance of the cell walls in these two preparations, as well as the slight haziness of the cytoplasm, is a consequence of the dark-ground illumination system used for these two preparations and from a cytochemical point of view should be ignored.

This becomes clear on examining Figs. 8 and 9 which show cells photographed by the incident-illumination method. In these cases the outline of the cell only becomes visible as a result of the refraction of light emitted by the fluorescent particles in the cell. The $mtDNA_0$ petite mutant (Fig. 8a) entirely lacks the small cytoplasmic bodies, and the cell outlines are barely visible in the photograph, although the latter were discernible in the original preparation. Figure 8 shows a $mtDNA_+$ petite mutant to illustrate the fact that these mutants, like the wild type (Fig. 8c), also show small fluorescent particles in the cytoplasm, generally in about the same number and size.

The number and disposition of these bodies in the wild-type and the mutant strains vary with physiological conditions and will be the subject of further communications. In the present context it is enough to note that the small cytoplasmic structures are only seen in strains believed to contain mtDNA, and in the following section we assemble evidence that they do in fact comprise mtDNA.

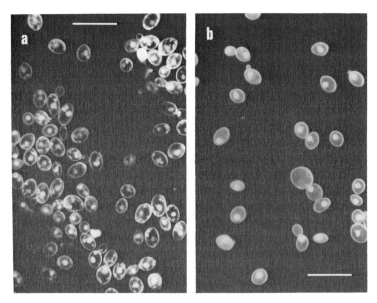

FIG. 7. Dark-ground fluorescence photomicrographs of cells of (a) a wild type and (b) a $mtDNA_0$ petite mutant. Reference bar, 10 μm.

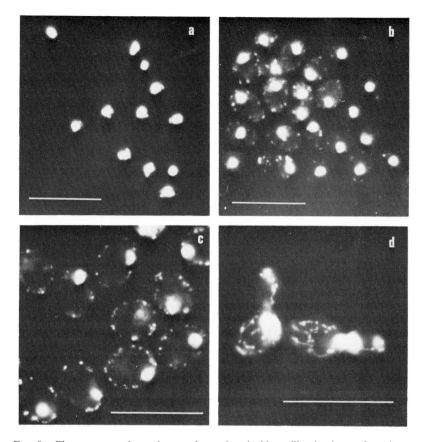

FIG. 8. Fluorescence photomicrographs, using incident illumination, of stationary-phase cells of (a) a $mtDNA_0$ petite mutant, (b) a $mtDNA_+$ petite mutant, (c) a wild type, and (d) growing cells of a wild type showing nuclear migration. In (a) the cell outlines were visible under the microscope, but only the nuclei can be seen in the photograph. Extra-nuclear fluorescent bodies are clearly visible in (b), (c), and (d). In (d) the characteristic "string-of-beads" appearance, sometimes seen in growing cells, is evident. Reference bar, 10 μm.

E.   Interpretation of Staining Results

It goes without saying that the conspicuous single body observed in the stained cells is the nucleus; its morphology and behavior during the growth of the cell are entirely characteristic (e.g., Fig. 8d), and this point needs no further comment. It may be noted, however, that in many vitally stained preparations the nucleus is not as clearly visible as in postvitally stained preparations, presumably because of poor penetration of the dye under living conditions.

Two main lines of evidence support the idea that the extranuclear stained bodies are in fact molecules (or aggregates of molecules) of mtDNA. The first is that treatment of suitably fixed cells with DNase was found to completely abolish the staining of both the cytoplasmic bodies and the nucleus. In order to achieve adequate penetration of the enzyme in this test, the ethanol-fixed cells had to be digested with 1.0 M NaOH at 37°C for 20 minutes. This procedure removed little DNA from the cells, and the characteristic staining pattern of the nucleus and cytoplasmic particles was unaffected. After digestion for 20 minutes at 30°C with DNase (50 μg/ml), however, the staining pattern was completely destroyed, and only weak background fluorescence throughout the cell was discernible (Fig. 9). Incubation of the alkali-treated cells with pronase or RNase was without effect.

mtDNA is the only DNA species clearly shown to exist outside the nucleus and to be absent in $mtDNA_0$ petite mutants. The 2-μm circles (Stevens and Moustacchi, 1971) have yet to be clearly assigned a cellular location, and they are in any case still present in $mtDNA_0$ petites (Clark-Walker, 1973). Coupled with the fact that mtDNA was intensely stained by DAPI in the cesium chloride gradients, there can be little doubt as to the mitochondrial identity of the extranuclear stained bodies.

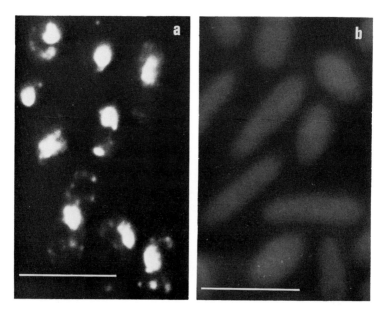

FIG. 9. DNase sensitivity of fluorescent DAPI-stained structures. (a) Cells of a wild-type strain treated with alkali as described in the text. (b) The same after DNase digestion. The photographic exposure for (b) was greatly increased in order to make the cell outlines just visible. Reference bar, 10 μm.

The argument is strongly reinforced by the results obtained on examining a large number of petite mutants, a selection of which is listed in Table I. Both $mtDNA_+$ and $mtDNA_0$ mutants were included, and the staining results correlated with other methods for assessing the presence of mtDNA. One which requires comment is the genetic property of suppressiveness. This is defined as the ability of a petite mutant to give rise to petite diploid progeny in crosses with respiratory-competent wild-type strains. Many petite mutants are, in varying degrees, suppressive, i.e., they give rise to varying proportions of petite progeny. So far none of the laboratories working in this field has reported a suppressive petite lacking mtDNA. However, nearly all nonsuppressive (neutral) petites are apparently entirely devoid of mtDNA, a correlation first pointed out by Nagley and Linnane (1970). This last correlation should in fact be treated with some caution. Quite apart from the difficulties of proving the absence of mtDNA, and of distinguishing weakly suppressive strains from true neutrals, it is also the case that rare neutral mutants containing mtDNA are known (Moustacchi, 1972; we also have isolated one such strain in our laboratory). However, provided these rare exceptions are borne in mind, the suppressiveness or neutrality of a petite mutant provides a useful guide to the presence or absence of mtDNA.

As shown in Table I, the staining procedure correlated perfectly with all the other evidence concerning the mtDNA content of the strains tested, thus providing further support for the nature of the stained cytoplasmic structures and also underlining the reliability of the staining test.

TABLE I

COMPARISON OF METHODS FOR DETECTING mtDNA IN CYTOPLASMIC PETITE MUTANTS

Strain	Suppressive-ness (%)	mtDNA detected by:[a]		
		Analytical cesium chloride	DAPI–cesium chloride	DAPI staining
N123/40	34	−	−	+
N123/39	40	−	−	+
N123 SUPP9	90	(+)	+	+
N123/N2	0	0	0	0
OX2 SUPP3	50	+	−	+
OX2 SUPP3 XEB	0	0	0	0
OX2 45EB2	23	+	−	+
239LUV2[b]	—	+	+	+
239HEB[b]	—	0	0	0
K3	40	+	+	+
K10	0	0	0	0
αDVUV15/1	0	0	0	0

[a] +, Present; (+), trace; 0, absent, −, not examined.
[b] Diploids; suppressiveness does not apply.

F. Sensitivity of the Staining Procedure

The sensitivity of the cytological procedure for the detection of mtDNA with DAPI seems to be very high. The yeast mtDNA molecule is believed to be 21–26 μm in length (ca. 40–50 $\times$ 10^6 daltons) (Hollenberg et al., 1970; Petes et al., 1973), and a diploid strain in which mtDNA comprises about 15% of the total DNA may have about 50 mtDNA molecules. In one such strain in stationary phase, we observed an average of 38 stained mtDNA aggregates per cell. Moreover, we found it possible to visualize DAPI-stained particles of T4 phage in suspension, and these particles carry a genome amounting to about 10^8 daltons, equivalent in size to about two mtDNA molecules. More rigorous tests are in progress, but at present it seems reasonable to suppose that DAPI should permit detection of a single full-sized mtDNA molecule, provided it is packaged in a compact form.

IV. Conclusion

The preparative application of DAPI provides a novel means of isolating mtDNA, potentially in a relatively intact state, albeit on only a moderate scale. A semipreparative application which should be of some use is under circumstances in which it is desired to demonstrate the selective replication of either nuclear DNA or mtDNA molecules. In these cases it is only neces-sary to use a radioactive label and then examine the DAPI–cesium chloride gradient fractions for radioactive DNA (as in Fig. 4).

However, its use in the detection of mtDNA, particularly at the cellular level, is likely to be just as valuable as either of the above applications. The microscopic test is simpler than all other techniques known for detecting mtDNA, and gives the correct answer even in the rare cases in which the correlation between suppressiveness and mtDNA content breaks down. It should also prove effective in cases in which, for some reason, extraction procedures fail or involve excessive nucleolytic degradation. However, the staining procedure itself might fail or be ambiguous in certain cases, for instance with a petite mutant carrying only a few small molecules and, in such circumstances, the DAPI–cesium chloride gradient procedure might be more successful. Thus the two procedures for detecting mtDNA may be regarded as complementary, and between them provide a useful new tool for the molecular geneticist.

ACKNOWLEDGMENTS

The authors are greatly indebted to Prof. Otto Dann of the Institut für Angewandte Chemie der Friedrich-Alexander-Universität, Erlangen, for generous gifts of DAPI. Prof. Dann has expressed his willingness to supply small samples of DAPI for research purposes. We also wish to thank M. R. Young for help with fluorescence microscopy and for Fig. 7, S. M. Fox for photographic advice, and S. G. Oliver for the preparation of some petite mutants. Dr. J. Williamson first introduced us to DAPI, and we are most grateful for his help and advice.

REFERENCES

Bernardi, G., Piperno, G., and Fonty, G. (1972). *J. Mol. Biol.* **65**, 173.

Bleeg, H. S., Bak, L., Christiansen, C., Smith, K. E., and Stenderup, A. (1972). *Biochem. Biophys. Res. Commun.* **47**, 524.

Clark-Walker, G. D. (1973). *Eur. J. Biochem.* **32**, 263.

Dann, O., Bergen, G., Demant, E., and Volz, G. (1971). *Justus Liebigs Ann. Chem.* **749**, 68.

Finkelstein, D. B., Blamire, J., and Marmur, J. (1972). *Biochemistry* **11**, 4853.

Hollenberg, C. P., Borst, P., and van Bruggen, E. F. J. (1970). *Biochim. Biophys. Acta.* **209**, 1.

Mahler, H. R., and Perlman, P. S. (1973). *Mol. Gen. Genet.* **121**, 285.

Moustacchi, E. (1972). *Biochim. Biophys. Acta* **277**, 59.

Nagley, P., and Linnane, A. W. (1970). *Biochem. Biophys. Res. Commun.* **39**, 989.

Newton, B. A. (1967). *Biochem. J.* **105**, 50P.

Newton, B. A., and Le Page, R. W. F. (1967). *Biochem. J.* **105**, 50P.

Petes, T. D., Byers, B., and Fangman, W. L. (1973). *Proc. Nat. Acad. Sci. U.S.* **70**, 3072.

Stevens, B. J., and Moustacchi, E. (1971). *Exp. Cell Res.* **64**, 259.

Wickerham, L. J. (1951). *U.S., Dep. Agr., Tech. Bull.* **1029**.

Williamson, D. H. (1970). *Symp. Soc. Exp. Biol.* **24**, 247.

Williamson, D. H., and Fennell, D. J. (1974). *Biochim. Biophys. Acta* (in press).

Williamson, D. H., Moustacchi, E., and Fennell, D. J. (1971). *Biochim. Biophys. Acta* **238**, 369.

Chapter 17

Cytoplasmic Inheritance and Mitochondrial Genetics in Yeasts

D. WILKIE

Department of Botany and Microbiology,
University College,
London, England

I. Introduction

The phenomenon of cytoplasmic inheritance relates to the transmission of genetic elements outside the control of the nucleus of eukaryotic organisms. The existence of extranuclear genetic units has been difficult to establish in the past, although genetic evidence has been accumulating over the last 60 years. These data describe segregation patterns not readily explainable

in terms of chromosomal transmission; during nuclear division chromosomes duplicate and segregate one to each daughter cell with a precision
which ensures that each cell inherits only one set of genetic factors in a
haploid and in meiotic products, and two sets in a diploid. The alternation
of haploid and diploid phases of the yeast *Saccharomyces cerevisiae* is illustrated in Fig. 1. Thus during vegetative growth mitotic division ensures that
each cell in the culture has precisely the same nuclear genetic complement.
Clearly, if genetic characteristics were seen to segregate among vegetative
progeny, this would provide grounds for suspecting that the controlling

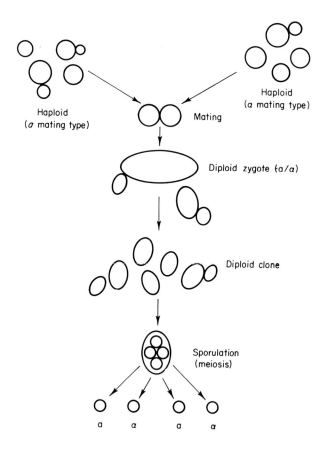

FIG. 1.   Life cycle of *Saccharomyces.*

factors were not carried in chromosomes. In genetic crosses haploid strains of opposite mating type are mixed and fuse pairwise to form diploid zygotes (Fig. 1). Diploid cultures are obtained from zygotes, and individual diploid cells can be induced to divide meiotically to give meiotic tetrads. If the two parental strains differed with respect to a nuclear gene, the derived diploids would all be heterozygous at the particular locus involved and would all be phenotypically similar. This would depend in turn on the dominance relationship of the pair of alleles under consideration. To be accurate, it must be pointed out that occasionally (in about 1 in 10^3 nuclear divisions) a heterozygous diploid cell may segregate two nuclei homozygous for the respective alleles as a consequence of mitotic crossing-over (see review of Hawthorne and Mortimer, 1969).

Meiosis in a heterozygous cell results in four haploid products (the ascus); two inherit one allele and two inherit the other. Again in the interests of accuracy, it may be noted that aberration of this strict 2:2 segregation of alleles occurs and about 1 ascus in 20 may show 3:1 or 4:0 ratios. These have been attributed to complications of the crossing-over event itself and come under the general heading of gene conversion, a subject of considerable controversy among microbial geneticists.

In situations in which meiotic tetrads regularly deviate from a 2:2 ratio, e.g., all showing 4:0 for respective parental phenotypes, this would substantiate a hypothesis of cytoplasmic inheritance of the controlling genetic factors.

II. The Petite Mutation

The earliest record of cytoplasmic control in yeasts and, until recently, the only one, was the case of the respiratory mutant petite. Isolated by Ephrussi and his collaborators, the petite phenotype of small colony on standard agar medium, inability to respire, and absence of membrane-bound cytochromes a and b has been well known for more than 20 years. It has been the subject of numerous investigations and reviews, but it must be admitted that the causal factors in spontaneous petite formation (which occurs with high frequency) remain obscure. The irreversible nature of the petite condition indicates a loss mutation, and this has been substantiated in large measure by the finding that the DNA in the mitochondria of petite strains frequently shows a loss of GC content and in some strains is apparently lost altogether. This indicates a flaw in the replication system of the mitochondrion. It is not surprising that DNA-reacting compounds such as acriflavine and ethidium bromide can induce the petite mutation with 100% efficiency without at the same time showing any detectable effect on the integrity of nuclear DNA replication.

III. Genetics of the Petite Mutation

When different petite mutants are crossed, the resulting diploid cultures are petite. Thus there is no complementation between petites, irrespective of the condition of their respective mitochondrial DNA (mtDNA), nor is there restoration of a wild-type genome following any possible recombination between these molecules. This indicates a common lesion in the mtDNA of all petites, generally referred to as loss of the ρ factor. When petites (ρ^-) and wild type (ρ^+) are crossed, diploid cultures contain both ρ^+ and ρ^- cells, the proportion of ρ^- depending on the petite strain used. In some $\rho^+ \times \rho^-$ crosses the frequency of ρ^- diploids is not significantly different from that found in the haploid parental ρ^+ strains or in diploids from the $\rho^+ \times \rho^+$ cross, in which case the petite involved is said to be neutral. When the frequency of ρ^- diploids is relatively high, the ρ^- parental strain is said to be a suppressive petite. The implication is that the ρ^- condition in some way assumes a dominant or repressive role in a proportion of the hybrid zygotes, this proportion being a characteristic of the petite strain used. The possible mechanism of suppressive action is discussed in Chapter 10 and later in this section.

Mutation of any one of a number of nuclear genes (the p series) results in the petite phenotype. In most of these mutants, the factor replicates normally as typified in the cross

$$pp^+ \qquad \times \qquad P\rho^-$$
(nuclear petite)   (cytoplasmic petite)

which gives rise to a proportion of ρ^+ diploids (depending again on the suppressiveness of the ρ^- strain). When these are sporulated, tetrads segregate 2 pp^+:2 $P\rho^+$. It is concluded that the p genes in this category specify components directly or indirectly involved in the differentiation of the respiratory chain.

Other genes of the p series, e.g., p_3, appear to be involved in the mtDNA replication system and, when mutant, lead to the ρ^- condition. When crossed with a ρ^- petite, these nuclear mutants produce only ρ^- diploids, but give ρ^+ diploids when crossed with a wild type. When sporulated, the latter show two petite: two wild type on tetrad analysis.

These findings demonstrate the complex interaction between nuclear and mitochondrial genetic information.

IV. Genetics of Mitochondrial Resistance to Antibiotics

The demonstration that the antibacterial antibiotic chloramphenicol blocked protein synthesis in isolated mitochondria, and the subsequent finding that the drug acted similarly in intact yeast cells, led to the isolation of antibiotic-resistant mutants in this organism (Thomas and Wilkie, 1968b; Linnane et al., 1968). Since the growth of yeast cells was arrested by the antibiotic only in media containing nonfermentable substrates such as glycerol, and not in the presence of glucose, this indicated specific inhibition of the mitochondria by the drug. In analogy with bacterial inhibition, it was assumed, and subsequently shown to be the case at least with regard to erythromycin, that the antibiotic reacted with mitochondrial ribosomes in blocking protein synthesis (Grivell et al., 1973). It was reasonable to expect, again in analogy with bacteria, that a proportion of resistant mutants would have an altered binding site on the mitochondrial ribosome as a mechanism of resistance, and that the site may be specified by mtDNA. Genetic analysis of several erythromycin-resistant mutants (E^R) establishes this last point in the following way. In the cross $E^R \times E^S$, segregation of E^R and E^S into diploid cells during vegetative growth of zygotes was noted; on plating samples of diploid cells resulting colonies were either E^R or E^S. When E^R diploids were sporulated, individual tetrads showed in general 4 E^R:0 E^S, while the tetrads of E^S diploids were of the type 4 E^S:0 E^R. Thus cytoplasmic inheritance of the E^R factor was indicated. Evidence for the location of the factor in mtDNA was provided by inducing the ρ^- condition in the E^R parental strain and demonstrating failure to transmit E^R to ρ^+ diploid progeny in the cross $E^R\rho^- \times E^S\rho^+$. In other words, destruction of mtDNA in the petite coincided with loss of the resistance factor. In the reciprocal cross $E^R\rho^+ \times E^S\rho^-$, all ρ^+ diploid progeny were E^R. These findings were independently reported by Thomas and Wilkie (1968b) and by Linnane et al. (1968). The latter group further reported that some ρ^- mutants could transmit the E^R factor, a point now generally established. Thus the criteria for assigning a mutation to the mitochondrial genome were clearly defined: (1) segregation of resistance versus sensitivity among vegetative diploid progeny from crosses between resistant and sensitive haploid strains, (2) non-Mendelian tetrad ratios (usually showing no meiotic segregation), and (3) loss of transmissibility of the resistance factor, in at least some of the ρ^- mutants of the resistant strain, in crosses to the wild type (ρ^+, sensitive).

V. Recombination of Mitochondrial Factors

The isolation of several mitochondrial mutants resistant to other anti-bacterial antibiotics with the inheritance characteristics of E^R opened the way for formal genetic analysis of mitochondria, i.e., the study of recombination of these factors in crosses (Thomas and Wilkie, 1968a). The first of these crosses involved three mitochondrial markers, resistance to erythromycin (E^R), spiramycin (S^R), and paromomycin (P^R). Some of the results are shown in Table I. Segregation and recombination of markers was seen in the vegetative diploid progeny of zygotes, and the frequency of each of the eight possible arrangements of the mitochondrial genes was recorded in samples of diploid cells. Plating of these cells gave colonies apparently individually pure with respect to mitochondrial phenotype. In this early work it was apparent that, in crosses involving certain strains, there was a significant deviation from equality of the complementary types, both parental and recombinant. These results indicated that genome selection was operating in these cases, although its mechanism and timing were obscure. Furthermore, from the analysis of individual zygote clones (obtained as colonies on agar minimal medium using different auxotrophic parental strains), it was apparent that in three-point crosses (Thomas and Wilkie, 1968a) the eight possible mitochondrial types were never observed in any one zygote clone. Also, the types that did appear in one zygote clone could be totally absent from another zygote clone. This could be explained if a restricted number of genome types was randomly selected to form a pool in zygotes, from which transmission to zygote daughter cells proceeded. If so, it seems to be a valid exercise to sample diploid cells from a mass zygote culture in order to obtain an overall picture of the relative frequency of mitochondrial types. In calculating recombination it also seems necessary to confine the genetic analysis to those crosses in which little or no asymmetry in the distribution of reciprocal types is seen (but see Section VII). Furthermore, a basic assumption must be made regarding the timing of the recombination events in the zygote. For example, if recombination is delayed until after the first two or three zygote daughter cells have been budded off, and evidence of this will be discussed presently, recom-

TABLE I

DISTRIBUTION OF MITOCHONDRIAL TYPES AMONG DIPLOID CELLS SAMPLED
FROM A MASS MATING OF STRAINS 41 ($O^S C^R E^R$) AND 4a ($O^R C^S E^S$)

$O^S C^R E^R$	$O^R C^S E^S$	$O^S C^S E^S$	$O^R C^R E^R$	$O^S C^R E^S$	$O^R C^S E^R$	$O^S C^S E^R$	$O^R C^R E^S$
24	128	13	18	8	33	4	11

binant classes in zygote clones will be very low in number, leading to a gross underestimation of the frequency of recombination events. In the meantime a considerable number of data has become available from several laboratories in which segregation patterns of mitochondrial types in zygote clones and diploids from mass matings has been explained in terms of transmission control of mitochondria by the zygote, peculiarities of the recombination process, and a combination of both.

VI. Aspects of Mitochondrial Transmission

An obvious next step in the study of mitochondrial inheritance was the analysis of zygote cell lineages, i.e., isolation of daughter cells as they were budded off zygotes in multifactorial crosses, cloning them and determining segregation patterns in the clones (Wilkie and Thomas, 1972). In a three-point cross between strain 4a carrying a mitochondrial resistance factor O^R (resistance to oligomycin) and strain 41 $C^R E^R$ (mitochondrial resistance to chloramphenicol and erythromycin), the first two or three buds usually inherited a mixture of parental and recombinant mitochondrial types, with O^R predominant among the parental types. Later-formed zygote buds usually inherited only one type, namely, the parental O^R.

Each of the cells sampled from a bud clone gave an apparently pure colony with regard to mitochondrial type when plated (but see Rank and Bech-Hansen, 1972), and all recombinant types were seen, but not altogether in any one zygote lineage. For example, the distribution of mitochondrial phenotypes in the clone from the first bud in one of the zygotes was:

Recombinant: $O^S C^S E^R$ (61%)   Parental: $O^R C^S E^S$ (6%)
$O^R C^S E^R$ (31%)
$O^S C^S E^S$ (2%)

The third and subsequent buds from this zygote gave clones homogeneous for $O^R C^S E^S$. In one extreme case a third bud from a different zygote gave a clone comprised entirely of the recombinant $O^S C^S E^S$. The fourth and subsequent buds from this zygote were again homogeneous for the parental type $O^R C^S E^S$. From these results it was difficult to draw any conclusions as to the linkage of the three markers, since there was a relative abundance of all recombinant classes emanating from young zygotes. Electron microscope preparations of the zygotes (Smith et al., 1972; Wilkie, 1972b) clearly showed that, even in the earliest stages following cell fusion, mitochondria had become disaggregated, while in older zygotes, fully differentiated, cristate mitochondria were seen. The hypothesis was put forward that, in young zygotes, mitochondrial genomes were effectively released from the mitochondria and engaged in multiple recombinational events, rather like bacteriophages, the results of which formed a pool of genomes available for transmission to daughter cells. In older zygotes this pool was no longer available but, coinciding with the reaggregation of mitochondria, only the (more or less) ever-present parental type of strain 4a was transmitted.

When a random sample of diploids from mass mating of these two strains was plated and the resulting colonies scored for mitochondrial phenotype, the results shown in Table I were obtained. The significant excess of parental type $O^R C^S E^S$ over that of parental strain 41 ($O^S C^R E^R$) was as expected in the knowledge that the latter type predominates to the extent of eventual exclusion of all other types in most zygotes. Whatever the mechanism of this selection, the findings clearly demonstrate a remarkable control by the zygote in the transmission of mitochondrial type. The parental selection seemed to be a characteristic of strain 4a, since in another cross between strain 41 and strain D22 predominance of a parental type did not occur. With regard to recombinant classes among random diploids (Table I), members of the reciprocal pair $O^S C^R E^S$ and $O^R C^S E^R$ were asymmetrical in distribution. On the basic assumption that all recombinant genomes, unlike parental ones, were randomly inherited, this skew distribution indicated an "aberration" at the level of recombination. This applied to recombination of the E^R factor, since the other two reciprocal pairs did not significantly deviate from equality, although some measure of skewness was apparent in the pair $O^S C^S E^R$ and $O^R C^R E^S$. The relatively high frequencies of all recombinant classes indicated either loose linkage or a high rate of multiple crossing-over.

In crosses with other strains, a different pattern of zygotic control of mitochondrial transmission was observed (Waxman *et al.*, 1973). Crosses were set up between strain D587-4b and E290 with the respective mitochondrial genotypes $E^R C^S$ and $E^S C^R$. These markers were independently isolated and not inherited from strains 4a and 41. Diploid cells were sampled from mass matings and from individual zygote clones and plated, and mitochondrial phenotypes of resulting colonies were determined. In one series of experiments a total of 5309 diploid cells were scored in this way. The frequencies of the four classes were: $E^R C^S$, 0.873; $E^S C^R$, 0.095; $E^R C^R$, 0.027; $E^S C^S$, 0.005. This asymmetric pattern of inheritance of both parental and recombinant types was a consistent feature of this cross. Control of the asymmetry was attributed to strain D587-4b, since in crosses between this strain and other strains (1381-17c and D585-11c) similar asymmetry was observed. This was so irrespective of the mitochondrial marker of strain D587-4b (a C^R marker was also isolated in this strain).

Zygote cell lineage analysis was then carried out in the cross D587-4b × E290. The results obtained were strikingly different from those described in the cross 41 × 40; the first bud of most zygotes inherited entirely the parental type $E^R C^S$, the second bud entirely $E^S C^R$, the third entirely $E^R C^S$, while later buds tended to inherit a mixture of parental and recombinants (Table II). When recombinant classes appeared in a bud clone, their frequency varied but could be as high as 38% of the diploids sampled, with the $E^R C^R$ type in significant excess of the reciprocal $E^S C^S$. This pattern of transmission of mitochondrial types to daughter cells again demonstrated the remarkable control by the zygote in this respect. Also, it was clear that the inheritance pattern was responsible for the asymmetry seen in random diploids from zygote clones and mass matings. Unlike cross 41 × 4a, it appeared that recombination events were either delayed in the zygote or, if they did take place, the products were not generally available for transmission to the first daughter cells.

As in the crosses already discussed, tetrad analysis of meiotic products from the various diploid segregants showed inheritance by all spores of the mitochondrial phenotype of the diploid from which the tetrad was derived. In the cross D587-4b × E290, further analysis of tetrads was undertaken. In appropriate crosses evidence was obtained to indicate that a single gene segregating in a Mendelian fashion, and presumably carried by strain D587-4b, controlled the mitochondrial inheritance pattern observed in the zygote cell lineage analysis.

The control exerted by this nuclear gene was overcome by blocking translation on cytoplasmic ribosomes with cycloheximide in zygotes. However, the duration of treatment with this drug was an important factor, and exposure of zygotes to cycloheximide (2 μg/ml) for 90 minutes, after their inception, was the most efficient procedure. Zygotes removed from the drug-containing medium recovered and produced clones in which the in-

TABLE II

AVERAGE DISTRIBUTION OF MITOCHONDRIAL TYPES IN ZYGOTE
CELL LINEAGES IN THE CROSS D587–4b $(E^R C^S)$ × E290 $(E^S C^R)^a$

	Frequency				Total
Clone	$E^R C^S$	$E^S C^R$	$E^R C^R$	$E^S C^S$	colonies
Bud 1	0.94	0.03	0.02	0.01	3263
Bud 2	0.01	0.93	0.05	0.01	2906
Bud 3	0.99	0.005	0.005	0	3163
Bud 4	0.90	0.01	0.08	0.01	3417
Bud 5	0.92	0	0.07	0.01	3643

^{a}Average of six zygote cell lineages.

heritance pattern of mitochondrial types was consistently different from that of untreated zygotes; the preferential transmission of the $E^R C^S$ parental type was abolished, as well as the normally asymmetric distribution of the two recombinant classes (Table III). This alteration in distribution was not apparent when cycloheximide treatment of zygotes was carried out for 30, 60, 120, or 150 minutes.

To obtain more information about the timing of events within zygotes leading to the control of mitochondrial transmission to buds, zygotes were allowed to mature for various times prior to exposure to cycloheximide. Results showed that a maturation time of 30 minutes followed by drug treatment for 60 minutes was an efficient combination in overcoming the asymmetry normally seen in zygote clones (Table III).

Thiolutin, a drug known to inhibit nuclear RNA polymerase activity in yeasts (Jimenez et al., 1973), was used to treat zygotes, and effects on mitochondrial transmission observed. When newly formed zygotes were transferred to medium containing 20 μg/ml of this drug, budding was prevented as in the case of cycloheximide treatment. Again, as in the case of cycloheximide treatment, a 90-minute exposure to thiolutin was the most efficient in altering the pattern of inheritance of mitochondrial types. Zygotes rescued after this time produced clones in which the distribution of both parental and recombinant classes was more or less symmetric. The conclusion may be drawn that a protein synthesized on cytoplasmic ribosomes has a controlling influence in the asymmetric distribution of mitochondrial types in crosses involving strain D587. The significance of the 90-minute treatment with cycloheximide and thiolutin may be related to the fact that this is the time after zygote formation when the first bud is initiated under standard conditions. Blocking the nuclear factor concerned at the level of either transcription or translation may interfere with the initiation of the first bud and thereby alter the pattern of inheritance. This alteration applies to the distribution of both parental and recombinant classes. Because of the

TABLE III

EFFECT OF CYCLOHEXIMIDE TREATMENT (90 MINUTES) OF INDIVIDUAL ZYGOTES ON THE DISTRIBUTION OF MITOCHONDRIAL TYPES IN ZYGOTE CLONES FROM THE CROSS D587-4b ($E^R C^S$) × E290 ($E^S C^R$)

	Frequency				Total colonies
	$E^R C^S$	$E^S C^R$	$E^R C^R$	$E^S C^S$	
Untreated[a]	0.891	0.074	0.034	0.001	4354
Treated[b]	0.385	0.499	0.050	0.066	2675

[a]Average of eight zygote clones.
[b]Average of four zygote clones.

total lack of information on the behavior of mitochondrial genomes in zygotes, it is difficult to formulate a hypothesis as to the nature of transmission control.

Protein synthesis in mitochondria prior to, during, and after zygote formation is apparently not a factor in mitochondrial inheritance in this case. Treatment of mating cells and zygotes with paromomycin, an antibiotic that specifically inhibits protein synthesis in yeast mitochondria (Wilkie, 1970), had no detectable effect on the asymmetric distribution of mitochondrial types in the clones of zygotes treated for 90 minutes. During this treatment budding is not arrested on standard agar medium (yeast extract, peptone, dextrose). These results indicated that mitochondrial protein synthesis was not a significant factor in the transmission pattern of mitochondria in this cross, at least during this period of treatment.

The selective transmission of the parental type of strain D587-4b in zygote clones was further investigated with respect to the suppressiveness of suppressive petites. Several highly suppressive petites of strain E290 were isolated ranging in suppressiveness between 66 and 99% when crossed to various strains other than D587-4b. These results were obtained from platings of random diploids from these crosses. When the suppressive petites were crossed with D587-4b, suppressive action was completely overcome and the highest frequency of petites found in zygote clones and random diploids was 2.6%. These findings seem to favor the hypothesis that suppressive action depends on recombination between ρ^+ and ρ^- genomes (see Chapter 10), if the assumption is made that recombination is generally delayed in zygotes from the cross D587-4b $\times$ E290 but not in the crosses between E290 and other strains. Thus selective transmission of the intact mitochondrial genome of D587-4b would still be possible. These experiments involving strain D587-4b are described in Waxman *et al.* (1973) and Waxman (1974).

In this section on mitochondrial transmission, it is clear that the zygote can exert considerable control over the pattern of inheritance, probably through the action of nuclear factors. In the analysis of segregation data obtained from zygote clones or random diploids, this fact must be taken into consideration. It is recommended that zygote cell lineage analysis be carried out as an adjunct to mitochondrial genetic analysis.

VII.   Aspects of Mitochondrial Recombination

Recombination data have already been discussed relating to the three-point cross 4a $(O^R C^S E^S) \times 41 (O^S C^R E^R)$. It was concluded, mainly from the analysis of random diploids, that recombination of the E alleles was asymetric, and that the recombinant containing E^R ($O^R C^S E^R$) was significantly in excess of the reciprocal class $O^S C^R E^S$. This suggests that the mechanism of recombination in the E region was largely nonreciprocal. Recombination in heteroallelic nuclear systems in yeasts is generally of this type, and a mechanism based on hybrid DNA formation has been proposed to explain this phenomenon (see Whitehouse, 1969). Before discussing this possibility further, some of the results obtained by the Slominski group should be considered.

These workers reported the finding of certain strains which when crossed to other strains resulted in marked asymmetry of mitochondrial types, both parental and recombinant, among random diploids. The controlling factor in these cases of asymmetry was cytoplasmically inherited, located in DNA because of association with ρ factor, and named the ω factor. In a typical cross between strain A of mitochondrial genotype $C^S E^R$ and strain B, $C^R E^S$, random diploids segregated 8% $C^R E^S$ and 54% $C^S E^R$ (parental types), and 38% $C^S E^S$ and 0.13% $C^R E^R$ (recombinant types) (data from Bolotin et al., 1971). Strain A in this case was arbitrarily labeled ω^+, and strain B was labeled ω^-. A series of strains was then designated ω^+ or ω^- in which crosses of the type $\omega^+ \times \omega^-$ showed asymmetry, whereas the crosses $\omega^+ \times \omega^+$ and $\omega^- \times \omega^-$ produced equality of parental types (usually about 45% of each) and recombinant types (about 5% of each). These investigators saw an analogy between the ω factor and the F factor of bacteria and put forward the concept of mitochondrial sex. Crosses of the type $\omega^+ \times \omega^-$ were termed "heterosexual," and other crosses "homosexual." This idea that the ω factor controls some form of sex-mediated transfer of genetic information from one mitochondrion to another was never convincing, and the Slonimski group favors the view that ω influences the mechanism of recombination, perhaps promoting the nonreciprocal type in its vicinity in a ω^+-ω^- situation. In the heterosexual cross previously cited (Bolotin et al., 1971), the ratio $C^R E^R / C^S E^S$ ($=290$), is high and gives a measure of the asymmetry of the recombinant classes or the degree of polarity for the pair of markers. Wolf et al. (1973) provide evidence that the polarity of recombination between two particular markers appears to depend on the distance from the ω factor, rather than reflecting the distance between the markers. Thus in a heterosexual, four-point cross involving mitochondrial markers O^R and P^R (resistance to oligomycin and paromomycin, respectively), as well as C^R and E^R, the polarity values for the pairs CE, CO, and CP are similar. The EO, EP pairs also show similar magnitudes of polarity, but this is less than that of the three pairs involving C. Since C is influenced to a greater extent than E by the ω factor with respect to polarity, C can be considered the nearer of the two markers to ω. It follows that markers that show no polarity in the ω^+-ω^- situation cannot be mapped in this system, except to say they are relatively far away from the ω region. Since the recombinants for the OP pair show no polarity, these markers are in the latter category. It is worth noting that the ω factor effect is not apparent in all crosses, since in a series of crosses between strain D587-4b and known ω^+ and ω^- strains (kindly provided by Prof. Slonimski), the characteristic pattern of mitochondrial inheritance seen in crosses involving D587-4b was repeated irrespective of the ω factor used (Waxman, 1974).

In homosexual crosses using these markers, the frequencies of recombinants in the total population for the pairs of markers CO, CP, EO, EP, and OP are similar and are about 20%. In the case of CE the frequency of the recombinants is lower at about 8% of the population, indicating a preference for C and E to segregate together compared with the other combinations of markers. When markers are considered in combinations of three, the frequency of all possible triplet arrangements is high. To explain these findings, Wolf et al. favor the already proposed hypothesis that the situation in zygotes resembles a bacteriophage cross in which there is unrestricted pairing and recombination between mitochondrial genomes. This could lead to a theoretical maximum of 25% recombinants after one round of exchanges, but this would be exceeded if there were subsequent rounds of exchanges. Unlike in the case of phage crosses, the experimenter cannot control the precise number of mitochondrial genomes in order to simplify the problem. However, attempts along these lines are being made by varying the carbon source in the culture medium of parental strains prior to mating; glucose-repressed cells apparently contribute fewer genomes to the cross than nonrepressed cells (Marmur personal communication).

VIII. The Petite Mutation and Mapping of Mitochondrial Genes

As discussed in detail elsewhere in this volume (Chapter 10), the mtDNA of petites shows progressive loss of genetic complexity on protracted treatment with acridines. Without knowing anything about the mechanism of deletion, it is a reasonable hypothesis that mitochondrial markers that tend to be deleted together are linked, as are those that are retained together. Having previously established that the markers RI, RII, and $RIII$ (chloramphenicol, spiramycin, and erythromycin) are linked in that order on the basis of proximity to the ω factor, Avner et al. (1973) used a petite strain to deduce the positions of two O^R markers (OI and OII). This petite strain had lost loci RI and RII but had retained $RIII$ and OI. Since OII was also deleted, it was concluded that this marker was not located between $RIII$ and OI on the basic assumption that the segment of DNA connecting $RIII$ and OI was intact and unchanged. Thus the order of the markers was deduced as *RI-RII-RIII-OI-OII*. There are obvious limitations to this approach to the location of mitochondrial markers, but the method has some merit and would repay more extensive investigation. It would be of interest, for example, to find out how repeatable this pattern of deletion and retention described by Avner *et al.* is in which some ω-linked markers are lost and others are retained.

It is clear from the work discussed so far that mitochondrial genetics is complex, with nuclear factors controlling transmission and mitochondrial factors influencing the recombination process in certain regions. A further complication may arise from the fact that mtDNA of yeast has a GC content of only 17%, which is concentrated in GC-rich regions (coding segments?) separated by extensive segments of poly dAT (Bernardi *et al.*, 1972). The kinetics of pairing and exchange within the AT regions where base sequences (at least short sequences) must be frequently repeated may not be the same as in normal DNA with four bases.

IX. Mitochondrial Markers

Mitochondrial genetics is still in an early stage of development, and much remains to be done to elucidate the system. A continuous search for new mitochondrial markers goes on. As well as the more obvious mutants showing resistance to drugs with primary antimitochondrial activity (see Wilkie, 1972a), a temperature-sensitive mutant has been described by Handwerker *et al.* (1973). A shift from 30°C to the nonpermissive 35°C resulted in the petite condition, more than 90% of cells becoming ρ^- within about five generations. Genetic analysis indicated that the mutation was mitochondrially located. It was tentatively concluded by these investigators that the controlling factor coded for a mitochondrial protein necessary for normal replication of mtDNA, a conclusion already arrived at by Williamson *et al.* (1971) in their studies on the effects of inhibition of mitochondrial protein synthesis by antibiotics.

A further source of mitochondrial markers is indicated from the work of Linstead *et al.* (1974) with the membrane-reacting drug chlorimipramine. The cellular toxicity, which correlates with the binding capacity of the drug to cell membrane, is much reduced in sensitive cells following induction of the ρ^- mutation, although this resistance is not manifest in all ρ^- mutants. The resistance segregated among diploid progeny in crosses with ρ^+ sensitive strains, indicating cytoplasmic control. It was concluded that mitochondrial transcription products were involved in specifying characteristics of the cell membrane. Supporting evidence for this conclusion has come from experiments in our laboratory in which acriflavine was used to block mitochondrial transcription. Cells budded off in the presence of acriflavine (2 μg/ml) were able to divide and form small clones of cells when transferred by micromanipulation to chlorimipramine-containing medium which totally inhibited budding of untreated cells.

X. Concluding Remarks on Mitochondrial Genetics

The general techniques being used to isolate and analyze mitochondrial mutants have been illustrated by some examples taken from the literature. These also illustrate the unique problems of mitochondrial inheritance both from the point of view of transmission and recombination. The term "mitochondrial sex" seems to be something of a misnomer based as it is on differences in the ω factor. These apparently affect the recombination mechanism in their vicinity, presumably in a gene-bearing, GC-rich segment. When more markers are found in other segments, perhaps they too will have their ω factors. It is useful to construct a model of mitochondrial recombination based on a preexisting system. In this case comparison with phage recombination is probably the best starting point, although it is likely that discrete differences between the two systems will be found.

It is more difficult to construct a model of mitochondrial transmission from zygote to daughter cell. The old idea of passive inheritance of these structures as cytoplasmic inclusions is probably wrong, and strict control by the zygote involving cytoplasmic protein synthesis is the more likely situation. In the case of the neutral petite lacking mtDNA crossed with ρ^+, for example, it is pertinent to ask why these petite mitochondria are not transmitted by zygotes to daughter cells, since they are transmitted during vegetative growth of the petite strain itself.

As well as the articles cited here, the reader may wish to consult the data listed in other publications in the field of mitochondrial genetics. These are not in any way insignificant, and each has something to add to the overall picture without contributing anything fundamentally new by way of analysis of segregation data. Among these are the following: Kleese et al. (1972), Lukins et al. (1973), Stuart (1970), and Wakabayashi and Gunge (1970).

XI.   Tetracycline Resistance

The antibacterial antibiotic tetracycline (TC) inhibits the mitochondrial ribosome of yeasts (Grivell, 1974) and shows similar inhibitory effects on yeast cells comparable to those of chloramphenicol, erythromycin, and paromomycin. As a result, the growth of most yeast strains is arrested at concentrations of about 0.1 mgTC/ml in nonfermentable medium.

Several spontaneous resistant mutants were isolated in the usual way and analyzed genetically by Hughes and Wilkie (1972). All but one showed nuclear control of resistance. The odd one was a high-level resistance mutant with a TC tolerance of 3 mg/ml. When crossed to a sensitive strain, cytoplasmic control was indicated by the segregation of resistance and sensitivity among random diploid cells. However, the resistance did not show the expected association with the ρ factor and was transmitted by all acriflavine-induced petites tested in appropriate crosses. When sensitive diploids were sporulated, all ascospores gave sensitive haploid cultures, i.e., tetrads were of the type 4 TC^S: 0 TC^R. The analysis proceeded to the next generation of diploids by crossing several of the F_1 sensitive haploids *inter se*. Only sensitive diploids were produced, leading to the conclusion that the first generation of diploids had failed to inherit the resistance factor, in which case a cytoplasmic location of the factor was indicated.

Resistant segregants of the first generation of diploids were then sporulated but, instead of the expected non-Mendelian pattern of inheritance, a strict 2:2 segregation of resistance to sensitivity was observed in the tetrads. Several of the F_1 haploids that had inherited the resistance factor were crossed; i.e., crosses of the type resistant × resistant were set up. A diploid clone was selected at random from each of three such crosses, and all were found to be resistant. Sporulation of each of the diploids gave tetrads that segregated once again 2:2, resistant to sensitive. There seems to be no precedent for this type of result, and a model was proposed in which resistance was vested in an episomelike cytoplasmic factor. In zygotes it was postulated that the factor had a cytoplasmic location and could be transmitted, or fail to be transmitted, to daughter cells. In diploids that inherited the factor, a chromosomal location was visualized during vegetative growth. At meiosis the factor would replicate in conjunction with the chromosome in which it was integrated, with subsequent transmission to two of the members of each tetrad. This would account for the 2:2 segregation in tetrads irrespective of the origin of the resistant diploid cell, whether it came from a resistant × sensitive or a resistant × resistant cross.

XII.   Omicron DNA (*o*DNA)

The presence of cytoplasmic DNA (*o*DNA) in yeast cells, as reported by Clark-Walker and Miklos (1974), may have a bearing on these findings with TC resistance. *o*DNA is a covalently closed circular molecule of buoyant density 1.701 gm/cm^3 with a major class size of 1.9 μm. It is located in the cytoplasm independently of mtDNA and, although similar to nuclear DNA in buoyant density, does not represent a nuclear gene amplification product. If *o*DNA is a normal constituent of all yeast strains, then it may be tacitly assumed that it has a genetic function.

XIII.   The Psi Factor

A cytoplasmically inherited factor (ψ) which modifies ochre-specific supersuppressors has been described by Cox (1971). The system is complex, and the maintenance of ψ requires the presence of a particular allele R of a nuclear gene.

These cases of cytoplasmic control, implying the presence of extrachromosomal genetic units, are strengthened by the finding of DNA molecules such as *o*DNA.

XIV.   The Killer Character in Yeasts

Certain strains of yeasts have been described by Bevan and Sömers (1969), which release a killer protein into the culture medium which brings about the death of sensitive cells. Strains that are neither killers or sensitives are termed neutral. Genetic analysis of the three phenotypes led these investigators to conclude that the killer character is under the control of two types of cytoplasmic genetic determinants, *k* and *n*. The presence of *k* determines the killer phenotype, while the neutral phenotype is conferred by *n*. A dominant nuclear allele *M* is required for the maintenance of the two factors. The genotype of killer cells (haploid) is thus *Mk*, and that of neutral cells *Mn*. The absence (0) of both types of determinant confers the sensitive phenotype, regardless of the nuclear allele *M* or *m*. The cytoplasmic factors could be transferred only by crossing, and diploids from killer and neutral

crosses had the killer phenotype. During vegetative growth killer cells segregated out neutral cells in varying proportions. When these cultures were
sporulated, a complete range of tetrad ratios from four killer: zero neutral
to zero killer: four neutral was obtained. Since these cytoplasmic units are
clearly not normal constituents of yeast cells, but assume rather the nature
of invasive elements (it has been claimed that they contain RNA), strictly
speaking they do not come under the general scheme of cytoplasmic inheritance. However, they may have a possible role as transducing agents both
for nuclear and mitochondrial factors and would be worth investigating in
this respect.

XV.   Cytoplasmic Mutagens

Acriflavine and ethidium bromide are highly specific mutagens in the
induction of the petite condition. Attempts have been made in several
laboratories using these mutagens to induce point mutations in mtDNA such
as E^R. Typical procedure is to expose cells for varying lengths of time to the
mutagen, rescue or otherwise select ρ^+ cells, and check for E^R. Resistance
to high concentrations of erythromycin (> 1 mg/ml) is usually attributable to
cytoplasmic control, whereas low-level resistance is usually nuclear (Thomas
and Wilkie, 1968b). Experiments of this type have been unsuccessful. A
variation in the method was tried by N. Eaton and J. Cohen (personal
communication) in which a few drops of a weak alcoholic solution of acriflavine (2–5 μg/ml) were introduced into a well cut out of the center of an
agar plate with a cork borer. Cells were transferred to this plate (which
contained erythromycin) by velvet-pad replication of a pregrown culture.
The advantage of this technique is that, on comparison with control replicates of cells untreated with acriflavine, any increase in the number of resistant colonies that grow is readily scored, but also those induced can be
located. Results were variable but encouraging, since in most cases more
resistant colonies grew on drug plates containing acriflavine than on control
plates. At these low concentrations of acriflavine (extremely low at the
periphery of the plate), the dye can be assumed to have no mutagenic effect
on nuclear factors.

More encouraging are the results of Putrament et al. (1973), which demonstrate the selective action of manganese as a mutagen of mitochondria.
Cells pregrown in the presence of manganese chloride (about 8 mM), and
plated on medium containing chloramphenicol or erythromycin, have a
higher proportion of viable cells resistant to chloramphenicol or erythromycin than untreated cells. Manganese also induces petites, but these are

eliminated on the antibiotic-containing medium which is nonfermentable. The nuclear mutations induced are relatively infrequent. If this method applies to yeast strains in general, it should facilitate the isolation of a wide range of mitochondrial mutants and other types of possible cytoplasmic mutants.

The nonspecific mutagenic agents N-methyl-N'-nitro-N-nitrosoguanidine (see Hardwerker *et al.*, 1973) and ultraviolet light (see Stuart, 1970) have also been used to induce mitochondrial mutations, but it is not clear that these agents have any selective action on mitochondrial factors other than the ρ factor.

XVI. Summary

The yeast cell is rare among eukaryotes in being a facultative anaerobe. This feature has permitted the isolation of mitochondrial mutants resistant to membrane-reacting drugs such as chlorimipramine and oligomycin, and to inhibitors of the mitochondrial ribosome. Mitochondrial genetics may be said to have begun when crosses were set up between strains carrying different mitochondrial markers and the progeny analyzed. A selection of the results obtained in this new field have been presented to give an account of the salient features of the system and to outline the problems. The problems arise mainly from our total lack of knowledge of the behavior of mitochondrial genomes in zygotes and of the mechanism of their transmission to daughter cells. Segregation data alone are clearly insufficient to explain these phenomena and, in constructing linkage maps, basic assumptions must be made. The analysis of the system has reached a most interesting stage in its development, and new information may be forthcoming soon. The recent finding of a TMP-permeable strain should allow autoradiography of zygotes and dividing cells, for example, and provide information on both mitochondrial and nonmitochondrial DNA molecules in the cytoplasm of the yeast cell.

REFERENCES

Avner, P. R., Coen, D., Dujon, B., and Slonimski, P. P. (1973). *Mol. Gen. Genet.* **125**, 9–52.
Bernardi, G., Piperno, G., and Fonty, G. (1972). *J. Mol. Biol.* **65**, 173–189.
Bevan, E. A., and Sömers, J. M. (1969). *Genet. Res.* **14**, 71–77.
Bolotin, M. D., Coen, D., Deutsch, J., Dujon, B., Netter, P., Petrochilo, E., and Slonimski, P. P. (1971). *Bull. Inst. Pasteur, Paris* **69**, 215–239.
Clark-Walker, G. D., and Miklos, G. L. G. (1974). *Eur. J. Biochem.* **41**, 359–365.
Cox, B. S. (1971). *Heredity* **26**, 211–232.
Grivell, L. A. (1974). In preparation.

Grivell, L. A., Netter, P., Borst, P., and Slonimski, P. (1973). *Biochim. Biophys. Acta* **312**, 358–367.

Handwerker, A., Schweyen, R. J., Wolf, K., and Kaudewitz, F. (1973). *J. Bacteriol.* **113**, 1307–1310.

Hawthorne, D. C., and Mortimer, R. K. (1969). *In* "The Yeasts" (A. H. Rose and J. R. Harrison, eds.), Vol. 1, pp. 386–453. Academic Press, New York.

Hughes, A. R., and Wilkie, D. (1972). *Heredity* **28**, 117–127.

Jiminez, A., Tipper, D. D., and Davies, J. (1974). *Antimicrob. Ag. Chemother.* (in press).

Kleese, R. A., Grotbeck, R. C., and Snyder, J. R. (1972). *J. Bacteriol.* **112**, 1023–1025.

Linnane, A. W., Saunders, G. W., Gingold, E. B., and Lukins, H. B. (1968). *Proc. Nat. Acad. Sci. U.S.* **59**, 903–910.

Linstead, D., Evans, I. H., and Wilkie, D. (1974). *In* "Biogenesis of Mitochondria" (A. M. Kroon and C. Saccone, eds.), pp. 179–193. Academic Press, New York.

Lukins, H. B., Tate, J. R., Saunders, G. W., and Linnane, A. W. (1973). *Mol. Gen. Genet.* **120**, 17–25.

Putrament, A., Baranowska, H., and Prazmo, W. (1973). *Mol. Gen. Genet.* **126**, 367–372.

Rank, G. H., and Bech-Hansen, N. T. (1972). *Genetics* **72**, 1–15.

Smith, D. G., Srivastava, K. C., and Wilkie, D. (1972). *Microbios* **6**, 231–238.

Stuart, K. D. (1970). *Biochem. Biophys. Res. Commun.* **39**, 1045–1051.

Thomas, D. Y., and Wilkie, D. (1968a). *Biochem. Biophys. Res. Commun.* **30**, 368–373.

Thomas, D. Y., and Wilkie, D. (1968b). *Genet. Res.* **11**, 33–41.

Wakabayashi, K., and Gunge, N. (1970). *FEBS (Fed. Eur. Biochem. Soc.) Lett.* **6**, 302–304.

Waxman, M. F. (1974). Ph.D. Thesis, City University of New York.

Waxman, M. F., Eaton, N., and Wilkie, D. (1973). *Mol. Gen. Genet.* **127**, 277–284.

Whitehouse, H. L. K. (1969). "Towards an Understanding of the Mechanism of Heredity." Arnold, London.

Wilkie, D. (1970). *In* "Control of Organelle Development" (P. L. Miller ed.), pp. 71–83. Academic Press, New York.

Wilkie, D. (1972a). *Med. Biol. Illus.* **22**, 119–124.

Wilkie, D. (1972b). *In* "Mitochondria/Biomembranes" (G. S. Van der Bergh, P. Borst, and E. C. Slater, eds.), pp. 85–94. North-Holland Publ., Amsterdam.

Wilkie, D., and Thomas, D. Y. (1972). *Genetics.*

Williamson, D. H., Maroudas, N., and Wilkie, D. (1971). *Mol. Gen. Genet.* **111**, 209–223.

Wolf, K., Dujon, B., and Slonimski, P. P. (1973). *Mol. Gen. Genet.* **125**, 53–90.

Chapter 18

Synchronization of the Fission Yeast Schizosaccharomyces pombe Using Heat Shocks

BIRTE KRAMHØFT AND ERIK ZEUTHEN

The Biological Institute of the Carlsberg Foundation,
Copenhagen, Denmark

I. Introduction

Synchronous yeast cultures have so far been prepared in several ways, e.g., by cyclic feeding and starvation (Williamson and Scopes, 1960; von Meyenburg, 1969), by x-ray irradiation (Spoerl and Looney, 1959), and by a gradient separation method (Mitchison and Vincent, 1965). In addition, unsuccessful attempts to synchronize yeast using heat shocks as described for the protozoon *Tetrahymena pyriformis* by Scherbaum and Zeuthen (1954) have been made. This method involved the application of a series of heat shocks at short intervals. Recently, one of us developed an alternative method to synchronize divisions in *Tetrahymena* (Zeuthen, 1971). This method also comprises a series of heat shocks, but with an interval between successive shocks equal to the time required for one generation at the same temperature as the one used between the shocks. On the basis of these results, we have developed a method to synchronize cultures of the fission yeast *Schizosaccharomyces pombe*, using heat shocks spaced one generation time apart (Kramhøft and Zeuthen, 1971). Details of this method are given in the following.

II.   Media and Cultures

The following complex medium was used in all experiments: glucose, 30 gm per liter; yeast extract (Difco), 5 gm per liter.

The ingredients are dissolved in distilled water, and the mixture is then autoclaved for 20 minutes. Stock cultures are kept on an agar medium of similar composition, except that in addition it contains 15 gm of Difco agar per liter. Results similar to the ones to be described can be obtained with cells grown in chemically defined medium (Polanshek, 1974).

Cultures of the haploid strain 972 h$^-$ were obtained from Jørgen Friis, Odense University. Stock cultures on slanted agar medium are stored at 4°C. Experimental cultures are produced in the following way. Twenty-four hours before starting a synchronous culture, a loop of cells from a slanted agar culture is inoculated into 25–50 ml of liquid medium and allowed to grow overnight at room temperature. The next day this culture, now in exponential growth phase, is used as inoculum for the experimental culture. This procedure induces no significant lag, and is therefore a convenient method for obtaining the desired cell density of the synchronized culture on the following day. Depending on the type of experiment to be conducted, cell densities from 2 to 6 $\times$ 10^6 cells/ml were intended at the end of the cyclic heat program.

The experimental cultures contain from 100 to 500 ml of culture in cotton-stoppered or screw-capped Erlenmeyer flasks. The culture is stirred well with a magnetic Teflon-coated stirring bar equipped with an equatorial ridge. The presence of this ridge is important, because grinding of the cells between the stirring bar and the glass must be avoided.

No measures are taken to ensure aeration, because the glucose concentration of this medium (3%) is so high that the cells' respiration is partially repressed (Heslot *et al.*, 1970). For the same reason the culture volume/flask volume ratio is not important.

Immediately after inoculation the culture is immersed in the cyclic thermostat, described in the following section, and is then subjected to the synchronizing treatment, usually overnight.

III.   The Cyclic Thermostat

The synchronizing apparatus consists in principle of an arrangement which controls and records the temperature of a cyclic water thermostat in which the culture flask is immersed. A schematic drawing of the equipment is shown in Fig. 1. The temperature of the thermostat is controlled electronically at two alternating predetermined levels. An electronic clock switches the control from one level to the other at preset time points. For further details see the legend for Fig. 1.

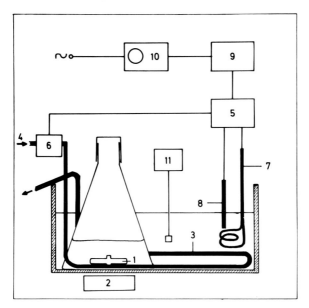

FIG. 1.   The cyclic thermostat. The culture vessel containing a ridged, Teflon-coated magnetic stirring bar (1) is immersed in a 5-liter water bath mounted on a magnetic stirrer (2). The flask is kept in position by means of lead rings (not shown). A motor-driven propeller (not shown) stirs the water. The bath is equipped with a cooling water coil (3) connected to a reservoir of cold water (4)—in our case tap water. The cold-water flow is intermittent and is controlled from the temperature-controlling unit (5) by means of a magnetic valve (6). If a cold-water bath is used as reservoir of cooling water, the magnetic valve is replaced by a pump. The temperature-controlling unit (5) energizes a 1000-W heater (7) regulated from a thermistor (8). The thermistor senses the temperature of the water bath and can be set (by 5) to regulate the temperature at two different levels—in our case 32° and 41°C. By means of an electronic clock (9), the temperature control can be shifted from one level to the other at preset time points. A signal watch (Electroboy, Firma Hugo Müller, Schwenningen am Neckar, West Germany) (10) starts the temperature cycles at a preset time point. Prior to this time the thermostat is regulated at the lower of the two temperature levels. The upward and the downward temperature shifts require 4 and 3 minutes, respectively. A temperature recorder is shown at 11.

IV.   The Synchronizing Procedure

A.   Choice of Parameters

We have adapted the heat shock procedure for *Tetrahymena* for use with *Schizosaccharomyces* (Zeuthen, 1971; Kramhøft and Zeuthen, 1971). In order to do this, five parameters are dealt with: (1) the time between successive heat shocks, (2) the temperature maintained between the shocks, (3) the shock temperature, (4) the duration of each temperature shock, and (5) the number of shocks necessary to induce good synchrony.

The principles of the *Tetrahymena* synchronizing heat treatment are based in short on the following. Parameter 1 shall equal the doubling time of the culture at the optimum temperature for cell multiplication. Parameter 2 is chosen as the temperature at which the multiplication rate is maximal. And parameter 3 shall equal the lowest temperature above which cell division is stopped. The said three parameters were fixed by the use of experiments in which exponential growth rates were measured as a function of temperature. They were fixed at 110 ± 13 minutes ($n = 13$), 32°C, and 41°C, respectively. The fourth parameter, the duration of each shock, was fixed at 30 minutes, again drawing on our experience with *Tetrahymena*. This program was tested in a few trial experiments, and five to seven heat shocks were decided on. The stated program of temperature treatment resulted in good synchrony, while less than five shocks resulted in poor synchrony or none at all.

B.   Results

Figure 2 shows cell multiplication and various biochemical parameters as a function of time in heat-synchronized cultures of *Schizosaccharomyces*. Each part represents the result of a typical experiment. Since no difference was observed between cultures treated either with six or with seven heat shocks, average values presented below are pooled from both groups of experiments.

Figure 2A shows cell multiplication. Following the sixth heat shock, the number of cells in the culture stays constant for about 70 minutes (71 ± 5 minutes, $n = 11$). Then a burst of division occurs during the next 30–40 minutes, during which the cell number doubles. During the next shock and for a further 70 minutes, cell division is again blocked, and then there is a new doubling. Mid cell division occurs 95 minutes (93 ± 8 minutes, $n = 11$) after the end of the shock. If the system is allowed to run freely, i.e., if the next shock is not applied, a second burst of division occurs about 90 minutes later.

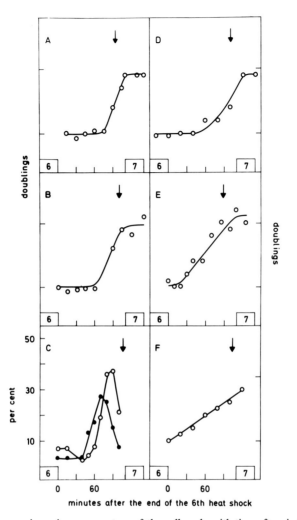

Fig. 2. Changes in various parameters of the cell cycle with time after six heat shocks. In each case the arrow indicates the time point of mid cell division in the respective experiment. (A) Cell division (number of cells per milliliter of culture). (B) DNA content (micrograms of DNA per milliliter of culture). (C) Proportion of the culture with binucleate cells (solid circles) and with cell plates (open circles). (D) Activity per milliliter of culture of aspartate transcarbamylase. (E) RNA content (micrograms of RNA per milliliter of culture). (F) Protein content (micrograms of protein per milliliter of culture). Cells were counted using a Bürker-Türk counting chamber. DNA was assayed by the diphenylamine method (Bostock, 1970; J. Creanor and J. M. Mitchison, personal communication). The cell plate index was measured on living cells, and the proportion of binucleate cells was determined after Giemsa-staining (Ganesan and Swaminathan, 1958; Ganesan and Roberts, 1959). The activity of the enzyme was assayed by the method of Gerhart and Pardee (1962), using borax buffer as suggested by Thorpe (1974). RNA was measured as described by Bostock (1970), and protein was assessed according to the method of Lowry et al. (1951). The abscissa rectangles represent the cyclic heat treatment.

Figure 2B shows the synthesis of DNA in a heat-shocked culture. DNA synthesis follows the same pattern as described for cell division. There is no net DNA synthesis until about 60 minutes (58 ± 13 minutes, $n = 4$) after the shock, and then the amount of DNA doubles and the doubling is finished before the initiation of the next shock. On the average mid-S occurs 80 minutes (81 ± 10 minutes, $n = 4$) after the end of the shock.

The synchrony can also be monitored by following the percentage of binucleate cells in the culture which expresses degrees of simultaneous nuclear division and the proportion of cells showing cell plates (*Schizosaccharomyces* divides by binary fission after the formation of a transverse cell wall, the cell plate). Figure 2C shows the result of one such experiment.

It is seen that a peak of binucleate cells and a peak of cells with cell plates appear at about 35 and 15 minutes, respectively, before the division. The peaks are somewhat higher (27 and 37%, respectively) than those obtained by Mitchison and Vincent (1965) and Polanshek (1974) using cultures of *Schizosaccharomyces* synchronized by a selection procedure.

The enzyme aspartate transcarbamylase (ATCase) is known as a step enzyme in *Schizosaccharomyces*, i.e., its synthesis is restricted to part of the cell cycle (Bostock *et al.*, 1966). Also in a heat-treated culture the activity of this enzyme increases in a stepwise manner. This is seen in Fig. 2D which shows the activity of the enzyme as a function of time after six shocks. No net increase in enzyme activity is observed until about 45 minutes (44 ± 12 minutes, $n = 6$) after the end of the shock; then the activity of the enzyme practically doubles during the next hour, with a midpoint at about 75 minutes (74 ± 16 minutes, $n = 6$) after the shock. During the next shock and until about 40 minutes after, the increase in ATCase activity is again blocked.

In the above paragraphs events have been dealt with that occur within parts of the normal cell cycle. These events are found to be discrete also in heat-synchronized cultures. RNA and bulk protein synthesis occur throughout the normal cell cycle, although at more-or-less smoothly increasing rates. The same holds true in the heat-synchronized system, except that synthesis is interrupted for some time by each shock. This is seen in Fig. 2E and F. A report by Kramhøft and co-workers, which discusses the details of the results presented above is in preparation (Kramhøft *et al.*, 1975).

C.   Discussion

It was stated in the previous section that the experimental conditions that led to the results presented here were based in part on data observed in preliminary growth experiments, and in part on information gained with a different cell system. Although the degree of synchrony thus obtained is satisfactory when compared to other synchronous systems, this is no guarantee that better synchrony would not be achieved if one or more of the parameters of the heat treatment were changed. To investigate this the following series of experiments was conducted.

In all cases the cultures received five standard heat shocks (41°C, spaced 110 minutes apart, and each lasting 30 minutes). A sixth shock was then varied according to the following scheme. (1) The time of initiation of the sixth shock was varied, while the shock temperature and the duration of the shock were kept constant; (2) the shock temperature was varied, and the interval between shocks and the duration of the shock were kept constant; and (3) the duration of the shock was varied, and the interval between shocks and the shock temperature were kept constant. No attempt was made to vary the temperature between the shocks.

In each experiment the degree of synchrony of the division following the sixth shock was measured, either as the slope of the division step as seen in Fig. 2A, or as the size of the peak fraction of cells with cell plates (Fig. 2C). Experiments conducted along the lines described above showed that the shock temperature and the interval between shocks selected initially proved to be optimal. (It should be noted, however, that the intershock interval must be adjusted as carefully as possible to the actual doubling time of the culture when cells are grown exponentially at the intershock temperature). However, the shocks chosen were probably too short. We are presently investigating this aspect, and preliminary results indicate that better synchrony may be obtained if the duration of each shock is increased. This change seems to increase the fraction of cells simultaneously with cell plates from about 35 to about 60% when the duration of the shocks is increased from 30 to 50 minutes.

V.   Conclusion

The current report has demonstrated the possibility of synchronizing populations of the yeast *S. pombe* with heat shocks spaced one normal generation time apart. Since this method has derived from results obtained with a completely different organism—the protozoon *Tetrahymena*—we venture the hypothesis that other cell cultures, including other yeasts, may be synchronized when working along the lines described here.

REFERENCES

Bostock, C. J. (1970). *Exp. Cell Res.* **60**, 16–26.
Bostock, C. J., Donachie, W. S., Masters, M., and Mitchison, J. M. (1966). *Nature (London)* **210**, 808–810.

Ganesan, A. T., and Roberts, C. (1959). *C. R. Trav. Lab. Carlsberg* **31**, 175–181.

Ganesan, A. T., and Swaminathan, M. S. (1958). *Stain Technol.* **33**, 115–123.

Gerhart, J. C., and Pardee, A. B. (1962). *J. Biol. Chem.* **237**, 891–896.

Heslot, H., Goffeau, A., and Louis, C. (1970). *J. Bacteriol.* **104**, 473–481.

Kramhøft, B., and Zeuthen, E. (1971). *C. R. Trav. Lab. Carlsberg* **38**, 351–368.

Kramhøft, B., Nissen, S. B., Thorpe, S. M., and Zeuthen, E. (1975). In preparation.

Lowry, H., Rosebrough, N. J., Farr, A. L., and Randall, R. J. (1951). *J. Biol. Chem.* **193**, 265–275.

Mitchison, J. M. (1970). *In* "Methods in Cell Physiology" (D. M. Prescott, ed.), Vol. 4, p. 131. Academic Press, New York.

Mitchison, J. M., and Vincent, W. S. (1965). *Nature (London)* **205**, 987–990.

Polanshek, M. (1974). Ph.D. Thesis (communicated by J. M. Mitchison). Edinburgh University.

Scherbaum, O., and Zeuthen, E. (1954). *Exp. Cell Res.* **6**, 221–227.

Spoerl, E., and Looney, D. (1959). *Exp. Cell Res.* **17**, 320–327.

Thorpe, S. M. (1974). Candidate Thesis, University of Copenhagen.

von Meyenburgh, K. (1969). *Vierteljahrschrift Naturf. Ges. Zürich* **114**, 113–222.

Williamson, D. H., and Scopes, A. W. (1960). *Exp. Cell Res.* **20**, 338–350.

Zeuthen, E. (1971). *Exp. Cell Res.* **68**, 49–60.

SUBJECT INDEX

CONTENTS OF PREVIOUS VOLUMES

Volume I

Volume IV

Volume V

Volume VI

Volume VII

Volume VIII

Volume IX